AF352900

Experimental and Numerical Modeling of Fluid Flow

Experimental and Numerical Modeling of Fluid Flow

Editor

Rüdiger Schwarze

MDPI • Basel • Beijing • Wuhan • Barcelona • Belgrade • Manchester • Tokyo • Cluj • Tianjin

Editor
Rüdiger Schwarze
TU Bergakademie Freiberg
Germany

Editorial Office
MDPI
St. Alban-Anlage 66
4052 Basel, Switzerland

This is a reprint of articles from the Special Issue published online in the open access journal *Applied Sciences* (ISSN 2076-3417) (available at: https://www.mdpi.com/journal/applsci/special_issues/Experimental_Numerical_Modeling_Fluid_Flow).

For citation purposes, cite each article independently as indicated on the article page online and as indicated below:

LastName, A.A.; LastName, B.B.; LastName, C.C. Article Title. *Journal Name* **Year**, *Volume Number*, Page Range.

ISBN 978-3-0365-5469-3 (Hbk)
ISBN 978-3-0365-5470-9 (PDF)

© 2022 by the authors. Articles in this book are Open Access and distributed under the Creative Commons Attribution (CC BY) license, which allows users to download, copy and build upon published articles, as long as the author and publisher are properly credited, which ensures maximum dissemination and a wider impact of our publications.

The book as a whole is distributed by MDPI under the terms and conditions of the Creative Commons license CC BY-NC-ND.

Contents

MDPI

Editorial

Experimental and Numerical Modeling of Fluid Flow

Rüdiger Schwarze

Institute of Mechanics and Fluid Dynamics, Lampadiusstraße 4, 09596 Freiberg, Germany;
ruediger.schwarze@imfd.tu-freiberg.de; Tel.: +49-3731-392486

Citation: Schwarze, R. Experimental and Numerical Modeling of Fluid Flow. *Appl. Sci.* **2022**, *12*, 9042. https://doi.org/10.3390/app12189042

Received: 5 September 2022
Accepted: 7 September 2022
Published: 8 September 2022

Publisher's Note: MDPI stays neutral with regard to jurisdictional claims in published maps and institutional affiliations.

Copyright: © 2022 by the author. Licensee MDPI, Basel, Switzerland. This article is an open access article distributed under the terms and conditions of the Creative Commons Attribution (CC BY) license (https://creativecommons.org/licenses/by/4.0/).

1. Introduction

Fluid dynamics is often related to complex flow conditions and systems, either in the context of fundamental research or in the context of industrial processes. Typically, the study of such flows requires experimental or numerical models which capture their complex, i.e., multiphase and turbulent, nature. Although such models have been the state of the art for a long time in areas such as aerodynamics or weather simulation, the field of the application of experimental and numerical flow models is constantly expanding. The goal of this Special Issue is to present an overview of applied experimental and numerical flow models, with a strong focus on the following topics: (i) new areas of application, (ii) advanced experimental or numerical models, (iii) innovative modeling approaches, and (iv) challenges in modeling techniques.

2. New Applications of Experimental and Numerical Flow Models

T. Zisis, K. Vasilopoulos, and I. Sarris [1] investigate the prevailing conditions in a railway tunnel after a train fire accident by means of CFD simulations. The goal is to analyze the ability of the ventilation system to create the proper conditions for safe passenger evacuation. Three scenarios with different fan activation times and different evacuation processes were examined. They found that the most important action in a tunnel fire is the time at which the ventilation system is activated after the start of a fire.

C. L. Pavlidis, A. V. Palampigik, K. Vasilopoulos, I. C. Lekakis, and I. E. Sarris [2] examine the airflow and pollutant dispersion around an isolated cubical building located in a warm Mediterranean climate with CFD simulations adapted to the local micro-climate conditions. The performance of the numerical model is tested with the help of measured data from the SILSOE cube experiment. In the simulations, the influence of the thermal boundary conditions of the building on the airflow and the pollutant dispersion is evaluated.

M. Zaboli, S. S. Mousavi Ajarostaghi, and M. Saffari Pour [3] analyze different configurations of a parabolic trough solar collector with inner helical axial fins as a swirl generator. After a successful validation of their CFD model, they use CFD simulations in order to resolve the inner flow and temperature fields. They show that the thermal performance of the collector can be significantly improved by using an innovative collector design.

K. Yousef, A. Hegazy, and A. Engeda [4] study complex flow behavior during gas entrainment in water in an inverted vertical U-tube with the addition of side air or water vapor. The results from both experiments and CFD simulations reveal a complex interplay between the water mass flow rate, pressure, and extracted air or vapor flow rates. Furthermore, critical states of the system are identified.

Z. Zhu, Y. Ju, and C. Zhang [5] investigate the obstructive sleep apnea–hypopnea syndrome, a highly prevalent respiratory disorder, in an experimental approach based on a computed tomography scanned extra-thoracic airway model. They use particle image velocimetry to study the air flow in their model. The in vitro measurements of oscillatory respiratory flow velocity show the temporal evolution of the complex flow patterns and the potential wall collapse of the extra-thoracic airway model with obstructive sleep apnea–hypopnea syndrome.

D. K. Kim [6] analyzes the performance improvement of refrigerators by different innovative methods. The effects of three methods, modifying the flow channel and the operating conditions and changing the positions of the fan in the mechanical chamber room, are analyzed in a combined experimental and numerical approach. The results show that the efficiency of the refrigerator can be improved by these methods.

T. Bano, F. Hegner, M. Heinrich, and R. Schwarze [7] study the fluid–structure interaction (FSI) involving flexible elastic structures in a water cross stream with an FSI-CFD model. Transient simulations are carried out for flexible flaps of different thicknesses in a laminar glycerin flow with Reynolds numbers ranging from $3 < \mathrm{Re} < 12$. The bending line and the maximum tip deflection of the flaps in the simulations agree well with findings from a previous experimental study.

F. Wang, L. Wang, G. Chen, and D. Zhu [8] examine the oil droplet size distribution in an aero-engine bearing chamber with CFD and population balance models. The CFD simulations reveal the correlations between the initial oil droplet size distribution, the oil droplets' coalescence and breakup, air, and oil mass flow, analyzed using the CFD and population balance models.

X. Qu, Q. Guo, Y. Zhang, X. Qi, and L. Liu [9] propose a miniaturized four-sensor electrical probe with a new signal processing method as an innovative multiphase flow measurement technique. Measurements were taken during a gas–liquid two-phase flow experiment in cap-bubble flow regime to test the performance of the proposed signal process scheme. The local flow parameters obtained by the probe measurement are in good agreement with the results from the visual measurement technique in the same flow conditions, proving the correctness of the proposed method.

Finally, Y.-B. Seo, S. S. Paramanantham, J.-Y. Bak, B. Yun, and W.-G. Park [10] use a CFD approach to investigate a boiling model for rod bundle flows related to reactor cooling in nuclear power plants. The comparison of the numerical results with the experimental values of the vapor volume fraction and vapor bubble–water interfacial area concentration show good agreement for different initial conditions.

Funding: This research received no external funding.

Acknowledgments: Thanks to all authors and peer reviewers for their valuable contributions to this Special Issue. I would also like to express my gratitude to all the staff and people involved in this Special Issue.

Conflicts of Interest: The author declares no conflict of interest.

References

1. Zisis, T.; Vasilopoulos, K.; Sarris, I. Numerical Simulation of a Fire Accident in a Longitudinally Ventilated Railway Tunnel and Tenability Analysis. *Appl. Sci.* **2022**, *12*, 5667. [CrossRef]
2. Pavlidis, C.L.; Palampigik, A.V.; Vasilopoulos, K.; Lekakis, I.C.; Sarris, I.E. Air Flow Study around Isolated Cubical Building in the City of Athens under Various Climate Conditions. *Appl. Sci.* **2022**, *12*, 3410. [CrossRef]
3. Zaboli, M.; Mousavi Ajarostaghi, S.S.; Saedodin, S.; Saffari Pour, M. Thermal Performance Enhancement Using Absorber Tube with Inner Helical Axial Fins in a Parabolic Trough Solar Collector. *Appl. Sci.* **2021**, *11*, 7423. [CrossRef]
4. Yousef, K.; Hegazy, A.; Engeda, A. Experimental and CFD Investigation into Using Inverted U-Tube for Gas Entrainment. *Appl. Sci.* **2020**, *10*, 9056. [CrossRef]
5. Zhu, Z.; Ju, Y.; Zhang, C. In-Vitro Experimental Modeling of Oscillatory Respiratory Flow in a CT-Scanned OSAHS Tract. *Appl. Sci.* **2020**, *10*, 7979. [CrossRef]
6. Kim, D.K. Experimental and Numerical Study on Performance Enhancement by Modifying the Flow Channel in the Mechanical Chamber Room of a Home Refrigerator. *Appl. Sci.* **2020**, *10*, 6284. [CrossRef]
7. Bano, T.; Hegner, F.; Heinrich, M.; Schwarze, R. Investigation of Fluid-Structure Interaction Induced Bending for Elastic Flaps in a Cross Flow. *Appl. Sci.* **2020**, *10*, 6177. [CrossRef]
8. Wang, F.; Wang, L.; Chen, G.; Zhu, D. Numerical Simulation of the Oil Droplet Size Distribution Considering Coalescence and Breakup in Aero-Engine Bearing Chamber. *Appl. Sci.* **2020**, *10*, 5648. [CrossRef]

9. Qu, X.; Guo, Q.; Zhang, Y.; Qi, X.; Liu, L. A New Vector-Based Signal Processing Method of Four-Sensor Probe for Measuring Local Gas–Liquid Two-Phase Flow Parameters Together with Its Assessment against One Bubbly Flow. *Appl. Sci.* **2020**, *10*, 5463. [CrossRef]
10. Seo, Y.-B.; Paramanantham, S.S.; Bak, J.-Y.; Yun, B.; Park, W.-G. CFD Analysis of Subcooled Flow Boiling in 4 × 4 Rod Bundle. *Appl. Sci.* **2020**, *10*, 4559. [CrossRef]

 *applied sciences*

Article

Numerical Simulation of a Fire Accident in a Longitudinally Ventilated Railway Tunnel and Tenability Analysis

Thomas Zisis, Konstantinos Vasilopoulos * and Ioannis Sarris

Laboratory of Thermo Fluid Systems (LTFS), Department of Mechanical Engineering,
Ancient Olive Grove Campus, University of West Attica, Thivon Str. 250, Egaleo, 12244 Athens, Greece;
msrtf20x03@uniwa.gr (T.Z.); sarris@uniwa.gr (I.S.)
* Correspondence: kvassil@uniwa.gr; Tel.: +30-2105381131

Abstract: The present study examines the prevailing conditions in a railway tunnel after a train fire accident and the ability of the ventilation system to create the proper conditions for a safe passenger evacuation. The examined scenario included an event of a 20-MW diesel pool fire on a suburban train, immobilized in the middle of a 1.5-Km long, linear shaped rectangular tunnel ventilated by a longitudinal jet fan system, and the people's movement during the evacuation was effectuated along walking platforms. More specifically, three scenarios with different fan activation times and different evacuation processes were examined. A Large Eddy simulation model (LES) was used for the simulation of the air flow in the railway tunnel. The evaluation of the ventilation system criteria considered the achievement of the air critical velocity inside the railway tunnel, and for the people's safe evacuation, the Fractional Effective Dose (FED) value was examined. It was found that the most important action in a tunnel fire is the time, after the start of a fire, the ventilation system is activated.

Keywords: railway tunnel fire; FDS; tunnel ventilation; tenability analysis; numerical simulation

Citation: Zisis, T.; Vasilopoulos, K.; Sarris, I. Numerical Simulation of a Fire Accident in a Longitudinally Ventilated Railway Tunnel and Tenability Analysis. *Appl. Sci.* **2022**, *12*, 5667. https://doi.org/10.3390/app12115667

Academic Editor: Rüdiger Schwarze

Received: 5 May 2022
Accepted: 31 May 2022
Published: 2 June 2022

Publisher's Note: MDPI stays neutral with regard to jurisdictional claims in published maps and institutional affiliations.

Copyright: © 2022 by the authors. Licensee MDPI, Basel, Switzerland. This article is an open access article distributed under the terms and conditions of the Creative Commons Attribution (CC BY) license (https://creativecommons.org/licenses/by/4.0/).

1. Introduction

Tunnels are complex infrastructures, usually of semicircular or rectangular cross-sections, constructed to establish connections between different sections of a road or railroad network. The growing reports of indoor fire accidents with high population mobility, such as tunnels, have created the need to consider the prevailing conditions in such cases as well as the best evacuation plan to ensure human life [1].

The consequences after a tunnel fire can be much more serious than an outdoor fire because of high smoke concentrations and confined movement space [2,3]. Accidents involving fire and hazardous material displacement can be studied for different urban scales and settings. Although the fire poses a threat near its vicinity, toxic substances can travel along the flow of the air [4]. After a fire accident, high temperatures and toxic fumes are produced, which have harmful consequences for passengers who are trapped inside the tunnel. In 2003, an arson fire in the subway tunnel in Daegu, South Korea resulted in 198 deaths. In 2017, the Jinji Road Tunnel Fire in Shanxi, China, 2017, caused 31 deaths [5]. A wrong evacuation plan can lead people in to the fire or smoke direction, with a negative effect on their health due to the presence of toxic gases, heat, and radiation [6,7].

To minimize human exposure to smoke during a tunnel evacuation, various types of ventilation systems are used to remove the smoke from the tunnel or to keep the smoke in the tunnel above the head [8,9]. These can be either mechanical or natural ventilation systems [10]. Forced ventilation systems use fans to extract or push smoke out of the tunnel. Natural ventilation with vertical wells in the roof of the tunnel take advantage of the buoyancy of hot smoke, which escapes through the shafts due to the stack effect and piston effect due to the movement of vehicles or trains inside the tunnel. This approach generally does not require mechanical fans, resulting in space-saving at the higher level of the tunnel, and it is suitable for overground [11].

One of the key criteria for the passenger's safety after a fire accident is the number of toxic products of combustion that the individual inhales during the event and the evacuation process. This is calculated using the FED (Fractional Effective Doses) index. The FED index was first introduced by Purser [12] and has since been adopted by various organizations, such as the NFPA [13]. The basic idea is that a person who inhales a certain dose of toxic products for a period of time is considered unconscious or dead. To assess how a person's life is threatened, FED is the main factor considered [1], alongside the thermal exposure and the low visibility due to smoke (hence the difficulty of accessing emergency exits).

Evacuation in underground public transport systems can be complicated and the outcome of an evacuation scenario depends on many different parameters. These parameters can be the people's walking speed, the visibility, the design of the train and tunnel, or the physical condition of the people [14].

Tunnels are designed with elevated platforms to assist a safe passenger evacuation in case of a fire accident. The reduced height difference between the level of the train and the platform, compared to the evacuation of a train directly to the rail area, is a method of increasing the safety of the population. The difficulty of evacuating a train to the rails, especially for people with mobility difficulties, has been analyzed in the METRO project of Ingason et al. [15]

The purpose of a tunnel ventilation system, besides for maintaining air quality, is to keep the longest part of a tunnel free of smoke for as long as possible. In the case of a fire accident inside a tunnel, passengers are exposed to smoke, even though the ventilation system exists. In such a situation, the activation of the ventilation system must be delayed slightly or operated at a slower speed so as not to disturb the natural stratification of the air and smoke layers [16,17]. This state will be maintained if the air velocity does not exceed the critical velocity [18]. The critical velocity is defined as the minimum longitudinal ventilation velocity to avoid the phenomenon of smoke reversal or backlayering in case of fire [19]. Depending on the air velocity, hot smoke and cold air are mixed, and thus the smoke is dispersed downstream of the fire throughout the tunnel [20]. Hot smoke also moves against the flow of air, upstream of the fire front, and as it cools it descends to the ground. This phenomenon is called smoke backlayering. The amount of smoke that will be inverted upstream of the fire front and the downstream distance of where it will remain depends on the ventilation conditions [19,21].

Hu et al. [22] suggested that smoke stops spreading when the horizontal inertia forces caused by the buoyancy of hot smoke are equal to the inertia forces of the ventilation air. Based on a series of small-scale experiments that studied the correlation between critical velocity and heat release rate (HRR—Heat Release Rate), for different geometries of fire sources, conducted by Oka and Atkinson [23] and Li et al. [24], they found that the dimensionless critical velocity is proportional to the dimensionless $HRR^{1/3}$ in the case of low and medium HRR and is independent in the case of large HRR. Riess and Bettelini [25] found that the stack effect should be considered in the design of the ventilation of sloped tunnels with angles of inclination of more than 1–2%. Wu and Bakar [26] observed that, with the same fuel and longitudinal ventilation speed, the length of the inverted smoke layer gradually increases as the tunnel cross-section increases when the tunnel cross-section reaches a certain value.

Lee and Ryou [27] investigated the critical velocities for various tunnel aspect ratios in a series of small-scale experiments, and they found that the critical velocity increases with the aspect ratio. The study of a fire accident with vehicles blocking the tunnel showed [28,29] that, as the distance from the blockage upstream of the source fire accident increases, the length of the inverted smoke layer and the critical velocity initially decreases until it reaches a certain value that is equal to the value that these two sizes would have without the presence of an obstacle. Tanaka et al. [30] conducted a series of small-scale experiments to study the characteristics of a fire's, smoke considering the escalation of heat conduction through the tunnel walls and presented some equations for estimating the critical velocity and length of the inverted smoke layer.

According to NFPA [13], the critical velocity is "the minimum steady-state velocity of the ventilation airflow towards the fire, in a tunnel or corridor, required to prevent smoke reversal in the fire area." In this study, three different scenarios (A, B, and C), regarding the jet fan activation time, were examined. The jet fans were activated simultaneously at t = +180 s, +300 s, and +480 s, after the fire ignition, respectively. Then, the calculated volume fractions of toxic gases from the numerical simulation were used to calculate the FED value of the passengers evacuating the train. Three different scenarios for the passengers' evacuation were examined, with evacuation initiation times being t = +60 s, +120 s, and +300 s after the fire ignition.

Section 2 describes the numerical model, the dimensions and boundary conditions of the simulation model geometry, the grid formulation, and the grid independence test. Section 3 describes the validation method used to validate the numerical simulation results using empirical equations that resulted from experiments. The tenability analysis takes place in Section 4, where the mean and max FED values are compared for each scenario.

The literature, in its majority, studies the evacuation times, choke points during the evacuation of subway systems and the comfort of passengers evacuating tunnels via walking platforms, or how the critical velocity is affected by different conditions [14,24,29,31,32]. The present study focused on the correlation between the two actions taken in a tunnel fire accident: ventilation system activation and evacuation process initialization. These two actions play a great role in the exposure of the population to harmful gases and hot temperatures. The results show that the most important action is the activation of the ventilation system as soon as possible from the ignition of the fire. When this is not possible, according to the results, the best strategy is to wait for the people to evacuate the tunnel and then activate the ventilation system to prevent passengers from inhaling smoke due to the entertainment of the smoke layer in the tunnel ceiling.

2. Numerical Model

The Fire Dynamics Simulation (FDS) is a code developed by the NIST in cooperation with the VIT research center of Finland. The code numerically solves the Navier-Stokes equations and finds application in low velocity thermally driven flows, smoke production, and mass and heat transfer. The FDS code is a low Mach number Large Eddy Simulation solver (the Mach number is under 0.3). To define the turbulent characteristics, the Smagorinsky model is applied.

The governing equations for mass and momentum conservation are:

$$\frac{\partial \rho}{\partial t} + \nabla \cdot \rho \vec{u} = \dot{m}_b''' \tag{1}$$

$$\frac{\partial}{\partial t}(\rho \vec{u}) + \nabla \cdot \rho \vec{u} \vec{u} + \nabla \cdot \vec{p} = \rho g + \vec{f_b} + \nabla \cdot \tau_{ij} \tag{2}$$

where ρ is the density, g is the gravity acceleration, u is the velocity, p is the pressure, t is the time, and $\tau_{ij} = \mu\left(2S_{ij} - \frac{2}{3}\delta_{ij}(\nabla u)\right)$ is the stress tensor, where μ is the dynamic viscosity of the fluid, S_{ij} is the strain tensor, and δ_{ij} is the Kronecker delta.

The coefficient of turbulent consistency in Smagorinsky's analysis is defined as:

$$\mu_{LES} = \rho(C_s \Delta)\left(2S_{ij} : S_{ij} - \frac{2}{3}(\nabla \overline{u})^2\right)^{\frac{1}{2}} \tag{3}$$

where $\Delta = (\delta x \delta y \delta z)^{\frac{1}{3}}$ is the dimension of the spatial filter proportional to the dimension of the grid.

The energy conservation equation is written as:

$$\frac{\partial}{\partial t}(\rho h_s) + \nabla \rho h_s u = \frac{Dp}{Dt} + q''' - \dot{q}_b''' - \nabla q'' + \varepsilon \tag{4}$$

where h_s is the sensible enthalpy, $\dot{q}_b'''$ is the energy transferred to particles during evaporation (change of phase), and q'' are the radiative and conductive heat fluxes.

The equation for species conservation is written as:

$$\frac{\partial}{\partial}(\rho Y_a) + \nabla \cdot \rho Y_a \vec{u} = \nabla \rho D_a \nabla Y_a + \dot{m}_a''' + \dot{m}_b''' \tag{5}$$

where $\dot{m}_b''' = \sum_a \dot{m}_{b,a}'''$ is the species production rate as particles or droplets.

The time step used is variable and adjusted by the FDS code. During the calculation, the time step is adjusted so that the CFL (Courant, Friedrichs, Lewy) condition is satisfied. The default value of the size of the time step is [31]:

$$DT = \frac{5(\delta x \delta y \delta z)^{1/3}}{\sqrt{gH}} \tag{6}$$

where δx, δy, and δz are the dimensions of the smallest mesh cell, H is the height of the computational domain, and g is the acceleration of gravity.

2.1. Model Details

The studied tunnel is made of rectangular cross-sections with dimensions typical for a double railway tunnel (y = 11 m, z = 7 m), and its length is 1.5 Km.

In all cases, the center of the fire source was located in the middle section of the tunnel, as shown in Figure 1 at (x = 795.5 m, y = 2.25 m, z = 4 m), and the intensity was 20 MW, which corresponds to a traction engine fire. The fire's surface was 1 m² and it was located on the train's roof. The surface of the fire was modeled as a burner type and was inserted into the model as an open vent. The fire growth followed a typical at^2 rule, where the growth rate a was 0.1876 kW/s² and achieved nominal HRR within 327 s. $C_{12}H_{23}$ was used as the fuel, which is the standard composition of diesel, with a yield of combustion products of 0.1 for CO and 0.09 for soot. Combustion efficiency is defined as the amount produced per gram of fuel diffused along with the flow, compared to the diffusion of fire into the environment.

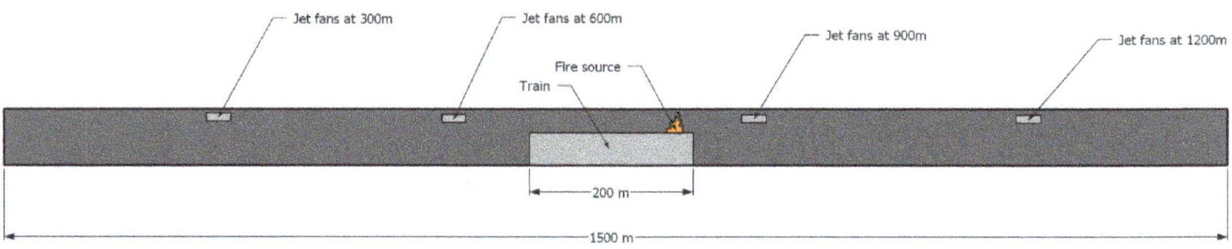

Figure 1. Side view of the simulation tunnel, tunnel dimensions, jet fans, train, and fire source location.

The jet fans used in the simulations are 3 m long with a surface area of 1 m². Four pairs of them were placed every 300 m at the height of 5 m and 1 m off the side walls

According to McGrattan et al. [32], the non-dimensional expression $D^*/\delta x$ was used to assess mesh resolution, where δx is the nominal size grid (in m), and the characteristic length of D^* is expressed as follows

$$D^* = \left(\frac{\dot{Q}}{C_p T_\infty \rho_\infty \sqrt{g}}\right)^{2/5} \tag{7}$$

where Q is the fire heat release rate (HRR) (W), ρ_∞ is the ambient density (kg/m^3), C_p is the specific heat capacity (J/kgK), T_∞ is the ambient temperature (K), and g is the gravitational acceleration constant (m/s^2). In the NUREG-1824 Directive published by the United States Nuclear Energy Regulatory Commission (USNRC) and the Electricity Research Institute (EPRI), it defines that D^* values should be in a range between 4–16 for an optimal analysis in FDS Code [33]. Unlike the number of cells, the grid resolution does not depend on the volume of the building, so the corresponding standard has the advantage of universal application regardless of the size of the building.

The grid independence test was conducted for three different meshes: coarse mesh with a mesh size of 0.75 m, medium mesh with a mesh size of 0.5 m, and fine mesh with a mesh resolution of 0.25 m. The number of cells for each type of mesh was 269,865, 924,000, and 4,254,149, respectively. As shown in Figure 2, the fine and medium meshes were compared. The grid errors using two computational grids were calculated using the following equation [34]

$$GCI = \frac{f_2 - f_1}{1 - r^p} \tag{8}$$

where f_2 is the numerical solution that results from the medium computational grid, and the corresponding from the finer is f_1, r is the refinement factor between the two computational grids, and p is the accuracy of the algorithm, which is 3 for the present study.

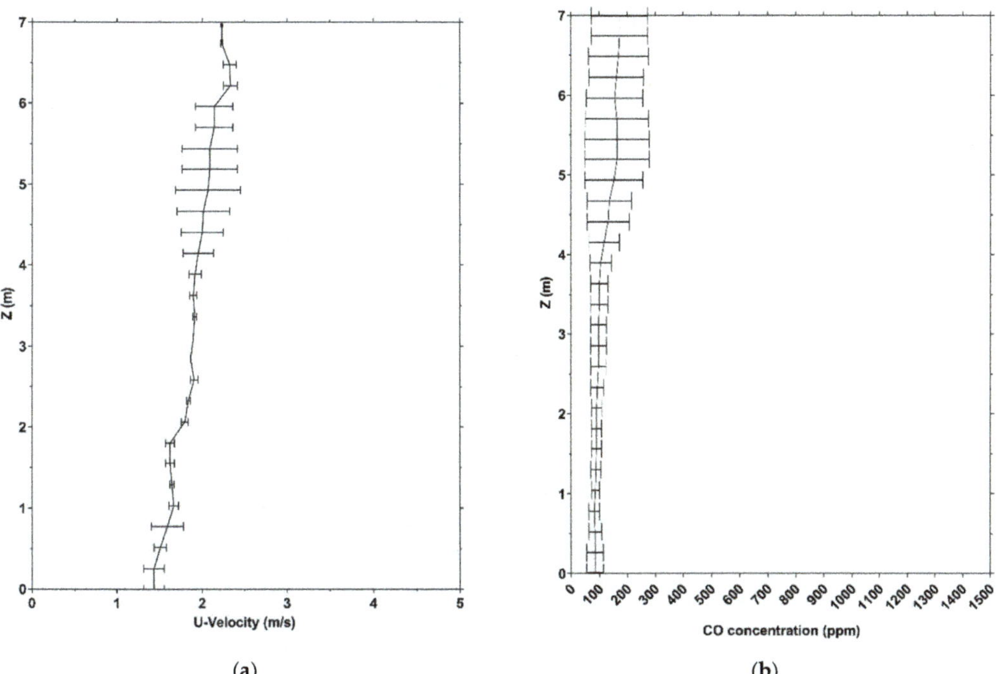

Figure 2. (a) Comparison of the grid convergence index (GCI) for the U mean velocity and (b) mean carbon monoxide concentration for the medium and fine mesh in the symmetry plane of the tunnel, with coordinates (x = 1100 m, y = 5.5 m, and z = from 0 to 7 m) and 28 sampling points.

The grid independence examined the U velocity and carbon monoxide concentration values. The values were extracted by a line, downwind of the fire source, in the symmetry plane of the tunnel with coordinates (x = 1100 m, y = 5.5 m, and z = from 0 to 7 m) with 28 sampling points.

The results did not show great deviation, so a combination of the fine and medium mesh was used as the final mesh for the examination of the three main scenarios. The fine mesh was used for the field around the fire, while, for the rest of the tunnel, the medium mesh was used. For the sake of convenience, a mesh size of 0.25 m was used in the area around the fire (50 m downstream and 50 m upstream), while, for the rest of the tunnel, a mesh size of 0.5 m was used, as shown in Figure 3. The total number of cells was 1,355,200.

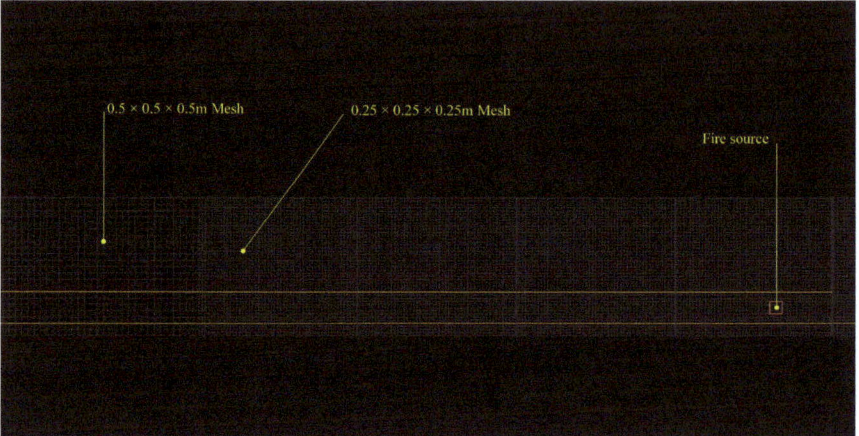

Figure 3. Grid meshes and fire source. The medium grid of 0.5 m was used for the majority of the tunnel, while the fine mesh of 0.25 was used near the field of the fire.

The mean value of errors of the U velocity for the coarse and medium meshes was approximately 14%, for the medium and fine mesh it was 5%, and for the medium and final mesh used it was 4%. For the CO concentrations, the mean error values for the coarse and medium mesh was 16%, for the medium and fine mesh was 8%, and for the medium and final mesh was 6% The mesh resolution, which was proposed in the UNSRC and EPRI directive, was used in many other tunnel fire simulations such as [35–37].

The work for the numerical simulation was conducted in the National HPC facility—ARIS, with computational time granted from the National Infrastructures for Research and Technology S.A. (GRNET S.A.). For each scenario, 22 cores and 56 GB of memory were used, and the typical time taken to complete each computation was 24 h.

2.2. Validation of Methodology

Kurioka et al. [38] performed experiments using three different tunnel scales, 1:10, 1:2, and a full scale, with rectangular and horseshoe cross-sections. The fire sources used were of square cross section. The aspect ratio of the tunnel cross-section dimensions, HRR, and forced longitudinal ventilation velocity was variable. From the experimental data of the 1/10 scale model, empirical formulas were exported to calculate the flame inclination, the flame height, the maximum temperature of the smoke layer, and its position. The effect of tunnel cross-section dimensions was incorporated into these models. The ratio $H^{3/2}/b^{1/2}$, where H is the height of the tunnel model and b is the width, was found to be representative for the investigation of fire phenomena in tunnels. It was also confirmed that these empirical formulas were sufficiently applicable to predict phenomena occurring during a fire accident, close to the square fire source field, by comparing the results of experiments in 1:2 and full-scale tunnels. Hu et al. [39] studied, experimentally and computationally, the maximum temperature on the tunnel roof. They first compared the FDS results with the experimental measurements, and then, with the values, they calculated with the model of Kurioka, Oka, Satoh, and Sugawa [38]. The temperature predicted by the

FDS was found to converge quite well with both the experimental measurements and the results of the empirical model.

The Li et al. [40] study provided a theoretical analysis of the maximum temperature in the tunnel's roof, based on the axon-symmetric fire plume theory, taking into account the heat release rate, the longitudinal ventilation velocity, and the geometry of the tunnel. They conducted scale tests to investigate the maximum temperature on the tunnel roof in case of fire. Li and Ingason [41] analyzed the effects of different ventilation systems, ventilation speeds, heat release rates, tunnel geometries, and fire sources on the maximum temperature under the tunnel roof for large fires. They used data from many model-scale tests and large-scale tunnel fire experiments around the world. They proposed some empirical formulas for the maximum temperature on the tunnel's roof nearby the fire source for low and high ventilation speeds. Rosignuolo et al. [42] merged the empirical relationships from the two studies above and defined a formula that exhibited non-linear regression and calculated temperature values as a function of distance from the fire source. They also distinguished between small and large fires because, in the event of a very large fire, the flame reaches the roof, so that the maximum temperature corresponds to the temperature of the flame, and it is difficult to estimate the mass flow rate. If a portion of the flame volume containing the combustion zone reaches the roof of the tunnel, the maximum temperature on the roof is reaching a constant value. Therefore, the maximum roof gas temperature is limited to a maximum of 1350 °C. The formula that calculates the temperature as a function of distance from the fire source is expressed as follows:

$$\frac{\Delta T(x)}{\Delta T_{max}} = 0.55 \; \exp\left(-0.143\frac{\chi - \chi_v}{H}\right) + 0.45 \; \exp\left(-0.024\frac{\chi - \chi_v}{H}\right) \tag{9}$$

where χ is the distance from the fire source, H is the height of the tunnel, and χ_v is the virtual origin, which is calculated as

$$\chi_v = \begin{cases} L_f - 10H, & L_f > 10H \\ 0, & L_f \leq 10H \end{cases} \tag{10}$$

where L_f is the flame length (m). The maximum temperature of the gas depends on its maximum value on the roof (ΔT_{max}), which in turn was calculated by defining two different areas, depending on the dimensionless ventilation speed V':

$$\Delta T_{max} = \begin{cases} 17.5\dfrac{\dot{Q}^{2/3}}{H_{ef}^{5/3}}, & V\prime \leq 0.19 \\[2ex] \dfrac{\dot{Q}}{u_0 b_{f0}^{1/3} H_{ef}^{5/3}}, & V' > 0.19 \end{cases} \tag{11}$$

where $\dot{Q}$ is the total heat release rate (kW), H_{ef} is the effective height of the tunnel (the height from the fire source to the ceiling) (m), u_0 is the ventilation velocity, b_{f0} is the radius of the fire source (m), and V' is defined as:

$$V' = \frac{u_0}{w^*} \tag{12}$$

where w^* is the characteristic velocity of the plume and is expressed by the following equation:

$$w^* = \left(\frac{g\dot{Q}}{b_{f0}\rho_0 C_p T_0}\right)^{1/3} \tag{13}$$

where g is the gravitational acceleration, (m/s^2), b_{f0} is the radius of the fire source (m), ρ_0 is the density of air (kg/m^3), C_p is the specific heat capacity (kJ/kgK), and T_0 is the ambient temperature (K).

The values calculated with the empirical model were compared to the numerical results for Scenario C, measured with thermocouples at the tunnel ceiling (z = 7 m) above and downwind the fire source (where x = 0 is directly above the fire source). A large difference in temperature values is shown in Figure 4. The figure shows the good agreement of the curves at 0 m during the initial stages of the fire. From this point on, the temperature decreases and follows a scattering trend. There is a consistent difference of around 400 °C between the empirical and the actual temperature values for every location, except for above the fire pool. This difference leads to the assumption that, since the flame length increases up to the point that it is equal to and greater than the effective height of the tunnel and starts crawling horizontally to the ceiling, the smoke layer surrounds the fire source. The results indicate that there was not enough oxygen for the fire to grow further, so the fire became ventilation controlled. Other possible reasons for these differences might be that there are limitations to the validity of parameters such as the geometry of the tunnel and the fire area (the FDS fire source has different dimensions than those used in the experiments).

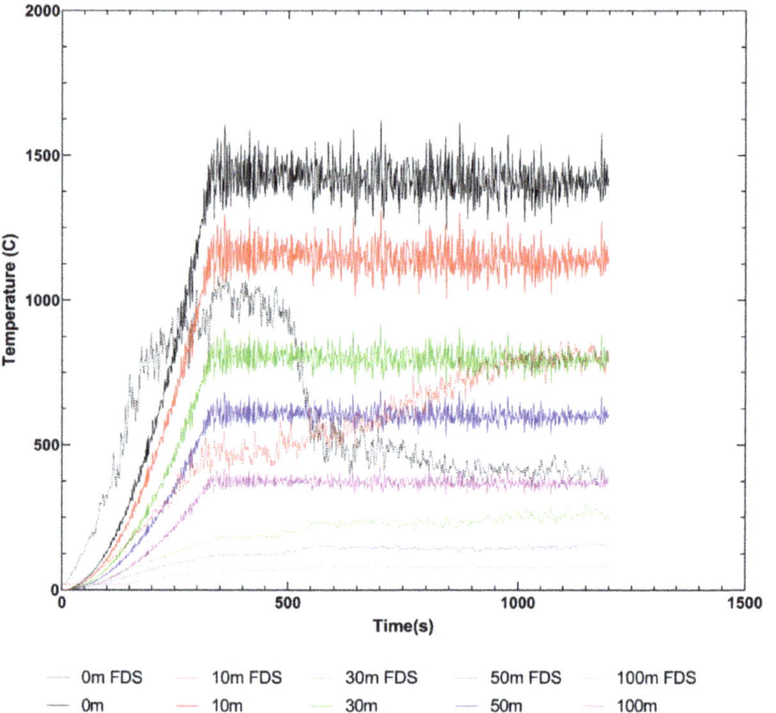

Figure 4. Temperatures at the tunnel ceiling (z = 7 m) above (x = 0 m) and downwind the fire source (x = 10, 30, 50 and 100 m) for Scenario C, calculated using the empirical model and the actual temperatures from the FDS output.

Additionally, for the validation of temperature, the results were compared with the hydrocarbon (HC) fire curve and modified hydrocarbon fire curve (HCM) used for fire resistance testing in structural designs and the RABT-ZTV train fire curve. The HC fire curve is used when there is a risk of liquid fires, e.g., in industrial plants (chemicals), on ships, pontoons, oil platforms, road, and railway tunnels, as well as for petrol and diesel pool fires. The HCM fire curve is derived from the HC curve and adjusted for French regulations. Both HC and HCM curves showed an abrupt temperature gradient in the first minutes, and their difference is the maximum temperature of 1100 and 1300 °C, respectively. The guideline for the equipment and operation of road tunnels (Richtlinie für die Ausstattung und den Betrieb von Straßentunneln) (RABT) defines the so-called RABT tunnel fire. This is often used for the fire protection design of tunnel structures. The RABT tunnel fire curve is an artificial fire curve and is intended to cover the maximum temperatures resulting from tunnel fires [19]. The curve can be used in tunnel design and testing. The HC and HCM fire curve equations are shown below in Equation (14) and Equation (15), respectively.

$$T = 20 + 1080 \cdot \left(1 - 0.325 \cdot e^{-0.167\,t} - 0.675 \cdot e^{-2.5\,t}\right) \tag{14}$$

$$T = 20 + 1280 \cdot \left(1 - 0.325 \cdot e^{-0.167\,t} - 0.675 \cdot e^{-2.5\,t}\right) \tag{15}$$

where t is time in minutes.

The RABT-ZTV train fire curve followed a linear growth for the first 5 min, where it reached a maximum temperature of 1200 °C. A comparison of the ceiling gas temperature from the FDS output for Scenario C and the HC, HCM, and RABT-ZTV curves is shown in Figure 5. The FDS output was in good agreement with the RABT-ZTV fire curve for the first minutes, until the activation of the ventilation system, which happened at 480 s. Scenario C was selected because of the delay in the activation of the jet fans, so the temperature rise was more in accordance with the fire curves.

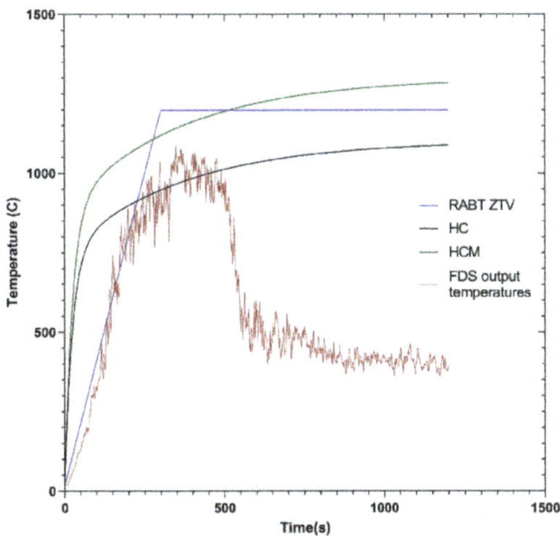

Figure 5. Tunnel ceiling temperature (z = 7 m) above the fire source (x = 795 m), and comparison of the FDS output for Scenario C and the HC, HCM, and RABT ZTV standard fire curves.

2.3. Evacuation and Tenability Analysis

For each scenario (A, B, and C) three different evacuation times were examined +60 s, +120 s, and +300 s after the start of the fire. The evacuation of passengers was effectuated along the walking platform on the side of the tunnel. The total number of passengers was 524 which corresponds to 80% capacity of an intercity train with 8 wagons, each of which can accommodate 80 passengers. The evacuation simulation and FED index calculation were conducted with Pathfinder, which is an interface for the FDS + EVAC code developed by the NIST [31].

For the three evacuation scenarios, eight occupant sources were created, each one simulating the door of a wagon. The total population number that exited the tunnel was 524 persons. This accounts for approximately 80% capacity of the train, assuming each wagon carries 80 passengers at full capacity. The profile of the passengers was set as the default profile provided, with a height of 1.8 m and a walking speed of 1.19 m/s. For the sake of convenience, only the entrance of the tunnel was marked as the exit point for the passengers to move towards to.

3. FDS Simulation Results

Three different scenarios were examined (A, B, and C) with three different fan activation times after a fire accident, at +180 s, +300 s, and +480 s, respectively. For all the cases, all the fans were activated at the same time. The three different scenarios for the different jet fan activation times were set up with the Pyrosim interface provided by Thunderhead Engineering. The extracted results were temperature, flow velocity, the volume fraction of gases, and heat release rate (HRR) values.

As shown in Figure 6, the input fire curve is in good agreement with the numerical results for the HRR curve.

Figures 7 and 8 show that the longer the activation of the ventilation system is delayed, the higher the temperatures appear in the tunnel ceiling. This rapid rise in high temperature can result in spalling, which is the detachment of structural material, such as concrete, often happening in an explosive way. This poses a great threat for the population in the tunnel.

Moreover, a time delay in the ventilation system activation creates a greater smoke dispersion, as shown in Figure 9, and thus a greater dispersion of the harmful combustion products. This can affect the evacuation process due to reduced visibility and soot inhalation.

As shown in Figure 10, the temperatures at the ceiling above the fire source are in good agreement with the RABT ZTV fire curve in the first minutes of the simulation. Then, when the ventilation system was activated, the temperature dropped and reached a steady value of around 450 °C.

Figure 11 shows that the development of ventilation flow started immediately after the activation of the jet fans. The air flow inside the tunnel needs about 300 s, after the jet fans achieve the nominal power, to reach the critical velocity value, but the thrust of the jet fans, even before they reach full power, is capable of pushing the smoke downwind of the fire source. Moreover, Scenario A showed some smoke layer entertainment, which was due to the pressure difference at the tunnel openings, which created a draft with a velocity around 0.5 m/s (Figure 12 at the 180 s time frame). As shown in Figure 12, the peak at 750 m (fire source) was due to the smoke layer and the presence of the train, which reduced the area of the tunnel in this section, resulting in higher velocities. This phenomenon can be seen in Figure 11, where, near the train location, a great volume of the flow had the critical velocity value. As depicted in Figure 13, the mean longitudinal velocity was in the range of 2.7 m/s and was achieved within 600 s from the jet fan activation time. Ventilation conditions are achieved as long as the jet fans reach their nominal output volume flow, but this does not mean the tunnel airflow velocity has reached the critical value.

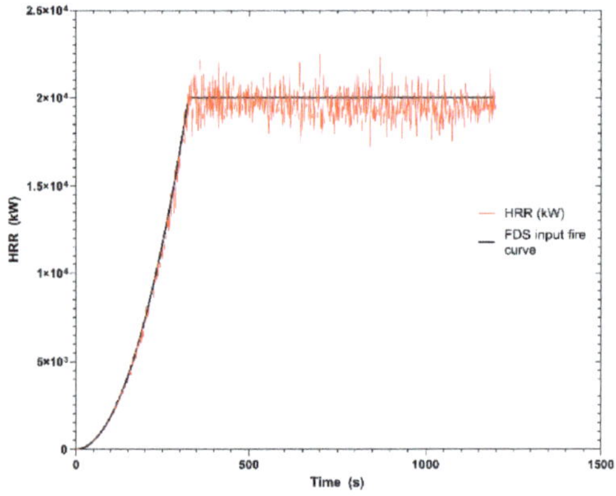

Figure 6. FDS input fire curve and FDS results fire curve.

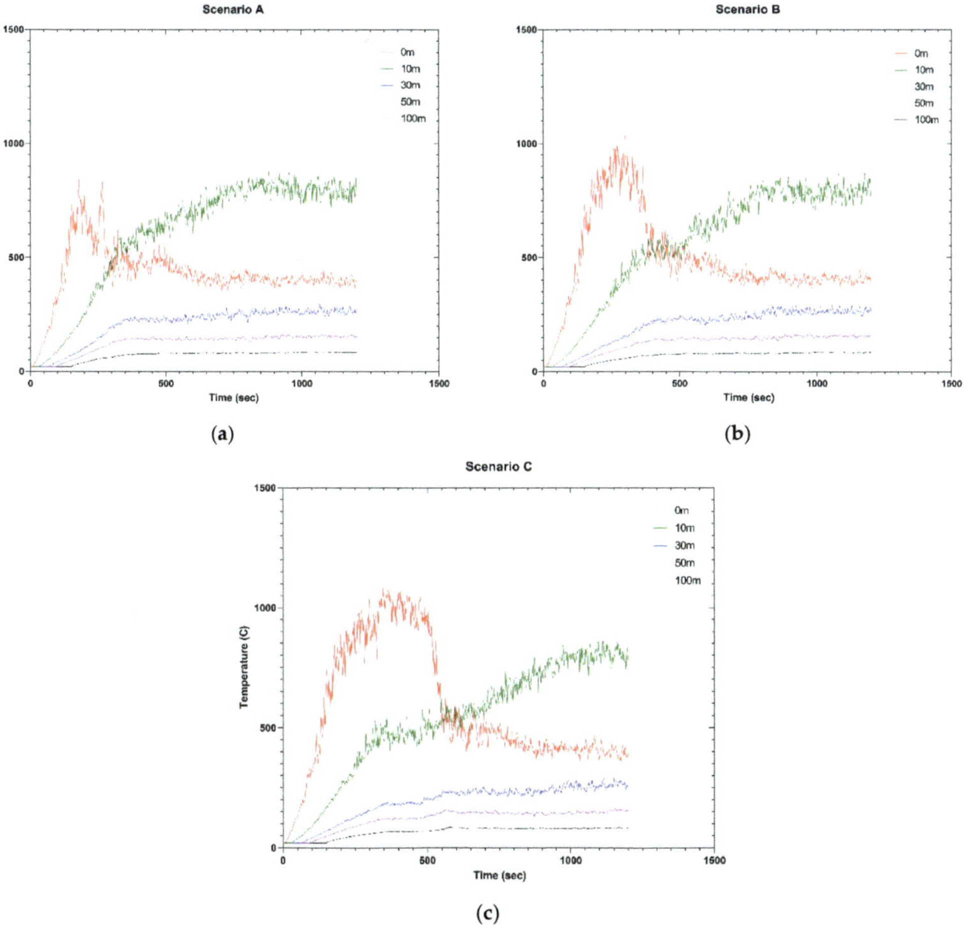

Scenario A

Scenario B

(**a**)

(**b**)

Scenario C

(**c**)

Figure 7. Tunnel ceiling (z = 7 m) gas temperature above (x = 0 m) and downwind (x = 10, 30, 50, and 100 m) from the fire source (**a**) for Scenario A, (**b**) Scenario B, and (**c**) Scenario C.

Figure 8. Tunnel ceiling (z = 7 m) temperatures for each scenario at the jet fan activation time frame: (**a**) Scenario A at +180 s, (**b**) Scenario B at +300 s, and (**c**) Scenario C at +480 s after the fire accident.

Figure 9. Smoke dispersion along the tunnel and backlayering at the jet fan activation time frame for (**a**) Scenario A at +180 s, (**b**) Scenario B at +300 s, and (**c**) Scenario C at +480 s.

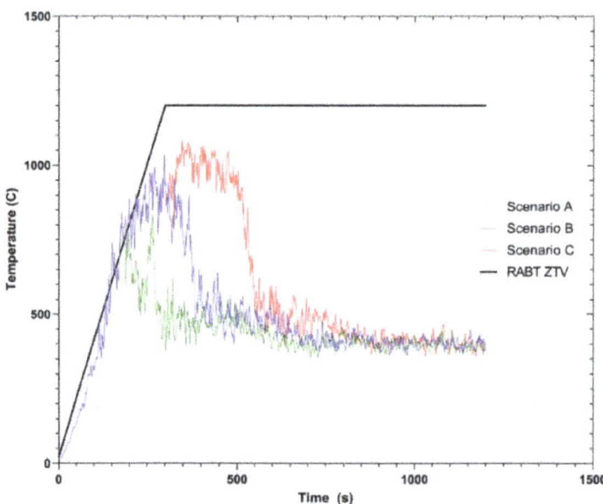

Figure 10. FDS ceiling temperatures above the fire source (x = 795 m, z = 7 m) for all three scenarios in comparison with the RABT ZTV fire curve.

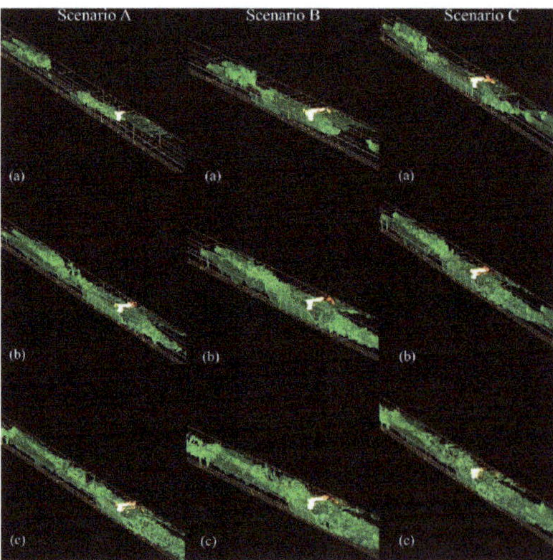

Figure 11. Development of the critical ventilation velocity of 2.7 m/s (isosurface with the green colour), at the train location, for (**a**) +90 s, (**b**) +180 s, and (**c**) +300 s after the activation of the jet fans. The jet fans reached their maximum output 90 s after their activation, while the development of the flow was reached approximately after 300 s.

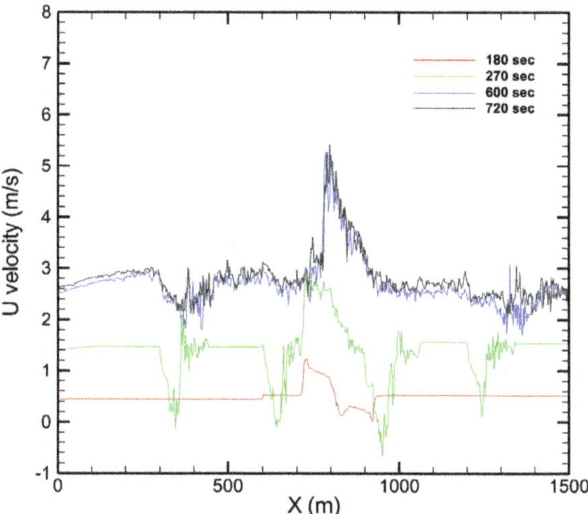

Figure 12. Lengthwise U velocity at symmetry plane y = 5.5 m and z = 3.5 m for Scenario A at different time frames (t = +180 s, +270 s, +600 s, and +720 s).

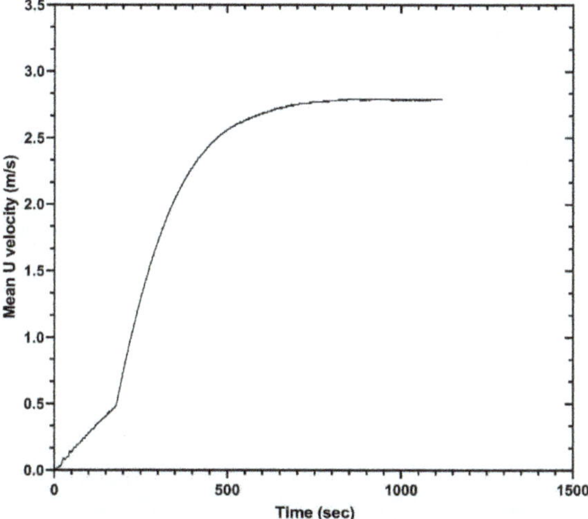

Figure 13. Lengthwise mean U velocity for Scenario A, upwind from the fire source.

4. Temperature and FED Index

Increased air temperatures may make it difficult for the people to move. This increase affects the train evacuation process. The critical temperature in a confined space, such as a tunnel, is the temperature where a person's movement and actions are impaired, and it is defined as 60 °C [43]. Figure 15 shows that, for every scenario, the temperature at the walking platform, at a height of 2 m, did not exceed 24 °C. In these conditions, the evacuation can be carried out normally.

The calculation of FED with Pathfinder used the equations described in the SFPE Handbook of Fire Protection Engineering, 5th Edition [44]. Only the concentrations of the narcotic gases CO, CO_2, and O_2 were used to calculate the FED value.

$$\text{FED} = \text{FED}_{\text{CO}} \cdot V_{\text{CO}_2} + \text{FED}_{\text{O}_2} \tag{16}$$

This calculation does not include the effect of hydrogen cyanide (HCN), and the effect of CO_2 is only due to hyperventilation. Carbon dioxide does not have toxic effects at concentrations of up to 5%, but it stimulates breathing, which increases the rate at which the other fire products are taken up.

$$\text{FED}_{CO} = 3.317 \cdot 10^{-5} \cdot CO^{1.036} \cdot V \cdot t / D \tag{17}$$

where CO is the concentration of carbon monoxide (ppm v/v 20 °C), V is the volume of air breathed each minute (L/min), with its value being 25 L/min for an activity level of light work (walking to escape), t is time in minutes, and D is the exposure dose (% COHb), with its value being 30% for an activity level of light work.

Hyperventilation due to carbon dioxide can increase the rate at which fire products are taken up. The multiplication factor is given by the equation

$$V_{CO_2} = exp(0.1903 \cdot \%CO_2 + 2.0004)/7.1 \tag{18}$$

where % CO_2 is the volume fraction of CO_2 (v/v). The fraction of an incapacitating dose of low O_2 hypoxia is calculated as:

$$\text{FED}_{O_2} = \frac{t}{exp\,[8.13 - 0.54 \cdot (20.9 - \%O_2)]} \tag{19}$$

where t is time in minutes, and % O_2 is the oxygen volume fraction (v/v). The measurement location for FED quantity sampling is 90% of occupant height. For example, FED data for an occupant with height = 1.8 m would be sampled 1.62 m above the occupant's location. When the occupant is not inside an FDS mesh, the FED calculation is paused until the occupant enters another mesh [31].

FED values never reached the critical value of 1, where a person is considered incapacitated, but some interesting results were extracted. The maximum values of FED are presented in Table 1 and compared in Figure 14. As shown, the longer the activation of the ventilation system was delayed, the more the uptake of combustion products and toxic gases increases. This is also shown in Table 2 and Figure 16, where the results for the mean fed values for each case are presented. It is important to be mentioned that, while maximum FED values for scenario A are about six times lower than the values for Scenario C, the mean FED values for Scenario A are an order of magnitude lower than Scenario C.

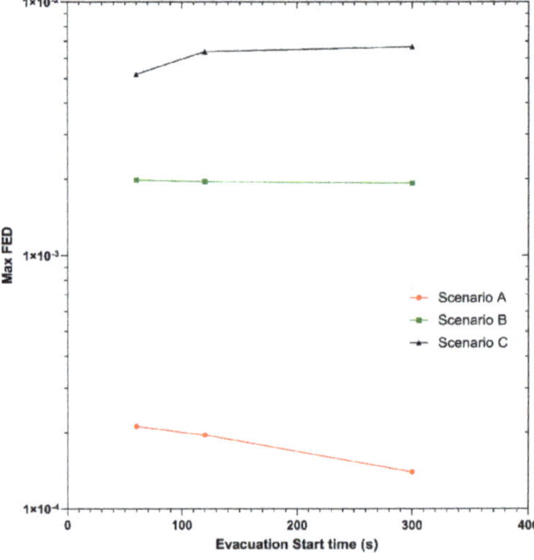

Figure 14. Max FED index for the three different scenarios and evacuation times.

Table 1. Max FED index for the three different scenarios and evacuation times.

	Max FED		
Evacuation Start Time (s)	**Scenario A**	**Scenario B**	**Scenario C**
60	2.12×10^{-4}	1.98×10^{-3}	5.20×10^{-3}
120	1.96×10^{-4}	1.96×10^{-3}	6.38×10^{-3}
300	1.40×10^{-4}	1.93×10^{-3}	6.67×10^{-3}

Table 2. Mean FED index for the three different scenarios and evacuation times.

	Mean FED		
Evacuation Start Time (s)	**Scenario A**	**Scenario B**	**Scenario C**
60	2.60×10^{-5}	4.14×10^{-4}	7.43×10^{-4}
120	2.60×10^{-5}	5.21×10^{-4}	1.37×10^{-3}
300	5.00×10^{-6}	4.23×10^{-4}	2.38×10^{-3}

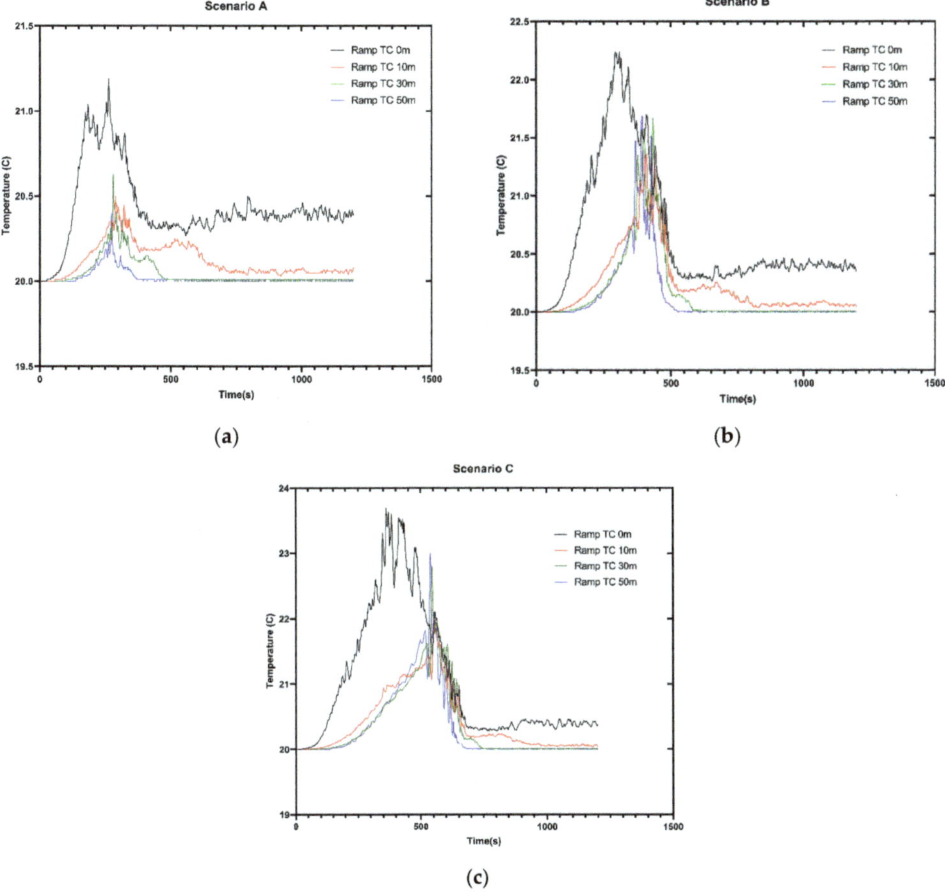

Figure 15. Temperature distribution upwind from the fire source at y = 0.5 m (middle of the walking platform) and at a height of 2 m for (**a**) Scenario A, (**b**) Scenario B, and (**c**) Scenario C. Temperatures do not exceed 24 °C.

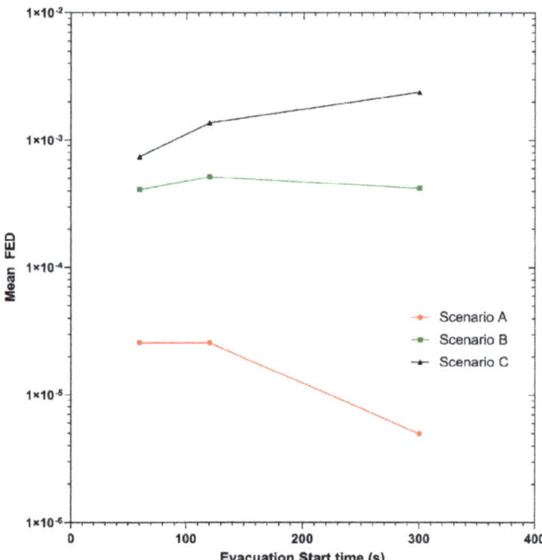

Figure 16. Mean FED index for the three different scenarios and evacuation times.

5. Discussion and Conclusions

Three different scenarios with different jet fan activation times, combined with three different evacuation scenarios, were studied. Firstly, validation of the temperature and critical velocity was conducted. The results showed a good agreement between the theoretical and simulation values, thereby the ventilation system was capable of providing sufficient air for smoke extraction.

The evacuation process showed that the passengers were not in danger, but important results regarding the importance of the jet fan activation time were extracted: From the FED values calculated for the nine cases in total, it was concluded that the most important action in a tunnel fire is the ventilation system activation time, after the start of a fire. If the ventilation system is delayed for any reason and the passengers have started evacuating the train, it is safer to wait for the population to travel some distance, possibly way past the train, and then the jet fans should be activated. If the ventilation system is activated in time, it seems safer to delay the initialization of the evacuation until the jet fans have reached their nominal output.

It is important to mention that there was no diversity in the population. Each simulated person had the same characteristics with each other as the movement speed and other factors that affect the evacuation process such as dimensions and comfort distance from each other. This affects results since, in real-life scenarios, the population is mixed (elderly people, people requiring assistance to move, people with baggage and families).

Author Contributions: Conceptualization, T.Z. and K.V.; methodology, T.Z., K.V. and I.S.; software, T.Z. and K.V.; validation, T.Z. and K.V.; formal analysis; T.Z.; resources, K.V. and I.S.; data curation, T.Z.; writing—original draft preparation, T.Z. and K.V.; writing—review and editing, K.V. and I.S.; visualization, T.Z.; supervision, K.V. All authors have read and agreed to the published version of the manuscript.

Funding: This research received no external funding.

Institutional Review Board Statement: Not applicable.

Informed Consent Statement: Not applicable.

Data Availability Statement: The data used in this study are available upon request from the corresponding author. The data can be easily reproduced from the theoretical analysis described in the study.

Acknowledgments: This work was supported by computational time granted from the National Infrastructures for Research and Technology S.A. (GRNET S.A.) in the National HPC facility—ARIS—under project ID pr011045-UrbanFirePlan2. The authors acknowledge and express particular appreciation to Thunderhead Engineering for providing licenses for Pyrosim and Pathfinder software, which was used for the designing, simulation scenarios and visualization of the results.

Conflicts of Interest: The authors declare no conflict of interest.

References

1. Ronchi, E. Evacuation Modelling in Road Tunnel Fires. Ph.D. Thesis, Polytechnic University of Bari, Bari, Italy, 2012.
2. Amundsen, F.H. Studies of driver behaviour in Norwegian road tunnels. *Tunn. Undergr. Space Technol.* **1994**, *9*, 9–15. [CrossRef]
3. Amundsen, F.H.; Ranes, G. Studies on traffic accidents in Norwegian road tunnels. *Tunn. Undergr. Space Technol.* **2000**, *15*, 3–11. [CrossRef]
4. Vasilopoulos, K.; Lekakis, I.; Sarris, I.E.; Tsoutsanis, P. Large eddy simulation of dispersion of hazardous materials released from a fire accident around a cubical building. *Environ. Sci. Pollut. Res.* **2021**, *28*, 50363–50377. [CrossRef] [PubMed]
5. Hong, W.-H. The progress and controlling situation of Daegu Subway fire disaster. *Fire Saf. Sci.* **2004**, *6*, s-5.
6. Zhang, T.; Zhang, Y.; Zhu, H.; Yan, Z. Experimental investigation and multi-level modeling of the effective thermal conductivity of hybrid micro-fiber reinforced cementitious composites at elevated temperatures. *Compos. Struct.* **2021**, *256*, 112988. [CrossRef]
7. Yan, Z.; Zhang, Y.; Guo, Q.; Zhu, H.; Shen, Y.; Guo, Q. Numerical study on the smoke control using point extraction strategy in a large cross-section tunnel in fire. *Tunn. Undergr. Space Technol.* **2018**, *82*, 455–467. [CrossRef]
8. Zhang, T.; Zhang, Y.; Xiao, Z.; Yang, Z.; Zhu, H.; Ju, J.W.; Yan, Z. Development of a novel bio-inspired cement-based composite material to improve the fire resistance of engineering structures. *Constr. Build. Mater.* **2019**, *225*, 99–111. [CrossRef]
9. Zhang, Y.; Zhu, H.; Guo, Q.; Carvel, R.; Yan, Z. The effect of technical installations on evacuation performance in urban road tunnel fires. *Tunn. Undergr. Space Technol.* **2021**, *107*, 103608. [CrossRef]
10. Buntelius, A.G.; Brickel, J.O.; Kuesel, T.R.; King, E.W. *Tunnel Engineering Handbook*; Chapman & Hall: Chapel Hill, NC, USA, 1996.
11. Zhang, Y.; Shen, Y.; Carvel, R.; Zhu, H.; Zhang, Y.; Yan, Z. Experimental investigation on the evacuation performance of pedestrians in a three-lane urban tunnel with natural ventilation in a fire scenario. *Tunn. Undergr. Space Technol.* **2021**, *108*, 103634. [CrossRef]
12. Purser, D. Assessment of Hazards to Occupants from Smoke, Toxic Gases, and Heat. In *SFPE Handbook of Fire Protection Engineering*; Springer: New York, NY, USA, 2008; pp. 2–96.
13. *NFPA 502*; Standard for Road Tunnels, Bridges and Other Limited Access Highways. NFPA: Quincy, MA, USA, 2011.
14. Carlson, E.-S.; Kumm, M.; Zakirov, A.; Dederichs, A. Evacuation tests with elevated platforms in railway tunnels. *Fire Saf. J.* **2019**, *108*, 102840. [CrossRef]
15. Ingason, H.; Kumm, M.; Nilsson, D.; Lönnermark, A.; Claesson, A.; Li, Y.Z.; Fridolf, K.; Åkerstedt, R.; Nyman, H.; Dittmer, T. *The METRO Project-Final Report*; Mälardalen University Press: Vasteras, Sweden, 2012.
16. Barbato, L.; Cascetta, F.; Musto, M.; Rotondo, G. Fire safety investigation for road tunnel ventilation systems—An overview. *Tunn. Undergr. Space Technol.* **2014**, *43*, 253–265. [CrossRef]
17. Brandt, R.; Haerter, H. On the Four Elements of Tunnel Safety: Fire, Air, Water and Earth. In Proceedings of the 7th International Conference Tunnel Safety and Ventilation, Graz, Austria, 12–13 May 2014; pp. 12–13.
18. Klote, J.H.; Ferreira, M.J.; Kashef, A.; Turnbull, P.G.; Milke, J.A. *Handbook of Smoke Control Engineering*; American Society of Heating Refrigerating and Air-Conditioning Engineers: Peachtree Corners, GA, USA, 2012.
19. Haukur, I.; Li, Y.Z.; Lönnermark, A. *Tunnel Fire Dynamics*; Springer: New York, NY, USA, 2015.
20. Gehandler, J. Road tunnel fire safety and risk: A review. *Fire Sci. Rev.* **2015**, *2*, 2–4. [CrossRef]
21. Carvel, R.; Beard, A. *The Handbook of Tunnel Fire Safety*; Thomas Telford: London, UK, 2005; 514p.
22. Hu, L.H.; Huo, R.; Chow, W.K. Studies on buoyancy-driven back-layering flow in tunnel fires. *Exp. Therm. Fluid Sci.* **2008**, *32*, 1468–1483. [CrossRef]
23. Oka, Y.; Atkinson, G.T. Control of smoke flow in tunnel fires. *Fire Saf. J.* **1995**, *25*, 305–322. [CrossRef]
24. Li, Y.Z.; Lei, B.; Ingason, H. Study of critical velocity and backlayering length in longitudinally ventilated tunnel fires. *Fire Saf. J.* **2010**, *45*, 361–370. [CrossRef]
25. Riess, I.; Bettelini, M. The Prediction of Smoke Propagation Due to Tunnel Fires. In Proceedings of the ITC Conference Tunnel Fires and Escape from Tunnels, Lyon, France, 5–7 May 1999; p. 999.
26. Wu, Y.; Bakar, M.Z.A. Control of smoke flow in tunnel fires using longitudinal ventilation systems—A study of the critical velocity. *Fire Saf. J.* **2000**, *35*, 363–390. [CrossRef]
27. Lee, S.R.; Ryou, H.S. An Experimental Study of the Effect of the Aspect Ratio on the Critical Velocity in Longitudinal Ventilation Tunnel Fires. *J. Fire Sci.* **2005**, *23*, 119–138. [CrossRef]

28. Lee, Y.-P.; Tsai, K.-C. Effect of vehicular blockage on critical ventilation velocity and tunnel fire behavior in longitudinally ventilated tunnels. *Fire Saf. J.* **2012**, *53*, 35–42. [CrossRef]

29. Gannouni, S.; Ben Maad, R. Numerical analysis of smoke dispersion against the wind in a tunnel fire. *J. Wind Eng. Ind. Aerodyn.* **2016**, *158*, 61–68. [CrossRef]

30. Tanaka, F.; Takezawa, K.; Hashimoto, Y.; Moinuddin, K.A.M. Critical velocity and backlayering distance in tunnel fires with longitudinal ventilation taking thermal properties of wall materials into consideration. *Tunn. Undergr. Space Technol.* **2018**, *75*, 36–42. [CrossRef]

31. McGrattan, K.; Hostikka, S.; McDermott, R.; Floyd, J.; Vanella, M. *Fire Dynamics Simulator User's Guide*; NIST Special Publication: Gaithersburg, MD, USA, 2019.

32. McGrattan, K.; Hostikka, S.; Floyd, J.; McDermott, R.; Vanella, M. *Fire Dynamics Simulator User's Guide*; NIST Special Publication: Gaithersburg, MD, USA, 2013.

33. U.S.NRC; EPRI. *Verification and Validation of Selected Fire Models for Nuclear Power Plant Applications: Fire Dynamics Simulator (FDS)*; U.S.NRC: Palo Alto, CA, USA, 2007.

34. Versteeg, H.K.; Malalasekera, W. *An Introduction to Computational Fluid Dynamics: The Finite Volume Method*; Pearson Education: New York, NY, USA, 2007.

35. Liu, C.; Zhong, M.; Tian, X.; Zhang, P.; Li, S. Study on emergency ventilation for train fire environment in metro interchange tunnel. *Build. Environ.* **2019**, *147*, 267–283. [CrossRef]

36. Król, A.; Król, M. Numerical investigation on fire accident and evacuation in a urban tunnel for different traffic conditions. *Tunn. Undergr. Space Technol.* **2021**, *109*, 103751. [CrossRef]

37. Cheng, C.H.; Chow, C.L.; Chow, W.K. A simulation study of tenability for passengers in a railway tunnel with arson fire. *Tunn. Undergr. Space Technol.* **2021**, *108*, 103679. [CrossRef]

38. Kurioka, H.; Oka, Y.; Satoh, H.; Sugawa, O. Fire properties in near field of square fire source with longitudinal ventilation in tunnels. *Fire Saf. J.* **2003**, *38*, 319–340. [CrossRef]

39. Hu, L.H.; Huo, R.; Peng, W.; Chow, W.K.; Yang, R.X. On the maximum smoke temperature under the ceiling in tunnel fires. *Tunn. Undergr. Space Technol.* **2006**, *21*, 650–655. [CrossRef]

40. Li, Y.; Lei, B.; Ingason, H. The maximum temperature of buoyancy-driven smoke flow beneath the ceiling in tunnel fires. *Fire Saf. J.* **2011**, *46*, 204–210. [CrossRef]

41. Li, Y.Z.; Ingason, H. The maximum ceiling gas temperature in a large tunnel fire. *Fire Saf. J.* **2012**, *48*, 38–48. [CrossRef]

42. Rosignuolo, F.; Souza, R.; Knaust, C.; Andreini, M. A Comparison Between Empirical Models and FDS Simulation to Predict the Ceiling Gas Temperature Distribution In a Tunnel Fire. In Proceedings of the World Tunnel Congress 2017—Surface Challenges—Underground Solutions, Bergen, Norway, 9–15 June 2017.

43. Shan-jun, M.O.; Li, Z.-R.; Liang, D.; Li, J.-X.; Zhou, N.-J. Analysis of Smoke Hazard in Train Compartment Fire Accidents Base on FDS. *Procedia Eng.* **2013**, *52*, 284–289. [CrossRef]

44. Hurley, M.J.; Gottuk, D.; Hall, J.R.; Harada, K.; Kuligowski, E.; Puchovsky, M.; Torero, J.; Watts, J.M.; Wieczorek, C. *SFPE Handbook of Fire Protection Engineering*, 5th ed.; Springer: New York, NY, USA, 2016; pp. 1–3493.

applied sciences

MDPI

Article

Air Flow Study Around Isolated Cubical Building in the City of Athens under Various Climate Conditions

Chariton L. Pavlidis, Anargyros V. Palampigik, Konstantinos Vasilopoulos, Ioannis C. Lekakis and Ioannis E. Sarris *

Laboratory of Thermo Fluid Systems (LTFS), Department of Mechanical Engineering, Ancient Olive Grove Campus, University of West Attica, Thivon Str. 250, Egaleo, 12244 Athens, Greece; cpavlidis@uniwa.gr (C.L.P.); msrtf21X03@uniwa.gr (A.V.P.); kvassil@uniwa.gr (K.V.); lekakis@uniwa.gr (I.C.L.)
* Correspondence: sarris@uniwa.gr; Tel.: +30-2105381131

Abstract: This study focuses on the airflow and pollutant dispersion around an isolated cubical building located in a warm Mediterranean climate, taking into account the local microclimate conditions (of airflow, albedo of building and soil, and air humidity) using a large-eddy simulation (LES) numerical approach. To test the reliability of computations, comparisons are made against the SILSOE cube experimental data. Three different scenarios are examined: (a) Scenario A with adiabatic walls, (b) Scenario B with the same constant temperature on all the surfaces of the building, and (c) Scenario C using convective and radiative conditions imposed by the local microclimate. For the first two cases the velocity and temperature fields resulting are almost identical. In the third case, the resulting temperature on the surfaces of the building is increased by 19.5%, the center (eye) of the wake zone is raised from the ground and the maximum pollutant concentration is drastically reduced (89%).

Keywords: LES; microclimate model; thermal radiation; pollutant dispersion; urban planning

Citation: Pavlidis, C.L.; Palampigik, A.V.; Vasilopoulos, K.; Lekakis, I.C.; Sarris, I.E. Air Flow Study Around Isolated Cubical Building in the City of Athens under Various Climate Conditions. *Appl. Sci.* **2022**, *12*, 3410. https://doi.org/10.3390/app12073410

Academic Editor: Rüdiger Schwarze

Received: 18 January 2022
Accepted: 21 March 2022
Published: 27 March 2022

Publisher's Note: MDPI stays neutral with regard to jurisdictional claims in published maps and institutional affiliations.

Copyright: © 2022 by the authors. Licensee MDPI, Basel, Switzerland. This article is an open access article distributed under the terms and conditions of the Creative Commons Attribution (CC BY) license (https://creativecommons.org/licenses/by/4.0/).

1. Introduction

The excessive population concentration in megacities has resulted in environmental pollution problems and overcrowded living conditions. The air quality in these cities is determined by the airflow in the complex urban terrain, the temporal and spatial conditions of pollutant sources, and the local urban microclimate parameters. The airflow and pollutant dispersion in an urban environment can be studied in different geometrical scales, such as blocks of buildings, street canyons, and isolated buildings. Several wind-tunnel, real-scale experiments and numerical studies examine the airflow and pollutant dispersion around blocks of buildings [1–5] and street canyons [6–12].

The city's local microclimate parameters are air and surface temperatures, humidity, direct and diffuse solar irradiation, and wind speed at different directions [13]. The knowledge of the urban microclimate is necessary to prevent and reduce pollution problems. The local microclimate conditions are playing an important role in the airflow and pollution distribution in an urban environment [14]. For this reason, complex models can be used to calculate mandatory variables such as the heat capacity of a building, the shortwave and longwave radiation according to the solar position, the emissivity, and the air humidity. According to these variables, the increment or decrement of the thermal comfort rate can be found to define the human comfort conditions. Another important issue is the values of the heating rate and of the thermal absorption of the surfaces of the building. Using complicated microclimate models, the distribution of the environmental temperature and their effect on pedestrians can be defined [15–19]. Small-scale meteorological models are widely used to define the comfort conditions inside the urban environment for winter and summer seasons [20–23].

A phenomenon by which warm layers of air are accumulated in densely populated centers is defined as an Urban Heat Island (UHI) creating uncomfortable thermal conditions for urban living [24]. A high air-temperature increment leads to the need for higher energy consumption for cooling in the summer, with a consequence of increasing air pollutants in the atmosphere [25].

It is also identified that the rapid augmentation of the air temperature depends on the airflow velocity. At calm wind conditions, where the airflow velocities are lower than 3 m/s, the density of hot gas masses is much higher than at heavy wind gusts [26]. As a result, wind gusts with high airflow velocities can remove the air hot masses from the urban environments, which leads to the decrement of the UHI phenomenon [27].

The effect of the microclimate in densely populated urban centers is also an important topic for study. Gronemeier and Raasch [28] studied the urban flow for the city of Hong Kong with PALM software. The numerical results showed a rapid increment in wind velocity at the pedestrian crossing areas and high changes of the airflow direction [28,29]. Another study by Pfafferott and Rißmann [30] with the PALM code examined the interaction between buildings and the urban microclimate using an energy balance solver. They found that the temperature increment of the urban environment was caused by the heated building, as the outcome of energy from the indoor model [30]. Resler and Eben [31] examined the airflow over the city of Prague in the Czech Republic with PALM code. Their study showed that the temperature and the concentration of NOx are in a good agreement with real field measurements during the summer season. In the winter season, the deviations between numerical results and experimental data pointed to inaccuracies in modeling the atmospheric boundary layer [31]. Thus, real field urban measurements can give important information for both real-scale meteorological phenomena and their impact on an urban boundary layer. To better understand the complex phenomena of the airflow in an urban environment, it is important to combine real field measurements and wind tunnel experiments so that these two methods can be interrelated [32,33]. Numerical methods can be also used to simulate the turbulence modelling of the airflow around both small- and large-scale cubical obstacles [34–38].

Several wind-tunnel experiments exist that describe the airflow characteristics around a cubical building. The wind-tunnel experiments are suitable for collecting detailed spatial experimental data [32,33,39–41] under stable conditions. However, experimental wind tunnel experiments are not efficient for the prediction of real-scale meteorological conditions.

The present study is focusing on the simulation of the flow and pollution distribution around an isolated building [42] and, more importantly, on the influence of a meteorological model on these results, i.e., what differences exist between the use of real boundary conditions and those employed for reasons of convenience. In addition, the use of an isolated building instead of a large urban area ensures good reliability of the numerical results for the available resources. In this respect, such studies do not seem to exist in the scientific literature.

The isolated building of this study is in the climatological area of Athens, Greece. Three different scenarios are examined. Scenario A studies the neutral flow and pollutant dispersion in the wake area of an isolated cubical building with adiabatic walls, without any buoyancy forces. Scenario B studies the same building with specified constant temperature of 50 °C on its surfaces. Scenario C studies the same building with convective and radiative conditions imposed by the local microclimate.

2. Configuration and Pollutant Dispersion Modelling

2.1. Problem Configuration

Figure 1 presents the geometry of the computational domain around the isolated cubical building of this study, with exactly the same dimensions as the SILSOE cube. The SILSOE cube field experiment for the study of the airflow around it [43], with a 6 m height cube placed in an open area, has been used as a test case in several other studies [44]. This geometry remains unchanged in all the different simulated scenarios of this work.

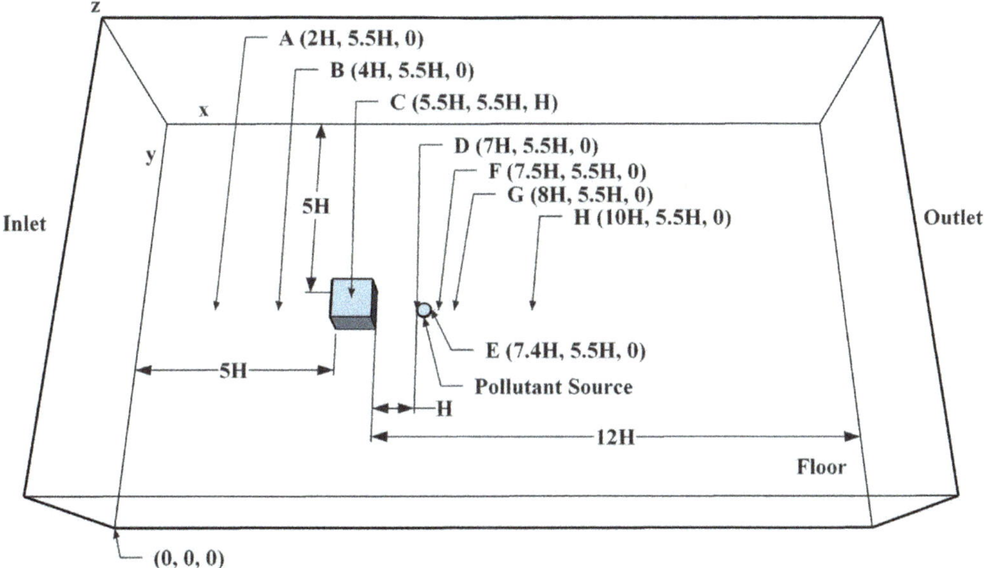

Figure 1. Computational domain and boundary conditions.

The global coordinate system origin is the frontal left edge of the computational domain and the distance between the frontal area of the computational domain and the building is 5 H. The distance between the lateral boundaries of the computational domain and the building's surface is also 5 H [45]. The distance from the rear surface of the building up to the outlet boundary of the computational domain is set at 12 H and the total height of the computational domain is 5 H. The resulting blockage effect value is approximately 1.81%. Under the recommendations of the German Association of Engineers (VDI), the blockage effect should be maintained below 10% during the simulations [46] and according to [46,47] should be even lower than 3%, which is satisfied in the present study.

The Reynolds number is kept constant at 4.03×10^6, based on the height of the cube and the free stream velocity. The specified flow field at the inlet of the computational domain is in the form of a logarithmic profile with the no-slip condition at the ground. A passive source of pollutants is located on the floor of the computational domain at a distance H from the rear surface of the building, using as a pollutant methane (CH_4) gas with a concentration and release of 600 ppm and 18.5 L/h, respectively. The dimensionless concentration coefficient K of the passive pollutant is defined as [48,49]:

$$K = \frac{(c_{Measured}/c_{Source})u_H H^2}{Q_{Source}} \tag{1}$$

where $c_{Measured}$ is the tracer concentration at any position, c_{Source} is the tracer concentration on the pollutant source, Q_{Source} is the release rate of the pollutant, and u_H is the velocity of the flow at the height of the building.

In Scenarios B and C, the temperature is affecting the flow field through the usual Boussinesq approximation. In these cases, the Grashof number, i.e., the ratio of thermal buoyancy forces over viscous forces is defined as [50]:

$$Gr_H = \frac{g\beta\left(T_{Surface} - T_\infty\right)H^3}{v^2} \tag{2}$$

where $T_{Surface}$ is the temperature on the surfaces of the building, T_∞ is the bulk temperature, β is the coefficient of thermal expansion, and v is the kinematic viscosity of the fluid. In Scenarios B and C, the Grashof number based on the height of the building is 2.43×10^{11} and 7.68×10^{11}, respectively. Thus, the buoyancy forces are intensive and the boundary layers close to the heated surfaces are highly turbulent.

2.2. Governing Equations

The PALM model system, an open-source software that simulates atmospheric and oceanic boundary layers, is used for the present simulations. The Navier-Stokes equations in a non-hydrostatic, filtered, and incompressible form under the Boussinesq approximation are solved. The basic equations for the conservation of mass, momentum, energy, and moisture are defined as:

$$\frac{\partial u_i}{\partial x_j} = 0 \tag{3}$$

$$\frac{\partial u_i}{\partial t} = -\frac{\partial u_i u_j}{\partial x_j} - \varepsilon_{ijk} f_j u_k + \varepsilon_{i3j} f_3 u_{g,j} - \frac{1}{\rho_0}\frac{\partial \pi^*}{\partial x_i} + g\frac{\theta_v - \langle\theta_v\rangle}{\langle\theta_v\rangle}\delta_{i3} - \frac{\partial}{\partial x_j}\left(\overline{u_i'' u_j''} - \frac{2}{3}e\delta_{ij}\right) \tag{4}$$

$$\frac{\partial \theta}{\partial t} = -\frac{\partial u_j \theta}{\partial x_j} - \frac{\partial}{\partial x_j}\left(\overline{u_j'' \theta''}\right) - \frac{L_v}{C_p \Pi}\Psi_{q_v} \tag{5}$$

$$\frac{\partial q_v}{\partial t} = -\frac{\partial u_j q_v}{\partial x_j} - \frac{\partial}{\partial x_j}\left(\overline{u_j'' q_v''}\right) + \Psi_{q_v} \tag{6}$$

$$\frac{\partial s}{\partial t} = -\frac{\partial u_j s}{\partial x_j} - \frac{\partial}{\partial x_j}\left(\overline{u_j'' s''}\right) + \Psi_s \tag{7}$$

where i, j, $k \in \{1, 2, 3\}$, the components of the velocity ($u_1 = u$, $u_2 = v$ and $u_3 = w$) are defined by the u_i variable at specific positions in the flow field x_i, where $x_1 = x$, $x_2 = y$ and $x_3 = z$, and t is time. q_v and s are the latent heat transfer and moisture, respectively. The gravitational acceleration is denoted by g, the density of the dry air is defined by ρ_0, and L_v is the latent heat of vaporization. The source term of the variable q_v, is defined as Ψ_{q_v} and the sink term of the variable s is defined as Ψ_s. The angle brackets correspond to the horizontal averages of the flow field and the double prime indicated the subgrid-scale variables.

The subgrid-scale turbulence kinetic energy is defined as:

$$e = \frac{1}{2}\left(\overline{u_i'' u_i''}\right) \tag{8}$$

The modified perturbation pressure based on the perturbation pressure p^* is defined as:

$$\pi^* = p^* + \frac{2}{3}\rho_0 e \tag{9}$$

The potential temperature is defined from the equation:

$$\theta = \frac{T}{\Pi} \tag{10}$$

where T is the instantaneous absolute temperature and the Exner function is calculated by the equation:

$$\Pi = \left(\frac{p}{p_0}\right)^{\frac{R_d}{C_p}} \tag{11}$$

where p is the hydrostatic pressure of the air, $p_0 = 1000$ hPa is the reference pressure, R_d is the specific gas constant for dry air, and C_p is the specific heat capacity under constant pressure.

The virtual potential temperature is defined by the equation:

$$\theta_v = \theta\left[1 + q_v\left(\frac{R_v}{R_d} - 1\right) - q_l\right] \tag{12}$$

where R_v is the specific gas constant for water vapors, and q_l is the liquid water mixing ratio.

2.3. Urban Surface Model (USM) of the PALM Model System

The urban surface model (USM) of the PALM model system is used for the energy balance in the computational domain. The energy balance solver is driven by three main procedures: (a) the solver predicts the temperature of the outer surfaces, (b) the turbulence dispersion of the sensible temperature is calculated at the surfaces near the walls, and (c) the calculation of the sublayer heat flux caused by convection. The first two procedures of the energy balance solver are executed simultaneously. The third procedure is executed under the calculations of a subgrid-scale model that predicts the thermal diffusion from the bluff body.

The energy balance solver is defined as:

$$C_0\frac{dT_0}{dt} = R_n - H - LE - G \tag{13}$$

where C_0 is the specific heat capacity, T_0 is the radiative temperature of the surface skin layer, R_n is the net radiation, H is the sensible heat, LE is the latent heat, and G is the heat flux at the surface of the ground. On the USM, the energy balance is calculated separately for each surface, and the three different types of radiations (sensible heat, latent heat, and ground heat flux) from the surface heat are combined.

The calculation of the convective heat transfer between the air and the outer surfaces is defined by the equation:

$$H = h(\theta_1 - \theta_0) \tag{14}$$

where θ_0 is the temperature of the outer surface of the building, θ_1 is the temperature of the air mass in contact with the outer surface of the building, and h is the coefficient of convective heat transfer and is parameterized only for the vertical surfaces inside the computational domain [51]. For the horizontal surfaces inside the computational domain, the coefficient of convective heat transfer is parameterized under the Monin-Obukhov similarity, which involves the calculation of the local friction velocity [52,53]. The friction velocity is used for the calculation of flow momentum near the surface, for each surface separately.

3. Initial and Boundary Conditions

The inlet velocity profile is described by the logarithmic law of the wall with a free-stream velocity value of 10.13 m/s. Figure 2 presents the normalized velocity profiles at both the inlet of the computational domain and in the X/H = 2 position from the inlet for Scenarios A–C, where A and B should be identical.

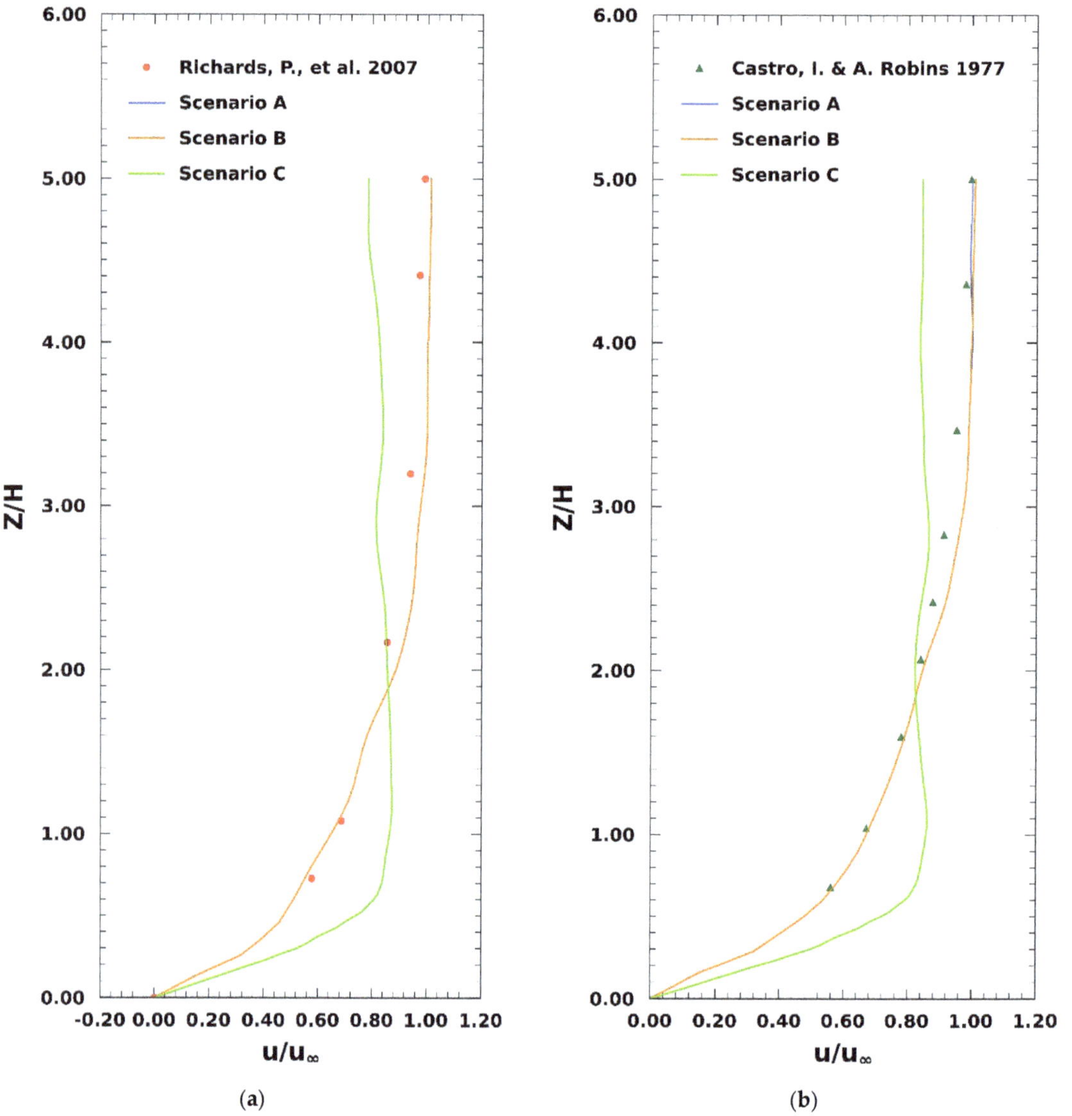

Figure 2. Normalized velocity profiles at the (**a**) inlet of the computational domain and (**b**) X/H = 2 position from the inlet for Scenarios A−C [44,54].

As observed in Figure 2a, an important deviation appears in Scenario C that corresponds to the application of the meteorological model in comparison with the other two scenarios. This difference appears also in the free-stream flow region and is caused by the strong buoyancy effects caused by the increased thermal heating of the air.

As shown in Figure 3, the present numerical results for the pressure coefficient are in good agreement with the corresponding experimental data of Scenarios A and B.

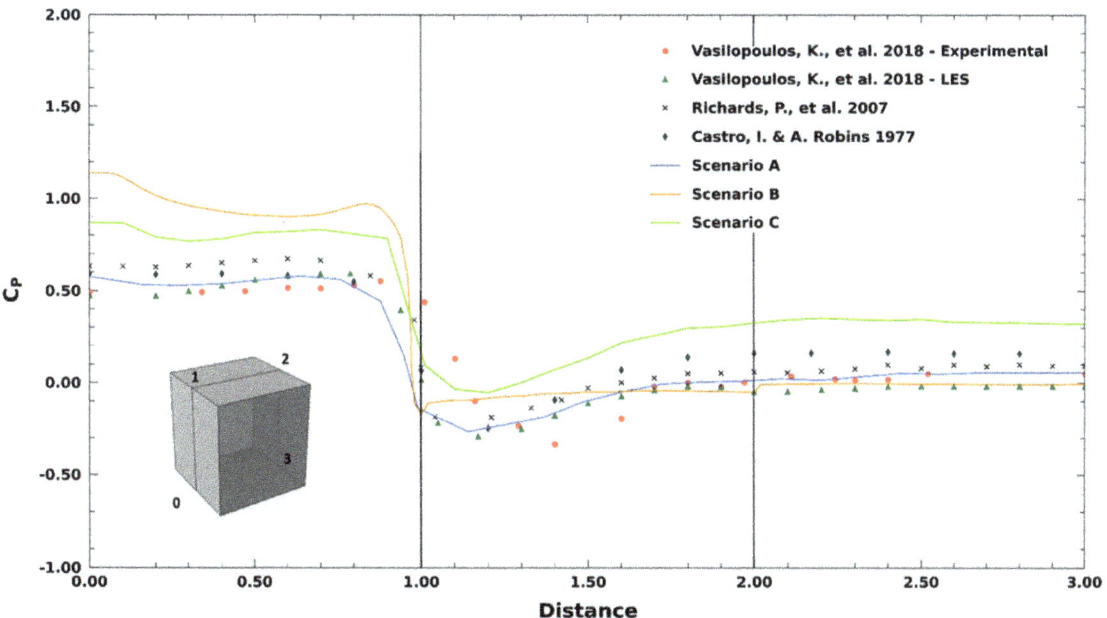

Figure 3. Pressure coefficient distribution at the building's symmetry plane for Scenarios A–C [4,44,54].

The pressure coefficient at the surfaces of the building for all scenarios is defined as:

$$C_P = \frac{p - p_{REF}}{\frac{1}{2}\rho_{REF}u_\infty^2} \tag{15}$$

where p is the static pressure of the fluid at a point, p_{REF} and ρ_{REF} are the static pressure and density of the fluid at the free-stream at the inlet of the computational domain, respectively, and u_∞ is the free-stream velocity at the inlet.

As observed in Figure 3, important differences exist between Scenarios A–C because of the heating of the surfaces of the building and the microclimate model application. The pressure coefficient distribution on the frontal surface of the building for Scenarios B and C is higher than the corresponding for Scenario A because of the influence of the mixed convection heat transfer, which causes reduction on the density of the streamlines in this region. In Scenario C, this influence is expected to be larger than for Scenario B. In Scenario B, the pressure distribution at the rear surface of the building significantly approaches the corresponding for Scenario A. In Scenario C, the pressure distribution at the rear surface of the building is the highest compared with other scenarios because of the domination of the buoyant heating forces there.

Figure 4 presents the vertical profile of the turbulent kinetic energy at the inlet of the computational domain for Scenarios A–C, as compared with the corresponding numerical data on the same normalized upstream position from the cube [18].

As described in Figure 4, the turbulent kinetic energy distribution at the inlet of the computational domain is almost identical for Scenarios A and B and different from Scenario C because of the effect of the buoyancy forces from the microclimate conditions.

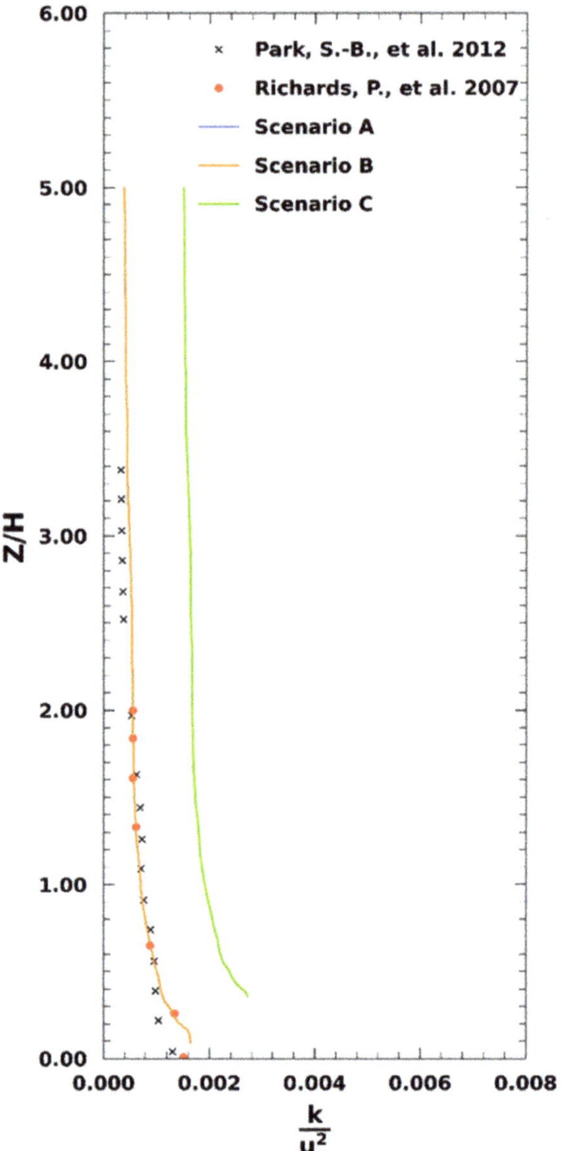

$$\frac{k}{u_\infty^2}$$

Figure 4. Turbulent kinetic energy at the inlet of the computational domain for Scenarios A–C [18,44].

On the lateral boundaries of the computational domain, periodic boundary conditions are employed because of the temporal and spatial periodicity of the flow. Neumann boundary conditions are used for the turbulent kinetic energy (e), temperature (θ), and the perturbation pressure (p^*) to calculate the following equations concerning the building's height:

$$\bar{e}\left(-\frac{\Delta y}{2}\right) = \bar{e}\left(+\frac{\Delta y}{2}\right) \tag{16}$$

$$\overline{\theta}\left(-\frac{\Delta y}{2}\right) = \overline{\theta}\left(+\frac{\Delta y}{2}\right) \tag{17}$$

$$\overline{p^*}\left(-\frac{\Delta y}{2}\right) = \overline{p^*}\left(+\frac{\Delta y}{2}\right) \tag{18}$$

The Dirichlet boundary condition is used for the velocity at the ground of the computational domain for the no-slip condition. Therefore, the velocity components at the ground are:

$$\overline{u}(z = 0) = 0 \tag{19}$$

$$\overline{v}(z = 0) = 0 \tag{20}$$

$$\overline{w}(z = 0) = 0 \tag{21}$$

On the staggered computational grid, the velocity components in the X and Y directions, respectively, are defined at the specific height of $z = \pm\frac{\Delta z}{2}$. Thus, the symmetry boundary condition is used as:

$$\overline{u}\left(-\frac{\Delta z}{2}\right) = -\overline{u}\left(+\frac{\Delta z}{2}\right) \tag{22}$$

$$\overline{v}\left(-\frac{\Delta z}{2}\right) = -\overline{v}\left(+\frac{\Delta z}{2}\right) \tag{23}$$

On the upper surface of the computational domain, the perturbation pressure (p^*) obeys the Neumann boundary condition. For the velocity to maintain the free-stream regime over the atmospheric boundary layer, Neumann boundary conditions are also applied to the velocity components of the velocity in the X and Y directions:

$$\partial_z \overline{u}\big|_{top} = constant \tag{24}$$

$$\partial_z \overline{v}\big|_{top} = constant \tag{25}$$

Additionally, on the upper surface of the computational domain, Dirichlet boundary conditions are used for both the turbulence kinetic energy (e) and the temperature (θ).

4. Nested Computational Grid

PALM provides the capability of computational grid self-nesting. In the self-nesting mode, there is a root/parent computational domain with the ability of nesting up to 63 levels of nested/child computational subdomains. Each subdomain may be root/parent to another nested subdomain. As a result, the second subdomain is simultaneously nested referring to the first parent, and also root, to the third nested. The nested computational subdomain receives all the appropriate and mandatory information for the three components of the velocity and all the prognostic variables of the vectors from its boundaries with the root computational domain. The flow field data are interpolated from the coarse to the finer computational grid. By the end of each time step, the corrected solution from the solver is reversely interpolated to the root/parent domain.

Figure 5 shows the grid arrangement used in the simulations. The root computational grid (coarse grid) is shown at full height of the computational domain in contrast with the nested (finer grid-darker parts), which is shown at two building heights.

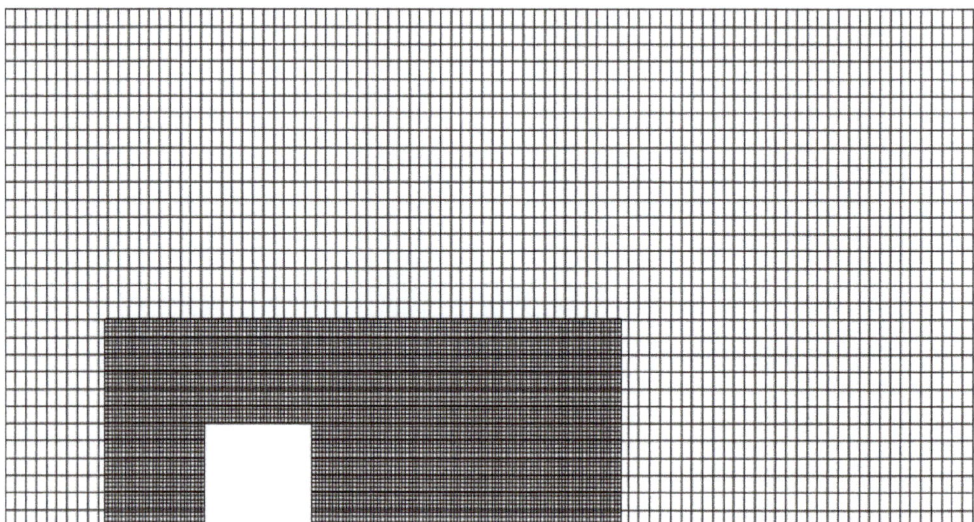

Figure 5. Root and nested computational domains of the flow field.

For the simulations, three different computational grids are used with an incremental resolution from the coarse to the finer: the coarse computational grid with 924,400 cells, the medium grid with 1,248,400 cells, and the fine grid with 1,842,400 cells. The grid errors using two computational grids are estimated from the following equation [4]:

$$GCI = \frac{f_2 - f_1}{1 - r^p} \tag{26}$$

where f_2 is the numerical solution that results from the medium computational grid and the corresponding from the finer is f_1. r is the refinement factor between the two computational grids and p is the accuracy of the algorithm, which is 3 for the present study. The refinement factor between the medium and finer computational grid is approximately 1.47.

Figure 6a presents the profile of the error bars of the normalized velocity, at the axial position 0.5 H from the rear surface of the building for scenario A. Additionally, Figure 6b shows the profile of the normalized concentration of the pollutant at the same axial position and scenario.

The mean value of the errors of the normalized velocity between the coarse and the medium computational grids is approximately 3.77% and the corresponding value between the medium and the fine grids is about 1.54%. The mean value of the errors of the pollutant between the coarse and the medium computational grids is approximately 2.77% and the corresponding value between the medium and the fine grids is about 1.24%. For both of the aforementioned variables, the decrement of the errors going from the coarse to the finer computational grids is clearly observed.

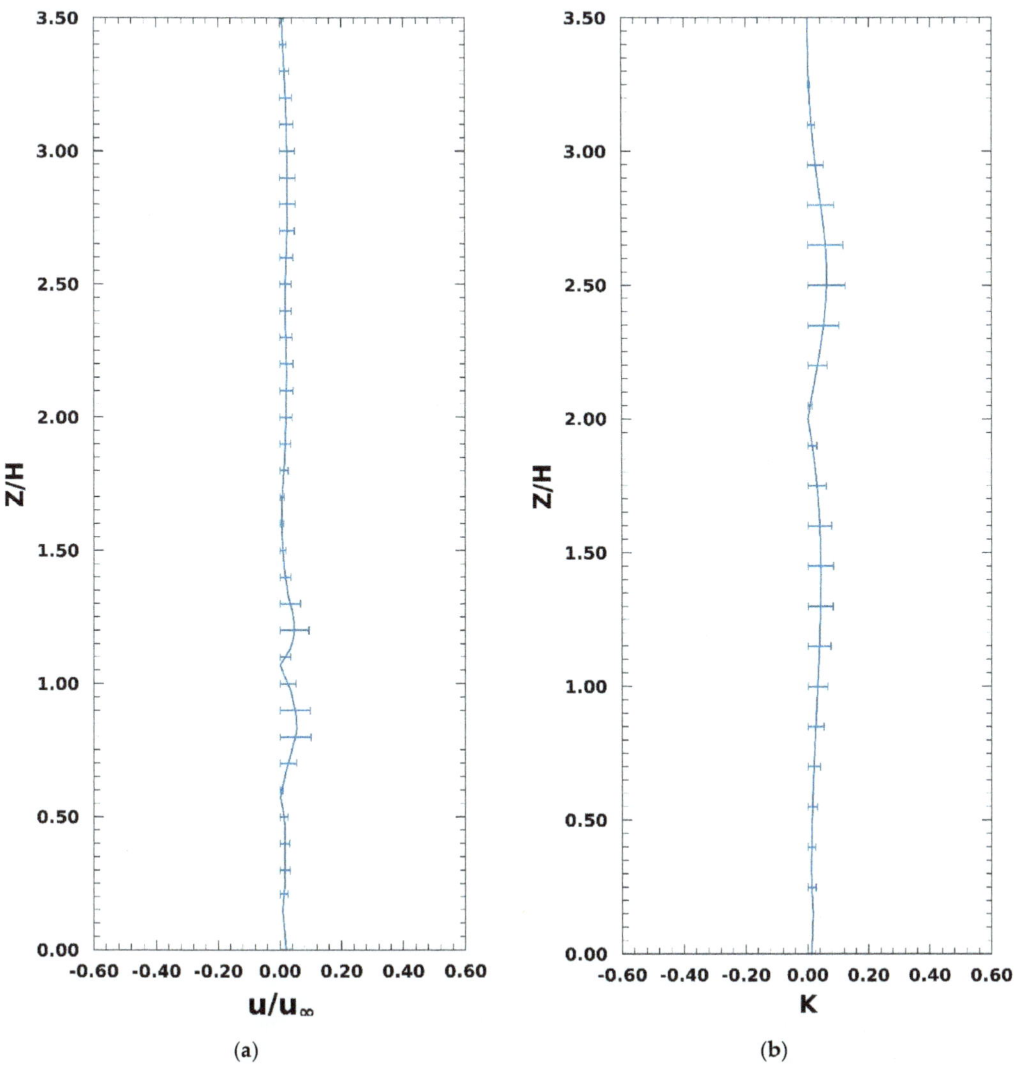

Figure 6. GCI error bars estimated from the fine to medium grids for Scenario A: (**a**) for the normalized velocities u_x/u_∞ and (**b**) for the normalized concentration of the pollutant at the axial position of 0.5 H from the rear surface of the building.

5. Numerical Details

The global discretization of the computational grid is accomplished using finite differences on a staggered Cartesian Arakawa-C grid [55]. The equations are spatially discretized on a fifth-order differential upwind scheme [56], while a third-order Runge-Kutta scheme is used on the temporal discretization [57]. The equations are implicitly filtered by the discretization of the computational grid and the subgrid-scale processes are calculated by the 1.5-order Deardorff numerical scheme [58]. Thus, it is assumed that the energy that is transported by the subgrid-scale vortices is proportional to the local gradients of the mean quantities [59,60].

The convergence criterion maintained on each simulation is kept below the 10^{-4} value for each computed variable based on the error. The time step is automatically adjusted for the CFL constant value of 0.9.

Every simulation is carried out until flow stationarity has been achieved. Figure 7 presents the normalized velocity components with respect to the free-stream velocity $(u_x/u_\infty, u_y/u_\infty, u_z/u_\infty)$, at the point with Cartesian coordinates of X: 10 H, Y: 5.5 H, and Z: 0.5 H and for the period from 600 to 1400 s for Scenario A.

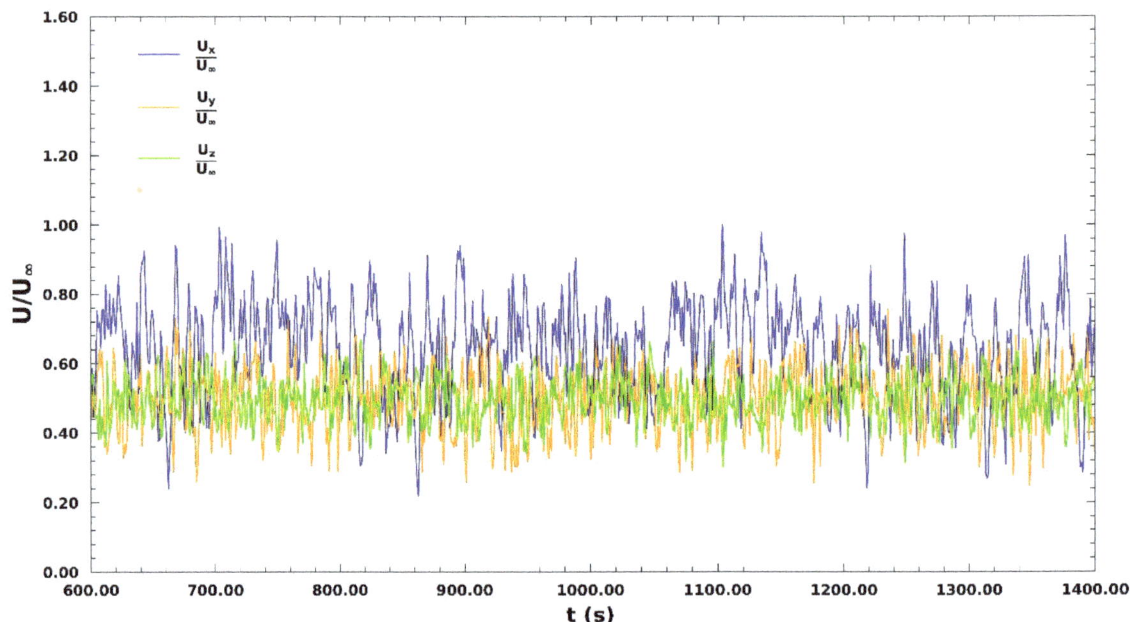

Figure 7. Velocity fluctuations of u_x/u_∞, u_y/u_∞, u_z/u_∞ at point X: 10 H, Y: 5.5 H, Z: 0.5 H from 600 up to 1400 (s) for Scenario A.

As it is shown in Figure 7, the flow seems to have attained stationarity conditions from the time point of 700 (s) onward. The attainment of stationarity conditions is confirmed by finding out that the mean value and higher order statistics are independent of the time of initiation of the measurements. If this is valid for the mean value and the autocorrelation function, the process is said to be weakly stationary. In this respect, the mean, the variance, and the autocovariance with a time delay of 2 (s) values are computed and given in tabulated form below (Table 1):

Table 1. Mean Value, Variance and Autocovariance for Specific Time Periods.

Averaging Period (s)	Mean Value	Variance	Autocovariance
600–800	1.0806	0.0473	0.3426
800–1000	1.0851	0.0538	0.3738
1000–1200	1.0879	0.0530	0.4316
1200–1400	1.0889	0.0524	0.4443

These computed values show that the flow field is at least weakly stationary.

However, although the flow field has reached stationary conditions it is possible the concentration field has not reached it yet.

Figure 8 presents the pollutant dispersion inside the computational domain for Scenarios A and C at the position with coordinates of X: 7.5 H, Y: 5.5 H, and Z: 0.5 H for the period of 200–1000 (s).

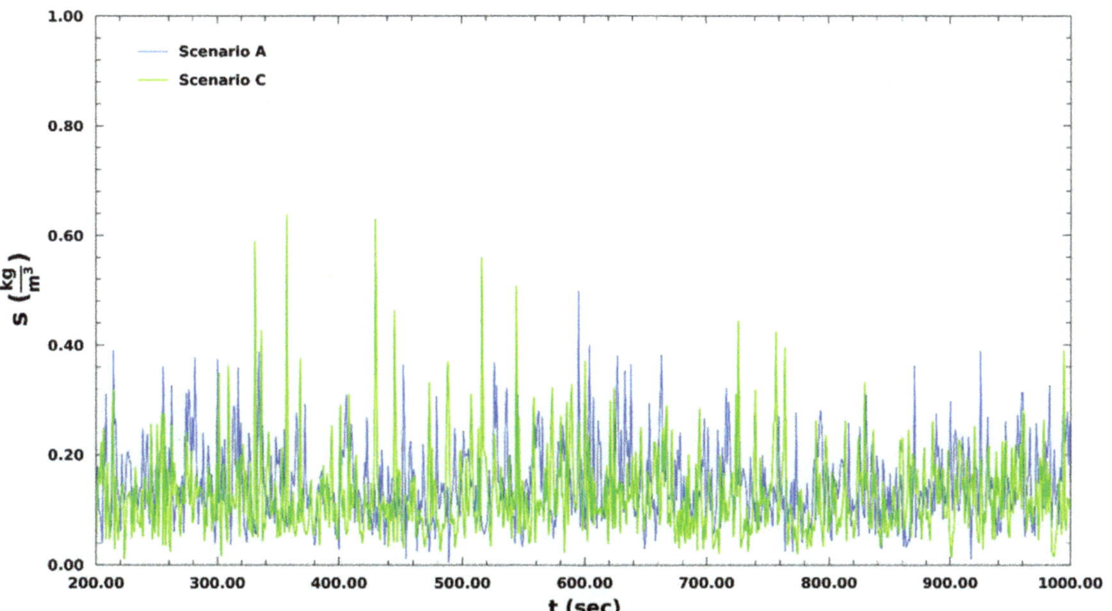

Figure 8. Concentration of the pollutant for Scenarios A and C at the position with coordinates of X: 7.5 H, Y: 5.5 H and Z: 0.5 H for the time period of 200 to 1000 (s).

The mean, the variance, and the autocovariance with time delay of 2 (s) values of the pollutant concentration are computed and given in tabulated form below (Table 2), which demonstrates that a passive scalar attains stationarity earlier in time than the velocity field:

Table 2. Mean Value, Variance and Autocovariance for Specific Time Periods.

Averaging Period (s)	Mean Value	Variance	Autocovariance
200–400	0.2449	0.0186	0.9584
400–600	0.2445	0.0198	0.9532
600–800	0.2480	0.0193	0.9548
800–1000	0.2463	0.0195	0.9524

6. Results and Discussion

The air flow field around the cubical geometry is presented in Figure 9 using normalized velocity profiles on the symmetry plane at different streamwise positions upstream, downstream, and at the middle of the roof of the cube (X/H = 5.5) for all three scenarios [44,54,61–65]. Comparisons are also made with Richards and Castro's experimental data [60] for Scenario A.

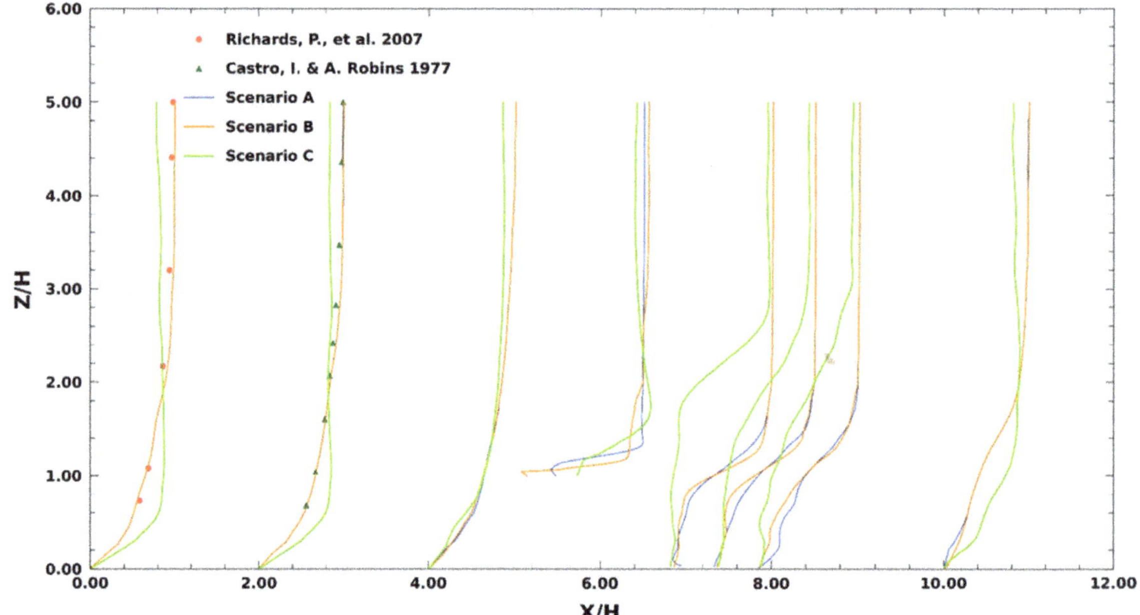

Figure 9. Normalized velocity profiles on the symmetry plane at positions X/H = 0, X/H = 2, X/H = 4, X/H = 5.5, X/H = 7, X/H = 7.5, X/H = 8 and X/H = 10 for Scenarios A–C [44,60].

Differences in the normalized velocity profile between Scenario C and Scenarios A and B are observed especially in the wake region of the cube, where the mixing layer for Scenario C has moved almost one cube height above the mixing layer of the other two scenarios. This is caused by the strong buoyant forces created in this case.

Figure 10 shows the streamlines of the flow for Scenario A. A detachment of the flow at the top surface of the building and a recirculation of the flow on both the frontal and the rear surfaces of the building are observed.

Different recirculation lengths X_f, X_b, and X_r are defined for the frontal, rear, and roof recirculation of the building, respectively.

Table 3 presents a comparison of the length of different recirculation zones in comparison with available experimental data.

Table 3. Recirculation length of the recirculation zones.

	X_f	X_b	X_r
Martinuzzi and Tropea [61]	1.04 H	1.61 H	-
Rodi [62]	0.651 H	2.182 H	0.432 H
Hoxey, Richards [63]	0.75 H	1.4 H	0.57 H
Richards and Norris [64]	0.9 H	1.4 H	0.9 H
Hu, Xuan [65]	-	1.31 H	0.94 H
Scenario A	0.79 H	1.56 H	0.56 H
Scenario B	0.56 H	1.97 H	0.64 H
Scenario C	0.86 H	2.15 H	0.73 H

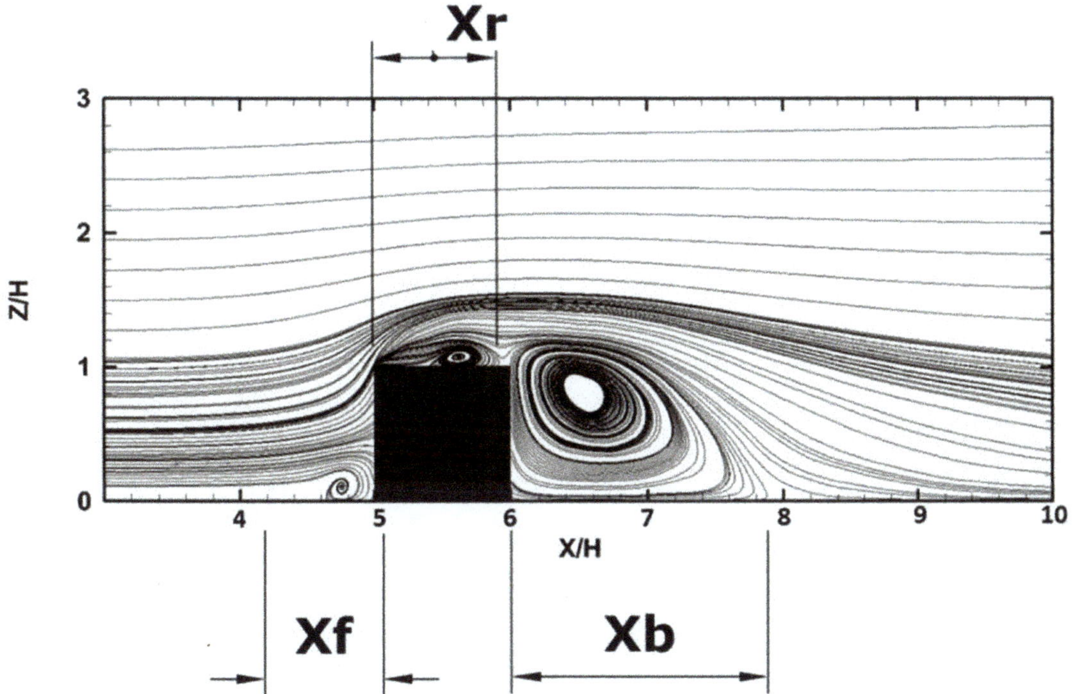

Figure 10. Velocity streamlines of the mean flow field on the symmetry plane of the computational domain for Scenario A.

The recirculation zone $\mathbf{X_f}$ in front of the upstream surface of the building for Scenario A is 41% higher than the corresponding for Scenario B and 8.14% lower than Scenario C. In addition, the recirculation zone on the roof $\mathbf{X_r}$ for Scenario A is 20.81% lower than the recirculation zone of Scenario B and 27.44% lower than for Scenario C. The recirculation zone in the wake region $\mathbf{X_b}$ of the building for Scenario A is 13% lower than the corresponding recirculation zone of Scenario B and 23.28% lower than the recirculation zone for Scenario C.

Figure 11 shows the normalized vertical temperature profile for Scenario C at the position with coordinates of X = 7 H and Y = 5.5 H, on the symmetry plane and two heights behind the building.

The present numerical results are in good agreement with the experimental data of Uehara and Murakami [41], who studied the stability of the atmospheric flow inside an urban street canyon placed normal to the wind direction. They found that inside the canyon exists a stable thermal stratification as shown above, which weakens the cavity eddy.

Figure 12 shows the temperature distribution on the vertical and horizontal surfaces of the building for Scenario C, along the intersection lines with the transverse symmetry plane (path No. 3) and two other planes parallel to it at distances 0.25 H (path No. 2) and 0.45 H (path No. 1), respectively.

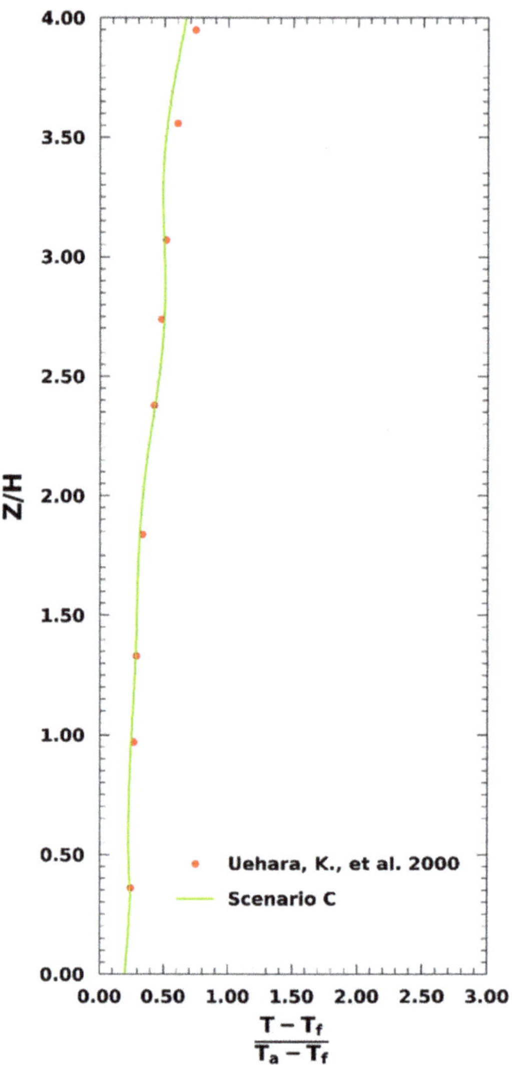

Figure 11. Normalized temperature profiles at the wake region of the flow for Scenario C, at X = 7 H and Y = 5.5 H, where T is the mean temperature, T_a is the ambient temperature, and T_f is the temperature at the floor of the computational domain [41].

The surface temperatures of the cube are the result of an energy balance between the incoming thermal radiation that is absorbed and reflected or emitted, and heat convection by the airflow. The amount of radiation that reaches the surfaces of the cube and the visibility of the solar path are determined by Athens's longitude and latitude. Albedo and emissivity of the surfaces control the shortwave and longwave radiation components reflected and emitted, respectively. The albedo and emissivity coefficients are 0.2 and 0.95, respectively, typical values for concrete material. The thermal heat capacity and thermal conductivity of the cube control its ability to store and conduct heat, respectively.

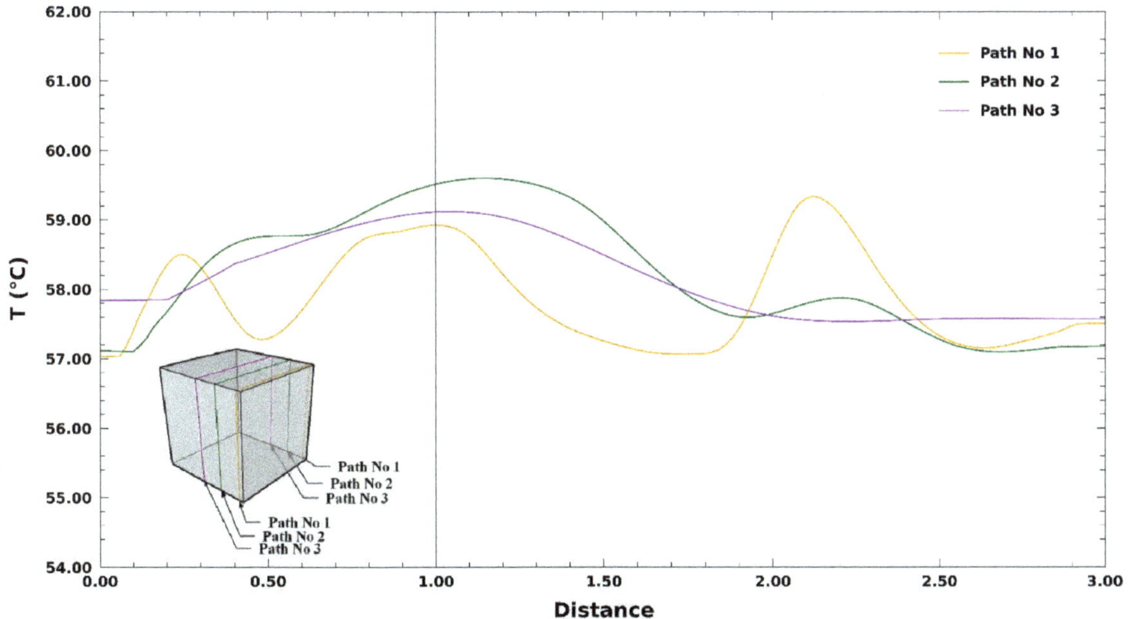

Figure 12. Temperature distribution on the vertical and horizontal surfaces of the building for the Scenario C along three different streamwise paths.

The temperature on the cube surfaces depends primarily on the convective thermal energy loss. In regions, where turbulence exists high heat exchange takes place. At the lower part of the frontal surface of the cube, where the horseshoe-shaped vortex appears, low surface temperatures appear. On the contrary, on the upper part of the frontal surface of the cube higher surface temperatures are shown. Similarly, on the wake zone of the cube where the arc vortex is present, low surface temperatures are shown because of the high heat exchange rate. On the first part of the roof surface where the flow separates the heat exchange is small, so the surface temperature is high, and on the second part it is reduced where the flow reattaches.

Figure 13 present the vertical mean pollutant concentration profiles computed for the time period from 800–1400 (s), at which all variables are statistically stationary on the symmetry plane for both Scenarios A and C at the Cartesian coordinate positions of X: 7 H, for a, X: 7.2 H for b, and 7.4 H, for c. The height Z extends from 0 to 2 H.

In Figure 13, the highest mean concentration of the pollutant is observed for Scenario A, where a large amount of the pollutant is trapped inside the wake region. In contrast, in Scenario C where because of the high buoyancy forces the maximum pollutant concentration is small in the above region as the pollutant is shifted almost one cube height higher away from the floor, thus creating a recirculation region of larger length and height than that corresponding for Scenario A.

As the pollutant in Scenario A is trapped inside the wake region of the cubical building, it becomes very harmful to the pedestrian's health. In Scenario C, this problem is exempted to a large extent, as the high levels of pollutant concentrations are at much higher heights than those of the pedestrians. This behavior is the result primarily of the buoyant forces acting on the roof and on the leeward surfaces of the building.

Figure 14 present the distribution of the flux concentration on the symmetry plane because of the x-velocity component for Scenarios A and C at the x-positions: (a) (X: 7 H, Y: 5.5 H), (b) (X: 7.2 H, Y: 5.5 H), and (c) (X: 7.4 H, Y: 5.5 H) for heights Z that extend from 0 to 2 H.

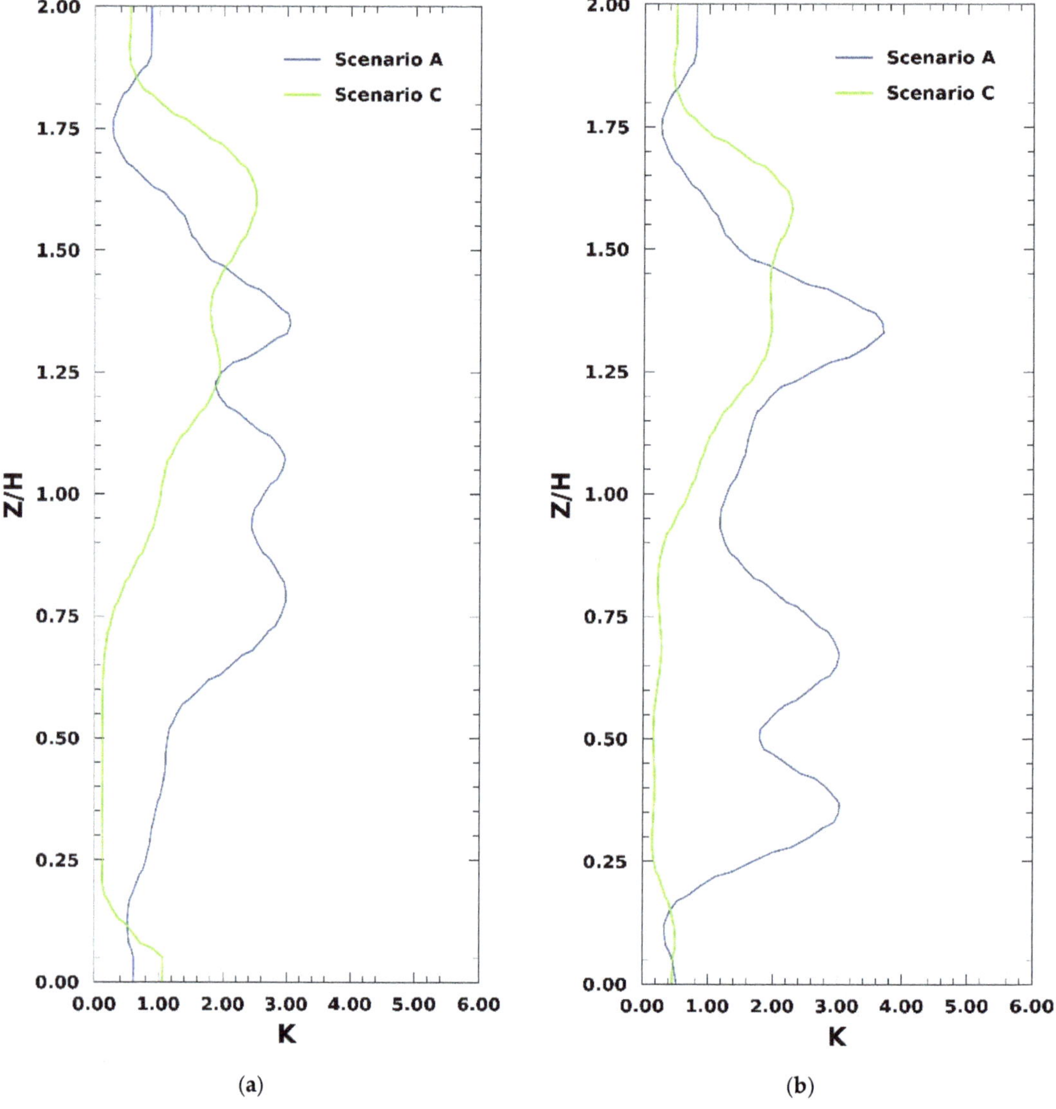

Figure 13. *Cont.*

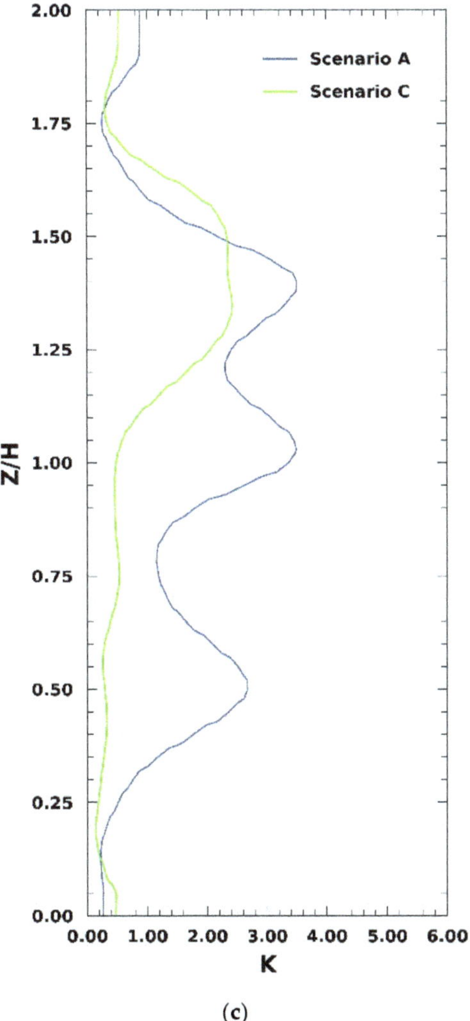

(c)

Figure 13. Vertical concentration profiles with the pollutant source at X = 7 H for Scenarios A and C at positions (**a**) (X: 7 H, Y: 5.5 H), (**b**) (X: 7.2 H, Y: 5.5 H), and (**c**) (X: 7.4 H, Y: 5.5 H).

In the wake region, up to the height of the cube, the flux concentration is similar for both cases A and C. At heights higher than the building, the concentration flux for Scenario A is larger of that of C because of the important influence of higher free stream air velocity. Additionally, the wake zone height in Scenario C exceeds the height of the building and the low values of air velocity produce low concentration fluxes.

Figure 15a–c present the distribution of the flux concentration on the symmetry plane because of the z-velocity component for Scenarios A and C at the x-positions: (a) (X: 7 H, Y: 5.5 H), (b) (X: 7.2 H, Y: 5.5 H), and (c) (X: 7.4 H, Y: 5.5 H) for heights Z that extend from 0 to 2 H.

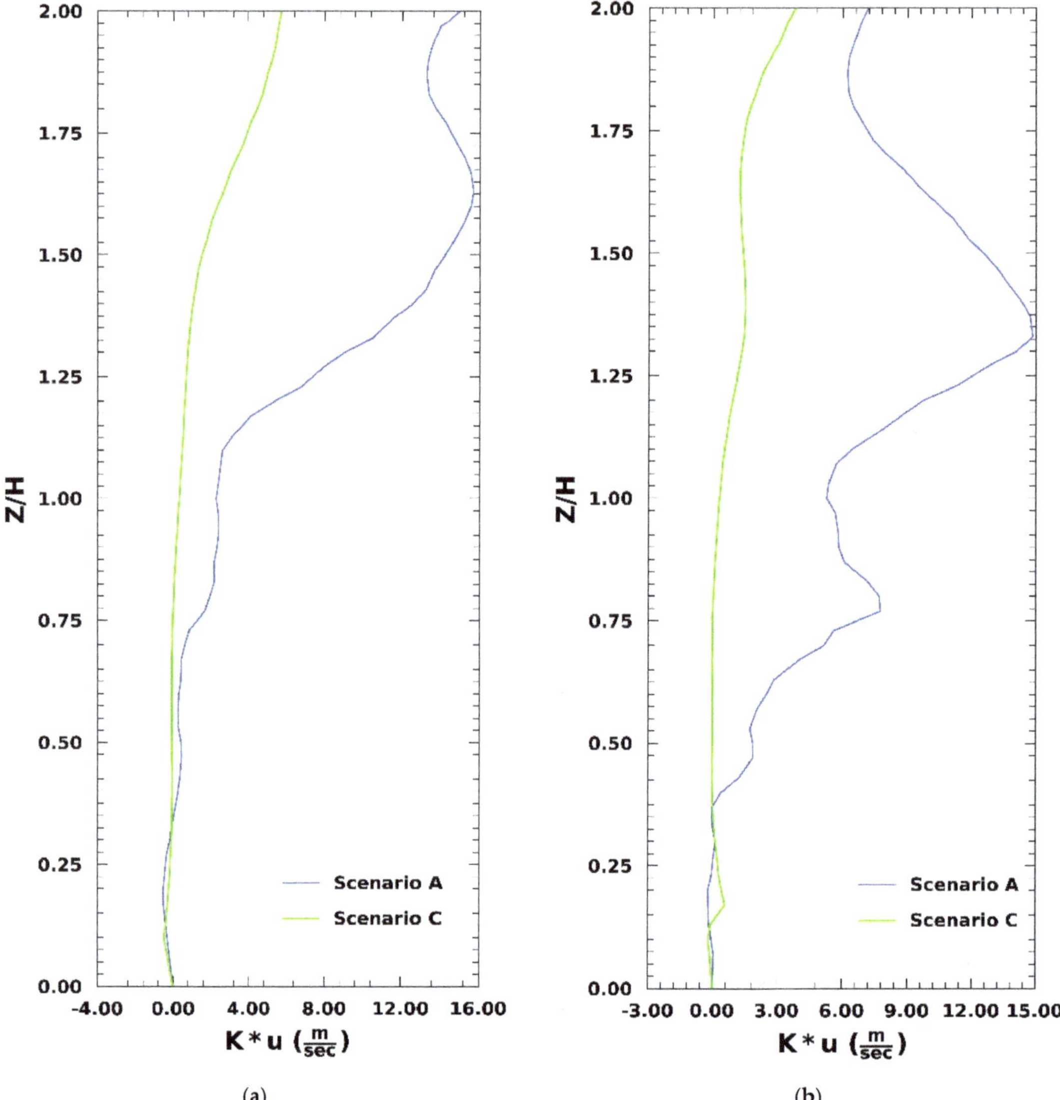

Figure 14. *Cont.*

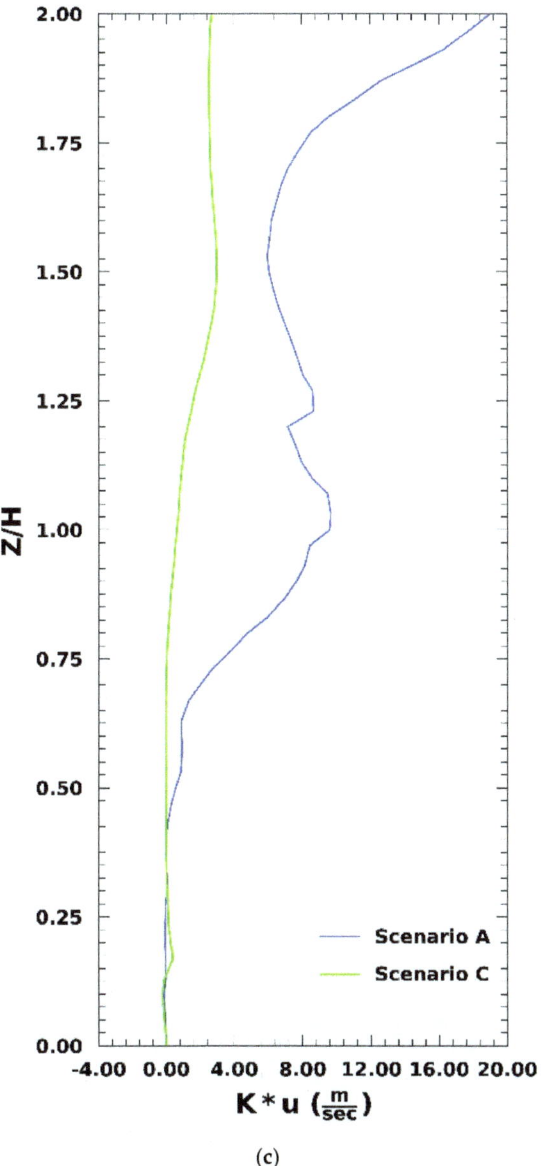

(c)

Figure 14. Concentration flux profiles that are due to the x−velocity component on the symmetry plane for Scenarios A and C at positions (**a**) (X: 7 H, Y: 5.5 H), (**b**) (X: 7.2 H, Y: 5.5 H), and (**c**) (X: 7.4 H, Y: 5.5 H) for heights Z that extend from 0 to 2 H.

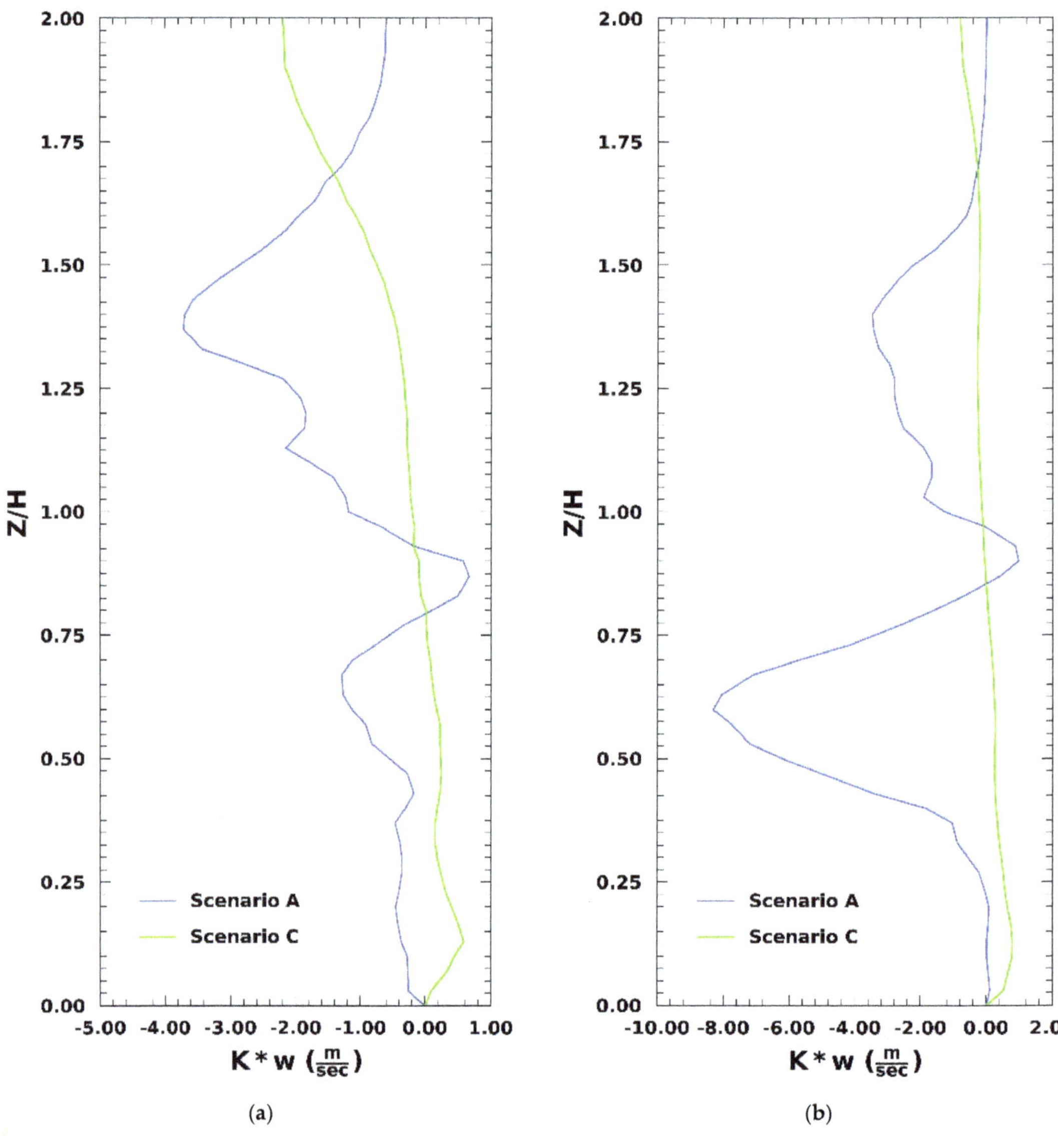

Figure 15. *Cont.*

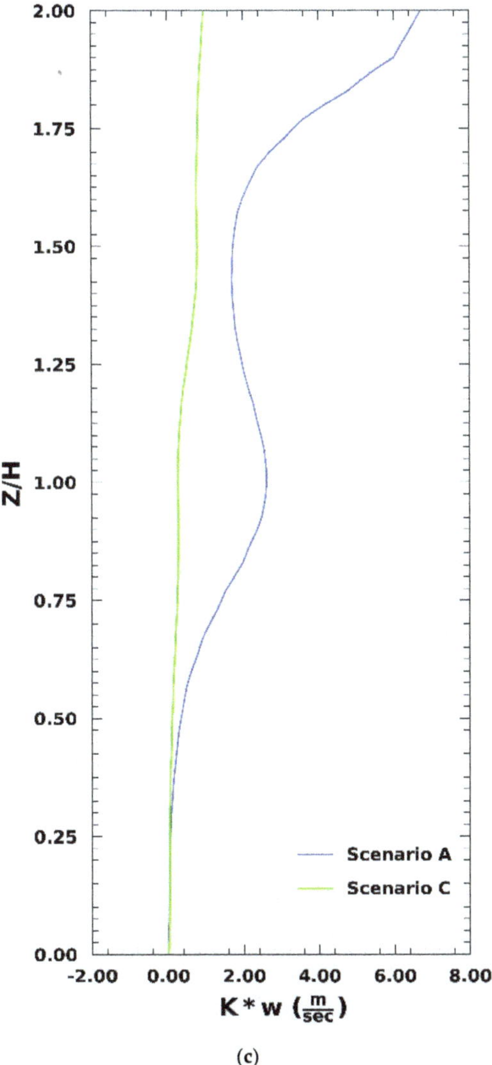

Figure 15. Distribution of the non-dimensional flux concentration at the z−component of the velocity on the symmetry plane for Scenarios A and C at positions (**a**) (X: 7 H, Y: 5.5 H), (**b**) (X: 7.2 H, Y: 5.5 H), and (**c**) (X: 7.4 H, Y: 5.5 H) for height Z that extends from 0 to 2 H.

For Scenario A, where the buoyancy forces do not affect the pollutant distribution, the concentration flux is toward the ground floor (Figure 15a,b) within the recirculation zone near the point of reattachment, in accord with the flow field of Figure 10, and away from the floor outside the recirculation zone. On the contrary, for Scenario C, the buoyancy forces lift the pollutants near position X = 7 H (left end of source) where at positions X = 7.2 H and X = 7.2 H (center and right end of source, respectively) the buoyancy forces are counterbalanced by momentum flux forces. It is expected that by placing the pollutant source closer to the leeward face of the building, the lifting forces that act on the concentration would be higher, resulting in higher concentration fluxes away from the ground. Thus, a higher fraction of the pollutants will escape the recirculation zone.

7. Conclusions

In the present work, three different scenarios are studied. The emphasis is on the flow and pollutant dispersion around an isolated cubical building in an open area and with a pollutant source placed on the ground and at a distance of one building height in the symmetry plane behind its rear surface. Scenario A with adiabatic walls of the building and without any buoyancy forces, Scenario B with the same constant temperature of 50 $^\circ$C on all the surfaces of the building, and Scenario C with convective and radiative conditions imposed by the local microclimate.

Each scenario is simulated using the large-eddy simulation (LES) approach. The numerical results that describe the nature and the behavior of the flow around an isolated building are in good agreement with the corresponding experimental and numerical data that were used for their validation.

It is observed that at higher flow field temperatures, the length and the corresponding height of the recirculation are increased. Thus, the recirculation region for Scenario B has higher length than that for Scenario A and lower than that for Scenario C, for which exist the highest temperatures from the three scenarios studied.

The pollutant in Scenario A is trapped inside the wake region of the cubical building with its highest concentration levels at a height close to what a pedestrian would breathe, thus being very harmful to his health. In Scenario C, the higher temperatures together with the associated buoyancy forces that are developed raise most of the pollutants to a height greater than that of the building, allowing them to escape the recirculation region of the building and be carried away by the wind, without causing any serious harm to the health of a pedestrian.

It is expected that by placing the pollutant source closer to the leeward face of the building, the lifting forces that act on the concentration would be higher, resulting in higher concentration fluxes away from the ground. Thus, a higher fraction of the pollutants will escape the recirculation zone.

Author Contributions: Conceptualization, C.L.P. and K.V.; methodology, C.L.P., K.V., I.E.S. and I.C.L.; software, C.L.P. and A.V.P.; validation, C.L.P. and K.V.; formal analysis; C.L.P.; resources, I.E.S.; data curation, C.L.P.; writing-original draft preparation, C.L.P. and K.V.; writing-review and editing, I.C.L. and I.E.S.; visualization, C.L.P. and A.V.P.; supervision, K.V.; All authors have read and agreed to the published version of the manuscript.

Funding: This research received no external funding.

Institutional Review Board Statement: Not applicable.

Informed Consent Statement: Not applicable.

Data Availability Statement: The data used in this study are available on request from the corresponding author. The data can be easily reproduced from the theoretical analysis described in the study.

Acknowledgments: This work was supported by computational time granted from the National Infrastructures for Research and Technology S.A. (GRNET S.A.) in the National HPC facility—ARIS—under project ID pr011045- UrbanFirePlan2. The authors acknowledge and express particular appreciation to the PALM group at the Institute of Meteorology and Climatology (IMUK) of Leibniz Universität Hannover, Germany that has developed the Parallelized Large Eddy Simulation Model (PALM).

Conflicts of Interest: The authors declare no conflict of interest.

References

1. Sini, J.-F.; Anquetin, S.; Mestayer, P.G. Pollutant dispersion and thermal effects in urban street canyons. *Atmos. Environ.* **1996**, *30*, 2659–2677. [CrossRef]
2. Barlow, F.J.; Harman, I.N.; Belcher, S.E. Scalar fluxes from urban street canyons. Part I: Laboratory simulation. *Bound. Layer Meteorol.* **2004**, *113*, 369–385. [CrossRef]
3. Rotach, M.W.; Gryning, S.E.; Batchvarova, E.; Christen, A.; Vogt, R. Pollutant dispersion close to an urban surface—The BUBBLE tracer experiment. *Meteorol. Atmos. Phys.* **2004**, *87*, 39–56. [CrossRef]
4. Vasilopoulos, K.; Mentzos, M.; Sarris, I.E.; Tsoutsanis, P. Computational Assessment of the Hazardous Release Dispersion from a Diesel Pool Fire in a Complex Building's Area. *Computation* **2018**, *6*, 65. [CrossRef]
5. Kim, D.J.; Lee, D.I.; Kim, J.J.; Park, M.S.; Lee, S.H. Development of a Building-Scale Meteorological Prediction System Including a Realistic Surface Heating. *Atmosphere* **2020**, *11*, 67. [CrossRef]
6. Kovar-Panskus, A.; Louka, P.; Sini, J.F.; Savory, E.; Czech, M.; Abdelqari, A.; Mestayer, P.G.; Toy, N. Influence of Geometry on the Mean Flow within Urban Street Canyons—A Comparison of Wind Tunnel Experiments and Numerical Simulations. *Water Air Soil Pollut. Focus* **2002**, *2*, 365–380. [CrossRef]
7. Dallman, A.; Magnusson, S.; Britter, R.; Norford, L.; Entekhabi, D.; Fernando, H.J. Conditions for thermal circulation in urban street canyons. *Build. Environ.* **2014**, *80*, 184–191. [CrossRef]
8. Madalozzo DM, S.; Braun, A.L.; Awruch, A.M.; Morsch, I.B. Numerical simulation of pollutant dispersion in street canyons: Geometric and thermal effects. *Appl. Math. Model.* **2014**, *38*, 5883–5909. [CrossRef]
9. Chatzimichailidis, A.E.; Argyropoulos, C.D.; Assael, M.J.; Kakosimos, K.E. Implicit Definition of Flow Patterns in Street Canyons—Recirculation Zone—Using Exploratory Quantitative and Qualitative Methods. *Atmosphere* **2019**, *10*, 794. [CrossRef]
10. Chatzimichailidis, A.; Assael, M.; Ketzel, M.; Kakosimos, K.E. Modelling the Recirculation Zone in street canyons with different aspect ratios, using CFD simulation. In Proceedings of the 17th International Conference on Harmonisation within Atmospheric Dispersion Modelling for Regulatory Purposes, Budapest, Hungary, 9–12 May 2016.
11. Ming, T.; Fang, W.; Peng, C.; Cai, C.; de Richter, R.; Ahmadi, M.H.; Wen, Y. Impacts of Traffic Tidal Flow on Pollutant Dispersion in a Non-Uniform Urban Street Canyon. *Atmosphere* **2018**, *9*, 82. [CrossRef]
12. Chatzimichailidis, A.E.; Argyropoulos, C.D.; Assael, M.J.; Kakosimos, K.E. Qualitative and Quantitative Investigation of Multiple Large Eddy Simulation Aspects for Pollutant Dispersion in Street Canyons Using OpenFOAM. *Atmosphere* **2019**, *10*, 17. [CrossRef]
13. Pisello, A.L.; Pignatta, G.; Castaldo, V.L.; Cotana, F. The Impact of Local Microclimate Boundary Conditions on Building Energy Performance. *Sustainability* **2015**, *7*, 9207–9230. [CrossRef]
14. Blanken, P.D.; Barry, R.G. (Eds.) Urban Climates. In *Microclimate and Local Climate*; Cambridge University Press: Cambridge, UK, 2016; pp. 243–260. [CrossRef]
15. Kim, J.-J.; Baik, J.-J. A Numerical Study of Thermal Effects on Flow and Pollutant Dispersion in Urban Street Canyons. *J. Appl. Meteorol.* **1999**, *38*, 1249–1261. [CrossRef]
16. Baik, J.-J.; Kim, J.-J.; Fernando, H. A CFD Model for Simulating Urban Flow and Dispersion. *J. Appl. Meteorol.* **2003**, *42*, 1636–1648. [CrossRef]
17. Kwak, K.H.; Baik, J.J.; Lee, S.H.; Ryu, Y.H. Computational Fluid Dynamics Modelling of the Diurnal Variation of Flow in a Street Canyon. *Bound. Layer Meteorol.* **2011**, *141*, 77–92. [CrossRef]
18. Park, S.B.; Baik, J.J.; Raasch, S.; Letzel, M.O. A Large-Eddy Simulation Study of Thermal Effects on Turbulent Flow and Dispersion in and above a Street Canyon. *J. Appl. Meteorol. Climatol.* **2012**, *51*, 829–841. [CrossRef]
19. Santiago, J.; Krayenhoff, E.; Martilli, A. Flow simulations for simplified urban configurations with microscale distributions of surface thermal forcing. *Urban Clim.* **2014**, *9*, 115–133. [CrossRef]
20. Nazarian, N.; Kleissl, J. CFD simulation of an idealized urban environment: Thermal effects of geometrical characteristics and surface materials. *Urban Clim.* **2015**, *12*, 141–159. [CrossRef]
21. Nazarian, N.; Kleissl, J. Realistic solar heating in urban areas: Air exchange and street-canyon ventilation. *Build. Environ.* **2016**, *95*, 75–93. [CrossRef]
22. Resler, J.; Krč, P.; Belda, M.; Juruš, P.; Benešová, N.; Lopata, J.; Kanani-Sühring, F. PALM-USM v1.0: A new urban surface model integrated into the PALM large-eddy simulation model. *Geosci. Model Dev.* **2017**, *10*, 3635–3659. [CrossRef]
23. Lee, D.-I.; Woo, J.-W.; Lee, S.-H. An analytically based numerical method for computing view factors in real urban environments. *Theor. Appl. Climatol.* **2018**, *131*, 445–453. [CrossRef]
24. Oke, T.R. The energetic basis of the urban heat island. *Q. J. R. Meteorol. Soc.* **1982**, *108*, 1–24. [CrossRef]
25. Tam, B.Y.; Gough, W.A.; Mohsin, T. The impact of urbanization and the urban heat island effect on day to day temperature variation. *Urban Clim.* **2015**, *12*, 1–10. [CrossRef]
26. Brown, M. Urban Parameterizations for Mesoscale Meteorological Models. In *Mesoscale Atmospheric Dispersion*; WIT Press: England, UK, 2000; pp. 193–255.
27. Fisher, B.; Kukkonen, J.; Piringer, M.; Rotach, M.W.; Schatzmann, M. Meteorology applied to urban air pollution problems: Concepts from COST 715. *Atmos. Chem. Phys.* **2006**, *6*, 555–564. [CrossRef]
28. Gronemeier, T.; Raasch, S.; Ng, E. Effects of Unstable Stratification on Ventilation in Hong Kong. *Atmosphere* **2017**, *8*, 168. [CrossRef]

29. Wang, W.; Xu, Y.; Ng, E. Large-eddy simulations of pedestrian-level ventilation for assessing a satellite-based approach to urban geometry generation. *Graph. Models* **2018**, *95*, 29–41. [CrossRef]

30. Pfafferott, J.; Rißmann, S.; Sühring, M.; Kanani-Sühring, F.; Maronga, B. Building indoor model in PALM model system 6.0: Indoor climate, energy demand, and the interaction between buildings and the urban climate. *Geosci. Model Dev. Discuss.* **2020**, *14*, 199. [CrossRef]

31. Resler, J.; Eben, K.; Geletič, J.; Krč, P.; Rosecký, M.; Sühring, M.; Belda, M.; Fuka, V.; Halenka, T.; Huszár, P.; et al. Validation of the PALM model system 6.0 in real urban environment; case study of Prague-Dejvice, Czech Republic. *Geosci. Model Dev. Discuss.* **2020**, *14*, 175. [CrossRef]

32. Allwine, K.J.; Leach, M.J.; Stockham, L.W.; Shinn, J.S.; Hosker, R.P.; Bowers, J.F.; Pace, J.C. Overview of Joint Urban 2003: An atmospheric dispersion study in Oklahoma City. In Proceedings of the 84th AMS Annual Meeting, Seattle, WA, USA, 11–15 January 2004.

33. Park, M.S.; Park, S.H.; Chae, J.H.; Choi, M.H.; Song, Y.; Kang, M.; Roh, J.W. High-resolution urban observation network for user-specific meteorological information service in the Seoul Metropolitan Area, South Korea. *Atmos. Meas. Tech.* **2017**, *10*, 1575–1594. [CrossRef]

34. Gowardhan, A.A.; Brown, M.J.; Pardyjak, E.R. Evaluation of a fast response pressure solver for flow around an isolated cube. *Environ. Fluid Mech.* **2010**, *10*, 311–328. [CrossRef]

35. Lim, H.C.; Thomas, T.G.; Castro, I.P. Flow around a cube in a turbulent boundary layer: LES and experiment. *J. Wind. Eng. Ind. Aerodyn.* **2009**, *97*, 96–109. [CrossRef]

36. Yazid, A.W.M.; Sidik, N.A.C. Prediction of the Flow around a Surface-Mounted Cube Using Two-Equation Turbulence Models. *Appl. Mech. Mater.* **2013**, *315*, 438–442. [CrossRef]

37. Mochida, A.; Tominaga, Y.; Murakami, S.; Yoshie, R.; Ishihara, T.; Ooka, R. Comparison of various k-ε models and DSM applied to flow around a high-rise building—Report on AIJ cooperative project for CFD prediction of wind environment. *Wind. Struct.* **2002**, *5*, 227–244. [CrossRef]

38. Dogan, S.; Yagmur, S.; Goktepeli, I.; Ozgoren, M. Assessment of turbulence models for flow around a surface-mounted cube. *Int. J. Mech. Eng. Robot. Res.* **2017**, *6*, 237–241. [CrossRef]

39. Brown, M.J.; Lawson, R.E.; Decroix, D.S.; Lee, R.L. Mean flow and turbulence measurement around a 2-darray of buildings in a wind tunnel. In Proceedings of the AMS 11th Joint Conference on the Applications of Air Pollution Meteorology, Long Beach, CA, USA, 9–13 January 2000. Report LA-UR-99-5395.

40. Brown, M.J. Comparison of Centerline Velocity Measurements Obtained around 1d and 3d Building Arrays in a Wind Tunnel. In Proceedings of the International Society of Environmental Hydraulics Conference, Tempe, AZ, USA, 5–8 December 2001.

41. Uehara, K.; Murakami, S.; Oikawa, S.; Wakamatsu, S. Wind tunnel experiments on how thermal stratification affects flow in and above urban street canyons. *Atmos. Environ.* **2000**, *34*, 1553–1562. [CrossRef]

42. Tseng, Y.-H.; Meneveau, C.; Parlange, M.B. Modeling Flow around Bluff Bodies and Predicting Urban Dispersion Using Large Eddy Simulation. *Environ. Sci. Technol.* **2006**, *40*, 2653–2662. [CrossRef]

43. Richards, P.J.; Hoxey, R.P.; Short, L.J. Wind pressures on a 6m cube. *J. Wind. Eng. Ind. Aerodyn.* **2001**, *89*, 1553–1564. [CrossRef]

44. Richards, P.J.; Hoxey, R.P.; Connell, B.D.; Lander, D.P. Wind-tunnel modelling of the Silsoe Cube. *J. Wind. Eng. Ind. Aerodyn.* **2007**, *95*, 1384–1399. [CrossRef]

45. Zheng, X.; Montazeri, H.; Blocken, B. CFD simulations of wind flow and mean surface pressure for buildings with balconies: Comparison of RANS and LES. *Build. Environ.* **2020**, *173*, 106747. [CrossRef]

46. Franke, J.; Baklanov, A. *Best Practice Guideline for the CFD Simulation of Flows in the Urban Environment: COST Action 732 Quality Assurance and Improvement of Microscale Meteorological Models*; Meteorological Institute, University of Hamburg: Hamburg, Germany, 2007.

47. Tominaga, Y.; Mochida, A.; Yoshie, R.; Kataoka, H.; Nozu, T.; Yoshikawa, M.; Shirasawa, T. AIJ guidelines for practical applications of CFD to pedestrian wind environment around buildings. *J. Wind. Eng. Ind. Aerodyn.* **2008**, *96*, 1749–1761. [CrossRef]

48. Li, W.-W.; Meroney, R. Gas dispersion near a cubical model building. Part II. Concentration fluctuation measurements. *J. Wind. Eng. Ind. Aerodyn.* **1983**, *12*, 35–47. [CrossRef]

49. Saathof, P.J.; Stathopoulos, T.; Dobrescu, M. Effects of model scale in estimating pollutant dispersion near buildings. *J. Wind. Eng. Ind. Aerodyn.* **1995**, *54–55*, 549–559. [CrossRef]

50. Bird, R.B.; Stewart, W.E.; Lightfoot, E.N. *Transport Phenomena*; Wiley: Hoboken, NJ, USA, 2006.

51. Krayenhoff, E.S.; Voogt, J.A. A microscale three-dimensional urban energy balance model for studying surface temperatures. *Bound. Layer Meteorol.* **2007**, *123*, 433–461. [CrossRef]

52. Monin, A.S.; Obukhov, A.M. Basic laws of turbulent mixing in the surface layer of the atmosphere. *Contrib. Geophys. Inst. Acad. Sci. USSR* **1954**, *24*, 163–187.

53. Obukhov, A.M. Turbulence in an atmosphere with a non-uniform temperature. *Bound. Layer Meteorol.* **1971**, *2*, 7–29. [CrossRef]

54. Castro, I.; Robins, A. The flow around a surface-mounted cube in uniform and turbulent streams. *J. Fluid Mech.* **1977**, *79*, 307–335. [CrossRef]

55. Arakawa, A.; Lamb, V.R. Computational Design of the Basic Dynamical Processes of the UCLA General Circulation Model. In *Methods in Computational Physics: Advances in Research and Applications*; Chang, J., Ed.; Elsevier: Amsterdam, The Netherlands, 1977; pp. 173–265. [CrossRef]

56. Wicker, L.J.; Skamarock, W.C. Time-Splitting Methods for Elastic Models Using Forward Time Schemes. *Mon. Weather. Rev.* **2002**, *130*, 2088–2097. [CrossRef]
57. Williamson, J.H. Low-storage Runge-Kutta schemes. *J. Comput. Phys.* **1980**, *35*, 48–56. [CrossRef]
58. Deardorff, J.W. Stratocumulus-capped mixed layers derived from a three-dimensional model. *Bound. Layer Meteorol.* **1980**, *18*, 495–527. [CrossRef]
59. Moeng, C.-H.; Wyngaard, J.C. Spectral Analysis of Large-Eddy Simulations of the Convective Boundary Layer. *J. Atmos. Sci.* **1988**, *45*, 3573–3587. [CrossRef]
60. Saiki, E.M.; Moeng, C.-H.; Sullivan, P.P. Large-Eddy Simulation of The Stably Stratified Planetary Boundary Layer. *Bound. Layer Meteorol.* **2000**, *95*, 1–30. [CrossRef]
61. Martinuzzi, R.; Tropea, C. The Flow Around Surface-Mounted, Prismatic Obstacles Placed in a Fully Developed Channel Flow (Data Bank Contribution). *J. Fluids Eng.* **1993**, *115*, 85–92. [CrossRef]
62. Rodi, W. Comparison of LES and RANS calculations of the flow around bluff bodies. *J. Wind. Eng. Ind. Aerodyn.* **1997**, *69*, 55–75. [CrossRef]
63. Hoxey, R.P.; Richards, P.; Short, J.L. A 6 m cube in an atmospheric boundary layer flow—Part 1. Full-scale and wind-tunnel results. *Wind. Struct.* **2002**, *5*, 165–176. [CrossRef]
64. Richards, P.; Norris, S. LES modelling of unsteady flow around the Silsoe cube. *J. Wind. Eng. Ind. Aerodyn.* **2015**, *144*, 70–78. [CrossRef]
65. Hu, J.; Xuan, H.B.; Kwok, K.C.S.; Zhang, Y.; Yu, Y. Study of wind flow over a 6 m cube using improved delayed detached Eddy simulation. *J. Wind. Eng. Ind. Aerodyn.* **2018**, *179*, 463–474. [CrossRef]

applied
sciences

Article

Thermal Performance Enhancement Using Absorber Tube with Inner Helical Axial Fins in a Parabolic Trough Solar Collector

Mohammad Zaboli [1], Seyed Soheil Mousavi Ajarostaghi [2], Seyfolah Saedodin [1] and Mohsen Saffari Pour [3,*

[1]   Faculty of Mechanical Engineering, Semnan University, Semnan 47148-71167, Iran;
    zaboli@semnan.ac.ir (M.Z.); s_sadodin@semnan.ac.ir (S.S.)
[2]   Faculty of Mechanical Engineering, Babol Noshirvani University of Technology, Babol 47148-71167, Iran;
    s.s.mousavi@stu.nit.ac.ir
[3]   Department of Mechanical Engineering, Faculty of Engineering, Shahid Bahonar University of Kerman,
    Kerman 76169-13439, Iran
*   Correspondence: mohsensp@kth.se

Abstract: In the present work, a parabolic trough solar (PTC) collector with inner helical axial fins as swirl generator or turbulator is considered and analyzed numerically. The three-dimensional numerical simulations have been done by finite volume method (FVM) using a commercial CFD code, ANSYS FLUENT 18.2. The spatial discretization of mass, momentum, energy equations, and turbulence kinetic energy has been obtained by a second-order upwind scheme. To compute gradients, Green-Gauss cell-based method has been employed. This work consists of two sections where, first, four various geometries are appraised, and in the following, the selected schematic of the collector from the previous part is selected, and four various pitches of inner helical fins including 250, 500, 750 and 1000 mm are studied. All the numerical results are obtained by utilizing the FVM. Results show that the thermal performance improvement by 23.1% could be achieved by using one of the proposed innovative parabolic trough solar collectors compare to the simple one. Additionally, the minimum and maximum thermal performance improvement (compare to the case without fins) belong to the case with P = 250 mm by 14.1% and, to the case with P = 1000 mm by 21.53%, respectively.

Keywords: solar energy; parabolic trough collector; thermal performance; turbulator; swirl generator; numerical simulation

---

Citation: Zaboli, M.; Mousavi Ajarostaghi, S.S.; Saedodin, S.; Saffari Pour, M. Thermal Performance Enhancement Using Absorber Tube with Inner Helical Axial Fins in a Parabolic Trough Solar Collector. *Appl. Sci.* **2021**, *11*, 7423. https://doi.org/10.3390/app11167423

Academic Editor: Rüdiger Schwarze

Received: 20 May 2021
Accepted: 2 August 2021
Published: 12 August 2021

Publisher's Note: MDPI stays neutral with regard to jurisdictional claims in published maps and institutional affiliations.

Copyright: © 2021 by the authors. Licensee MDPI, Basel, Switzerland. This article is an open access article distributed under the terms and conditions of the Creative Commons Attribution (CC BY) license (https://creativecommons.org/licenses/by/4.0/).

1. Introduction

Heat exchangers, which transfer thermal energy through direct and indirect contact between fluids, are considered an indispensable part of several industries, from pharmaceutical to petrochemicals. Indirect contact heat exchangers are extensively used in solar systems. Due to the increasing importance of solar energy, nowadays, improving the performance of solar systems is one of the most important challenges for human beings and researchers. One of the most efficient types of solar collectors is a parabolic trough solar collector (PTSC), which is employed in both domestic and power plant applications. Recently, many efforts have been made by scientists to increase the efficiency of this type of collector. Actually, the PTSC component is a heat exchanger in which heat transfer fluid (HTF) flows in the receiver tube and absorbs the radiated solar energy.

In order to improve the thermal performance of this type of heat exchanger, various methods have been proposed and studied by researchers. Berger et al. [1] categorized heat exchangers' augmentation procedures into passive and active subdivisions. One of two categories of ameliorating heat exchange methods is passive procedures [2]. It means that there is no requirement for any type of additional force. This method includes techniques such as nano particle [3], helical tubes [4], treated area [5], vortex generator [6], displaced increase devices [7], extended surfaces [8], and jagged surfaces [9].

Tang et al. [10] checked the heat transfer in the heat exchanger. According to their results, an increase of length enhances the performance of the vortex-generator fin. Azari et al. [11] checked the heat exchange in a heterogeneous circular microchannel, the results show the maximum heat exchange rates are got when the boundary conditions are symmetrical. Darzi et al. [12] have performed an experimental probing in a corrugated tube with nanofluid, they deduced that nanofluid with corrugated tubing could increase heat exchange by 330% compared to net liquid. Du et al. [13] numerically have worked on tube layouts, they demonstrated that the comprehensive efficiency of the optimized heat transfer device is ameliorated by about 50%. Bahiraei et al. [14] have checked on a triple-tube that their outcomes show that, based on heat exchange efficiency, an increase in height and reduced pitch are suggested. Abolarin et al. [15] have checked the impact of various geometries in twisted tapes, and the results pointed out that connection angle increases heat transfer. They found that when the wavelength is smaller, the temperatures near the walls increased. Zhang et al. [16] checked the thermal performance in corrugated pipes. According to the outcomes, the corrugated pipe ameliorated the heat transfer rate. Therefore, the other passive methods containing turbulator, etc., are usually better in the turbulent flow [17]. Several of the new works related to the utilization of inserts and the other passive methods to improve the thermal performance of the heat exchangers are displayed and listed in Table 1.

Table 1. Overview of articles to improve the heat transfer.

Author	Year	Method (Exp/Num)	Inserts	Result
Kareem et al. [18]	2015	Experimental and Numerical	Spirally corrugated tube	The tube of higher severity index has the best thermal performance.
Lu et al. [19]	2016	Numerical	Vortex generator	Their results pointed out that holes close to the leading-edge give better efficiency.
Mashoofi et al. [20]	2017	Experimental	Helically coiled	Findings showed that the use of a turbulator increases Nusselt number around 8–32%.
Aghaei et al. [21]	2018	Experimental and Numerical	Horizontal and vertical elliptic baffles	The heat exchange attains the utmost at the arrangement angle of 15°.
Noorbakhsh et al. [22]	2019	Numerical	Twisted-tapes	The enhancement of fin number goes up the heat exchange.
Kwon et al. [23]	2019	Numerical	static mixers	Static mixers registered heat transfer coefficient an increase of about 100%.
Ho et al. [24]	2019	Experimental	Mini-channel	Conforming outcomes, the friction factor reduced as the temperature of the inlet rises.
Liu et al. [25]	2019	Experimental	Heat sinks	Heat exchange had a peak at $\theta = 20°$ and $\beta = 0.8$.
Ramalingam et al. [26]	2020	Experimental	Automotive radiator	Enhancement heat dissipation was got for nanofluid at 0.8.
Qi et al. [27]	2020	Experimental	Triangle tube	Triangular tubes have the most positive impact on the rate of increase of heat exchange.
Nakhchi et al. [28]	2020	Numerical	Perforated hollow cylinders	Outcomes saw that the resistance of flow can be decreased up to 86.2% with go upping the perforated index.
Rahbarshahln et al. [29]	2020	Numerical	Microchannels	The use of hydrophobic models could decline the required power fluid pump up to 69% and go up heat flux up to 15%.

Table 1. *Cont.*

Author	Year	Method (Exp/Num)	Inserts	Result
Jamesahar et al. [30]	2020	Numerical	Flexible fins	Due to the oscillation of the fins, the heat exchange rate goes up.
Benabderrahmane et al. [31]	2020	Numerical	Central corrugated insert	The overall heat transfer performance was achieved in the range of 1.3–2.6.
Akbarzadeh and Valipour [32]	2020	Experimental	Helically corrugated tube	The maximum obtained thermal performance (2.29) belongs to the case with pitch length and roughness height of 3 mm and 1.5 mm, respectively.
Chakraborty et al. [33]	2020	Numerical	Helical absorber tube	Thermal efficiency and exergy rise by 1–10% and 0.2–3.2%, correspondingly.
Amani et al. [34]	2020	Numerical	Conical strip inserts	The overall thermal–hydraulic performance was obtained in the range of 0.679–1.107.
Khan et al. [35]	2020	Numerical	Absorber tube with twisted tape insert and tube with longitudinal fins	Thermal efficiency of the cases with twisted tape insert, tube with internal fins, and plain tube are by 72.26%, 72.10% and 71.09%, respectively.
Saedodin et al. [36]	2021	Numerical	Turbulence-Inducing elements	The maximum achieved thermal efficiency was by 29%.
Vasanthi and Jaya Chandra reddy [37]	2021	Experimental	Angular twisted strip inserts	The obtained enhancement efficiency was in the range of 145–215%.
Chakraborty et al. [38]	2021	Numerical	Helical absorber tube	Thermal efficiency and exergy rises by 4–10%, and 4–5%, correspondingly.
Akbarzadeh and Valipour [39]	2021	Numerical	Helically corrugated tube	The thermal performance shows a growth of 26–176%.
Peng et al. [40]	2021	Experimental and Numerical	Semi-annular and fin shape metal foam hybrid structure	Performance evaluation criteria augmentation, total entropy generation decrease in addition to exergetic efficiency growth maximally reach 360, 93.3 and 10.2%, correspondingly.

Solar energy is considered as a heat flux for solar collectors. The survey of a solar heat transfer device is trusted for increasing the thermal performance of the available system to obtain an outcome [41]. Moghaddaszadeh et al. [42] used two passive techniques to ameliorate the efficiency of heat exchangers in solar collectors. The outcomes illustrate that the use of nanofluid went up the Nusselt number up to 4%. Ghasemi and Ranjbar [43] in the numerical verification on solar parabolic collector demonstrated that adding nanoparticles increase thermal performance and using Al_2O_3 or CuO as nanofluid at 3% volume fraction enhances the heat exchange up to 28 and 35 percent, respectively. Reddy et al. [44] investigated the thermal analysis of parabolic trough collectors for various geometrical parameters such as porosity and fin aspect ratio. Their results pointed out that the porosity will increase the heat exchange by about 17%.

According to the presented literature review of previous works (Table 1) related to heat transfer enhancement of the PTSC by various passive methods, it can be concluded that among different kinds of utilized methods, employing helical axial fins has not been considered and analyzed previously. In the present work, one parabolic trough solar collector with inner helical axial fins as swirl generator or turbulator is considered and analyzed. This work consists of two sections. In the first section, four various schematics of the collector are appraised. In the second section, the selected geometry of the collector from the previous part (first section) is selected, and four various pitches of inner helical fins including 250, 500, 750 and 1000 mm are studied.

2. Materials, Methods and Boundary Conditions

Parabolic trough collector systems, as demonstrated in Figure 1a, in the studied geometry, a parabolic concentrator is used, which is due to the reduction of heat loss and increase of the produced temperature. Figure 1 shows the schematic of the checked geometry, and also Table 2 displays the geometrical constants and parameters. According to the check accomplished, for turbulent flow the entry region length is five times more than the diameter of the inner tube [35]; however, some researchers are of the opinion that it has a slight impact [45,46]. The three-dimensional numerical simulations have been done by finite volume method using commercial CFD code, ANSYS FLUENT 18.2. The geometrical parameters of the considered heat exchanger are illustrated in Figure 1c. The 30 °C inflow enters the tube at different velocities, namely 0.2, 0.24, 0.28 and 0.32 m·s^{-1}. Velocity–inlet and pressure–outlet boundary conditions were considered at the inlet and outlet ports, respectively. A heat flux of 60,000 W·m^{-2} (is calculated by dividing the power consumption (6000 W) to the side surface area (0.10 m^2) is transferred and applied to the outer part of the geometry made of an internally threaded steel layer. Moreover, the thermos–physical properties of tube material and fluid are noted in Table 3. The following situations are assumed in simulations:

- The inlet fluid at all velocities is 30 °C.
- The flow is turbulent and in a steady mood.
- The heat flux applied to the surface of the outer tube is 60,000 W·m-2.

Table 2. Physical parameters of the analyzed collector.

Parameters		Value
Diameter of inner the tube	D_1	30 mm
Diameter of outer the tube	D_2	60 mm
Length of the tube	L	2000 mm
Height of the fins	H	10 mm
Thickness of the fins	th	5 mm
Helix pitch of the fins	P	500 mm
Helical angle of the fins	α	180 °

Table 3. The thermos–physical properties of materials [46].

Fluid (Water)	Pipe (Steel)	Property
4182	502	Specific Heat [J/(kg·K)]
0.6	16	Thermal Conductivity [W/(m·K)]
998.2	7881.8	Density [kg/m^3]
0.001003	–	Viscosity [kg/m·s]

The present work includes two sections. In the first part, three geometries with various schematics of the inner helical axial fins are considered and investigated, and the obtained results are compared with the simple collector without any insert (fins). The schematics of the evaluated geometries in the first section of present work are illustrated in Figure 2. In the second section, the selected model according to the first section obtained results is considered with various pitch of the inner helical axial fins (P) including 250, 500, 750 and 1000 mm which the evaluated cases are depicted in Figure 3.

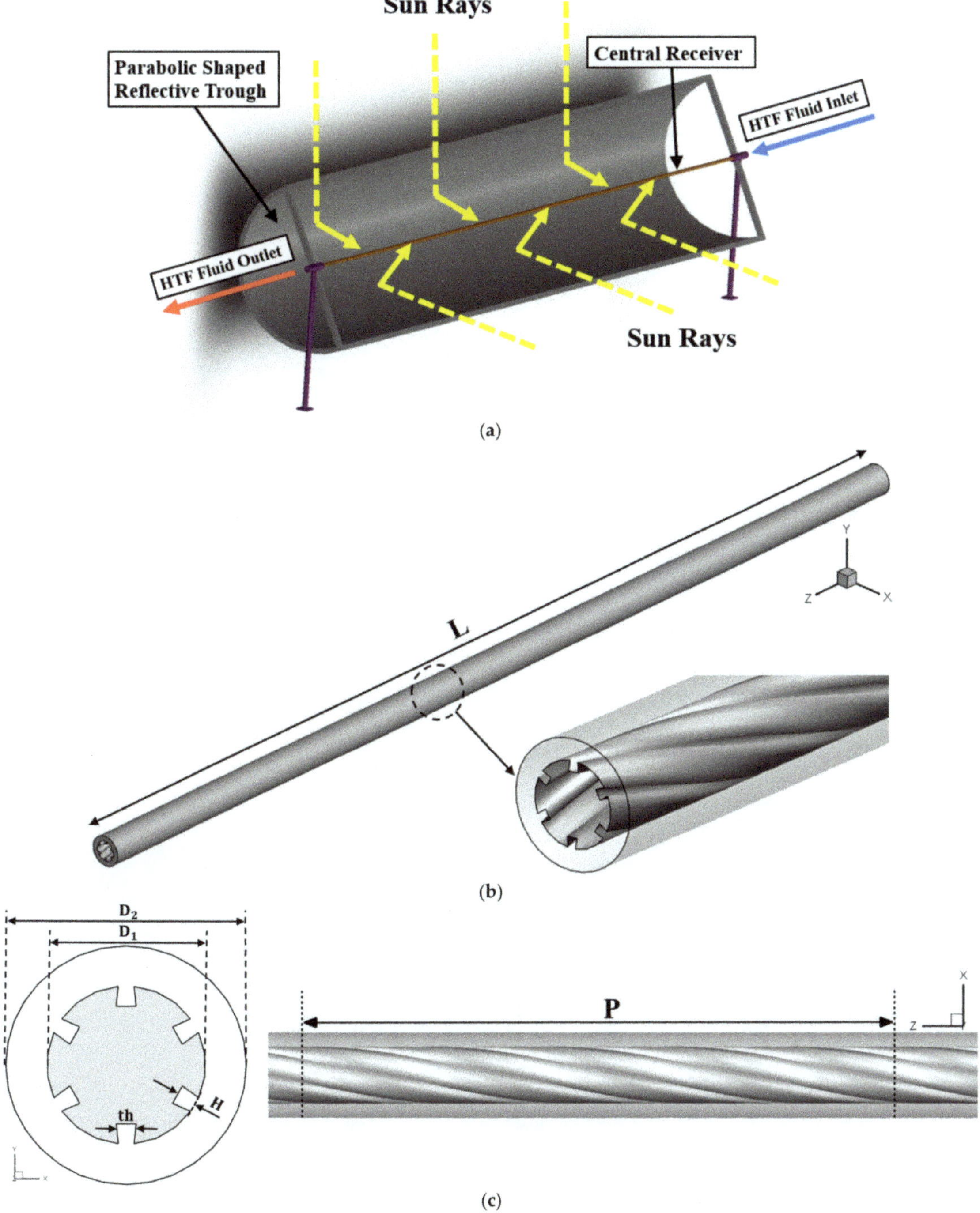

Figure 1. (**a**) A schematic of the PTC structure, (**b**) the schematic of the considered pipe and the inner helical axial fins, and (**c**) the geometrical parameters.

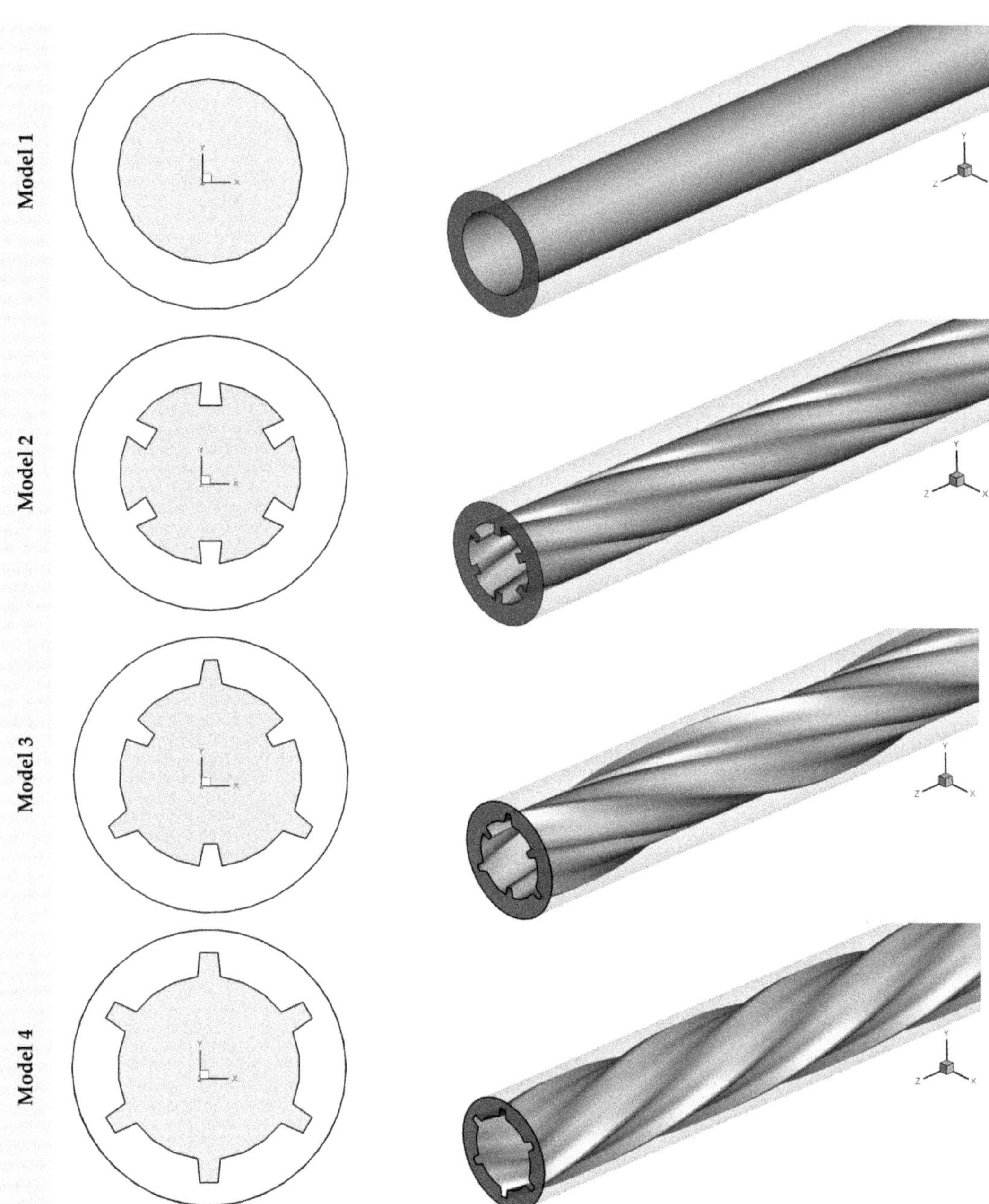

Figure 2. The schematics of the evaluated geometries in the first section of present work.

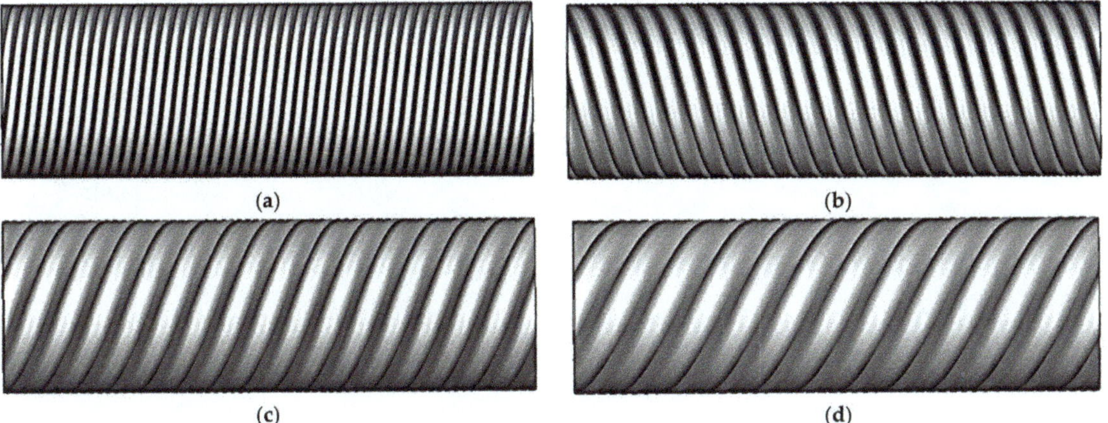

Figure 3. The schematics of the evaluated geometries in the second section of present work. (**a**) P = 250 mm, (**b**) P = 500 mm, (**c**) P = 750 mm and (**d**) P = 1000 mm.

3. Governing Equations and Dimensionless Parameters

The actual system is governed by the continuity, momentum, and energy equations that are defined, respectively, as [47,48]:

$$\frac{\partial \rho}{\partial t} + \nabla \nabla . \left(\rho \vec{v} \right) = S_m \tag{1}$$

$$\frac{\partial \left(\rho \vec{v} \right)}{\partial t} + \nabla . \left(\rho \vec{v} \vec{v} \right) = -\nabla p + \nabla \left(\overline{\overline{\tau}} \right) + \rho \vec{g} + \vec{F} \tag{2}$$

$$\frac{\partial (\rho E)}{\partial t} + \nabla . \left(\vec{v} (\rho E + p) \right) = \nabla \left(k_{eff} \nabla T - \sum_j h_j \vec{J}_j + \left(\overline{\overline{\tau}}_{eff} . \vec{v} \right) \right) + S_h \tag{3}$$

The thermal performance coefficient (η) can be exerted to quantify the yield of a heat exchanger that contains the Nusselt number and the friction factor (*f*), and is defined as [49–58]:

$$Nu = \frac{h_m d_h}{k} \tag{4}$$

$$f = \frac{2 d_h \Delta P}{\rho u^2 L} \tag{5}$$

$$\eta = \left(\frac{Nu}{Nu_0} \right) \left(\frac{f_0}{f} \right)^{\frac{1}{3}} \tag{6}$$

The three-dimensional numerical simulations have been done by finite volume method using commercial CFD code, ANSYS FLUENT 18.2. The spatial discretization of the mass, momentum, turbulence kinetic energy, turbulence dissipation rate and energy equations has been achieved by a second-order upwind scheme. The velocity-pressure coupling has been overcome by the SIMPLE algorithm. To calculate the gradients, Green-Gauss cell-based method has been utilized. The convergence criteria were set to 10^{-6} for the residuals of all equations except energy equation (10^{-8}).

4. Results

4.1. Grid Independency Study

The generated grids for the proposed collector geometry is displayed in Figure 4. Four variant grids were checked to certify the accuracy of the numerical outcomes. The analysis outcomes are offered in Table 4. It is vivid that the two middle mesh sizes do not vary substantially and in the interest of minimizing the computational time whilst attaining trust outcomes, demonstrate that a third option is a good selection.

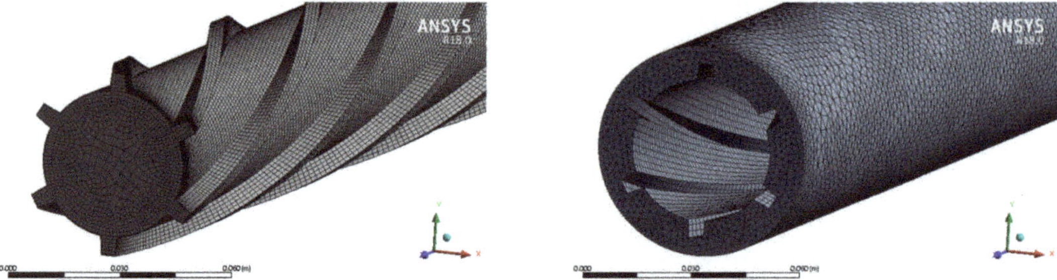

Figure 4. Considered grid for proposed geometry as PTC.

Table 4. Outcomes of the grid independence check.

Nusselt Number Error	Grid Number
17.2%	65421
9.04%	121470
2.13%	164957
1.98%	235712

4.2. Validation Study

The Dittus-Boelter correlation [59] provides the Nusselt number as a function of the Re and Pr according to Equation (7):

$$\mathrm{Nu} = 0.023\ \mathrm{Re}^{0.8}\ \mathrm{Pr}^{0.4} \tag{7}$$

Additionally, the Seban-Shimazaki correlation [59] (Equation (8)) is employed to compare with the simulation outcomes. A high correlation is suggested for $\mathrm{Pe_D} \geq 100$. The outcomes of the validation study are demonstrated in Figure 5. Therefore, the available work presents good accuracy. Additionally, to validate the present numerical model with the results of the performance of a real PTC system, the experimental results of Zou et al. [60] are used, and the comparison is presented in Figure 6. It can be concluded that present numerical method has good capacity for modeling the fluid and thermal characteristics in a PTC system. The uncertainty in measurement of the experimental parameters of the Zou et al. work [60] was as follows. The uncertainty in HTF velocity measurement was ± 0.05 m/s for HTF velocity up to 0.28 m/s and ± 0.2 m/s for HTF velocity between 0.28 and 0.32 m/s. The uncertainty of HTF temperature was $\pm 0.3°$C in this study.

$$\mathrm{Nu} = 5 + 0.025\ \mathrm{Pe_D}^{0.8} \tag{8}$$

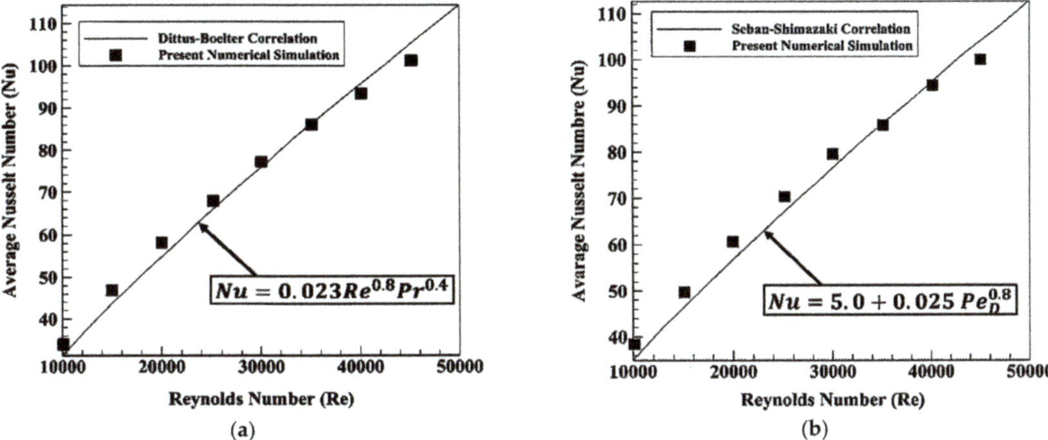

(a)

(b)

Figure 5. Validation betwixt the available simulation and (**a**) Dittus–Boelter correlation [41]; (**b**) Seban–Shimazaki correlation [59].

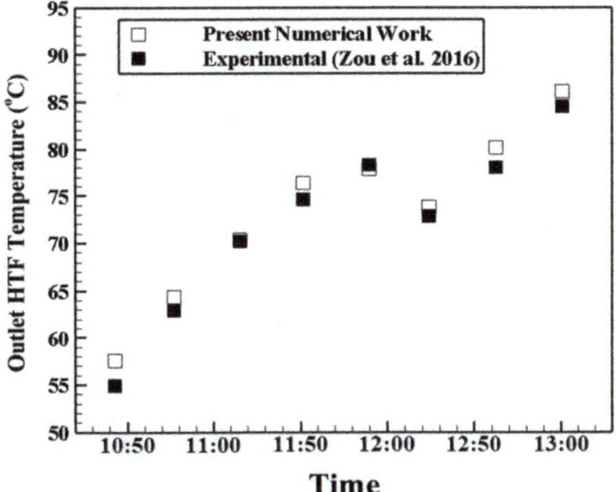

Figure 6. Results of the validation analysis between the present numerical work and experimental work of Zou et al. [60].

4.3. The Impact of the Schematic of the Proposed Collector's Fins

In this section, the influence of the collector shape on thermal efficiency is investigated numerically. Three various models are considered, and the obtained results have been compared with the simple shaped collector. The considered models here are shown in Figure 2. The variation of heat transfer coefficient versus different inlet velocity for various models is illustrated in Figure 7. Hence, it can be concluded that firstly, all the proposed models show higher heat transfer coefficient more than the simple collector (Model 1). Secondly, it is depicted that among the studied models, the maximum heat transfer coefficient belongs to Model 2 in all studied inlet velocities. Additionally, the minimum heat transfer coefficient belongs to Model 4 in all considered inlet velocities.

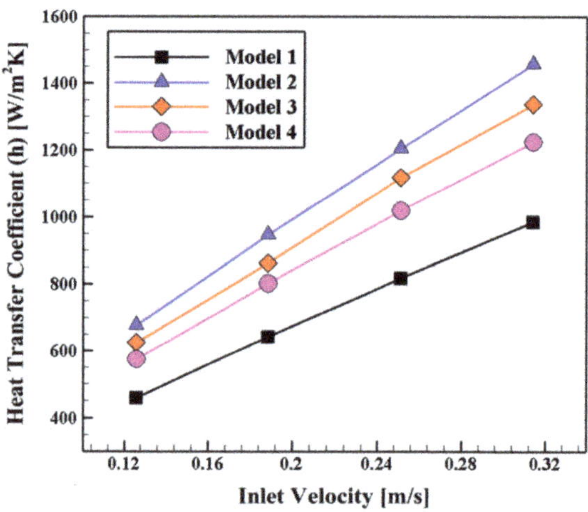

Figure 7. Heat transfer coefficient variation with the inlet velocity for various models at P = 500 mm.

To realize better the effect of geometry on the thermal performance of the proposed system, the streamline with the contour of temperature for all four studied models is shown in Figure 8. Hence, it can result that the schematic of the geometry and also the attendance of the inner axial helical fins have a significant impression on the rotation flow production and finally more heat transfer rate. Furthermore, contours show that Model 4 could enhance the heat transfer rate more than the other models. Additionally, all models depict better thermal performance in comparison with the collector equipped with simple channel (Model 1) because of the presence of swirl flows generated by the helical fins.

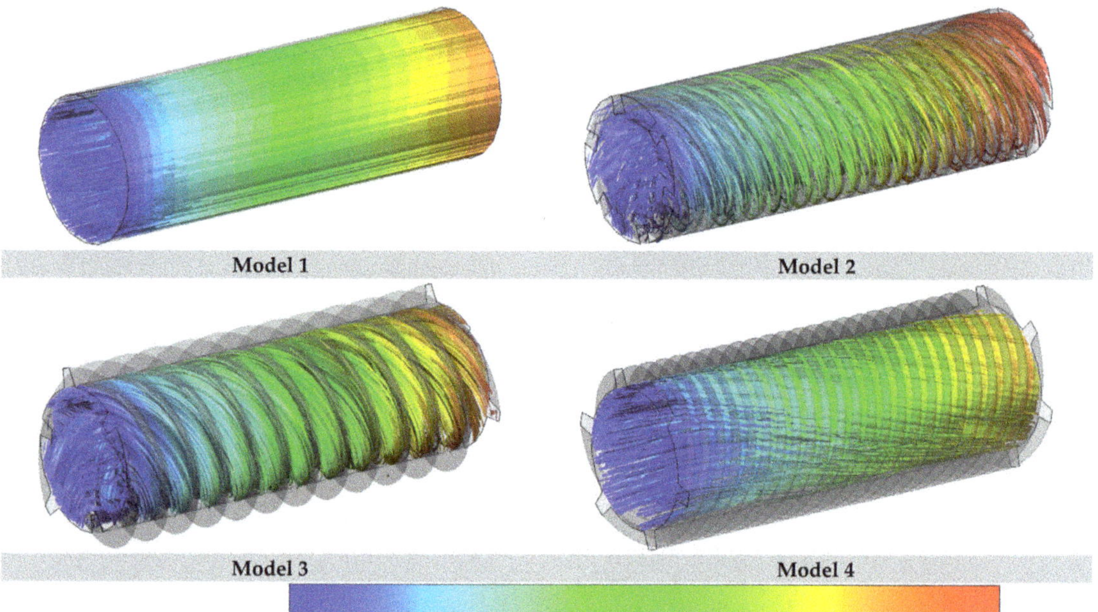

Figure 8. The streamline with contour of temperature for various models at P = 500 mm.

The variation of pressure drop and friction factor (f) versus various inlet velocities for various models are illustrated in Figure 9a,b, respectively. Figure 9a depicts that all models have more pressure drop compare to Model 1 (collector with simple channel, without helical fins) obviously because of presence of swirl flows. The maximum pressure drop belongs to Model 2 and the Model 4 shows lowest pressure drop among these four models. Additionally, as the inlet velocities rises, the pressure drop increases due to the effect of forced convection in the channel. This trend is the same for all models. Moreover, by increasing the inlet velocities, the differences between the studied models rises. This point could be important considering the pressure drop factor, as the differences between Model 4 and Model 1 (as base model) are very low.

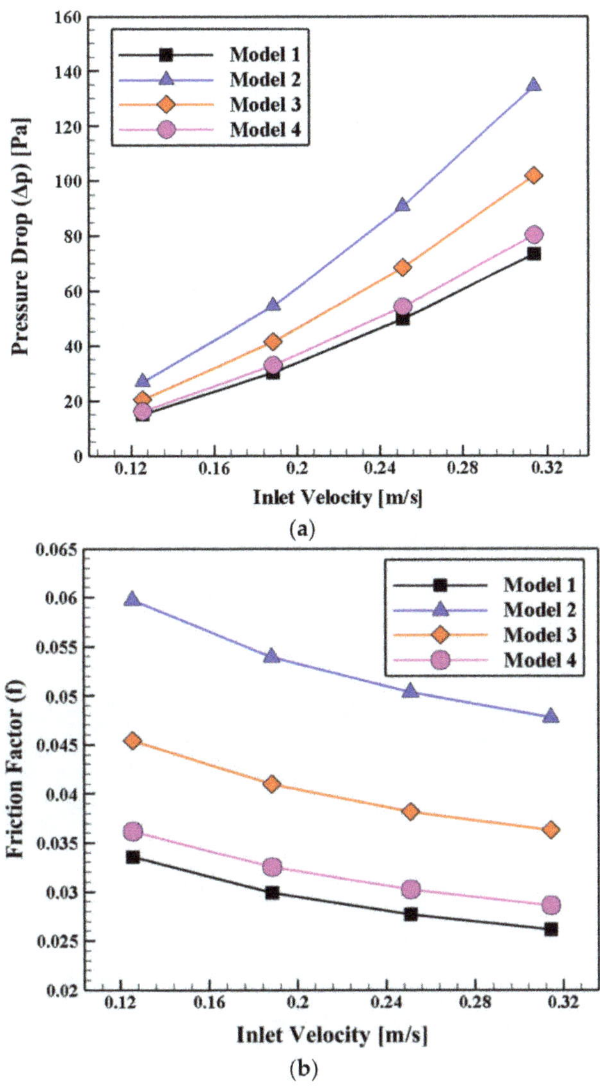

Figure 9. (**a**) Pressure drop and (**b**) friction factor variation with the inlet velocity for various models at P = 500 mm.

Figure 9b shows that the trend of variation of fraction factor among the models is the same with pressure drop (Figure 9a). However, it should be focused that the variation of friction factor with inlet velocities is not the same with pressure drop, and they are exactly the opposite. By increasing the inlet velocity, the friction factor declines. According to Equation (5), the friction factor is directly related to the pressure drop; however, the squared velocity is placed at the denominator of the equation.

Contours of temperature at outlet for various models and P = 500 mm are presented in Figure 10. The impact of changing the schematic of the collector on temperature distribution in the outlet port and also the heat transfer rate is presented clearly in this figure. Accordingly, all models show better temperature distribution than the Model 1 (simple one, without fins).

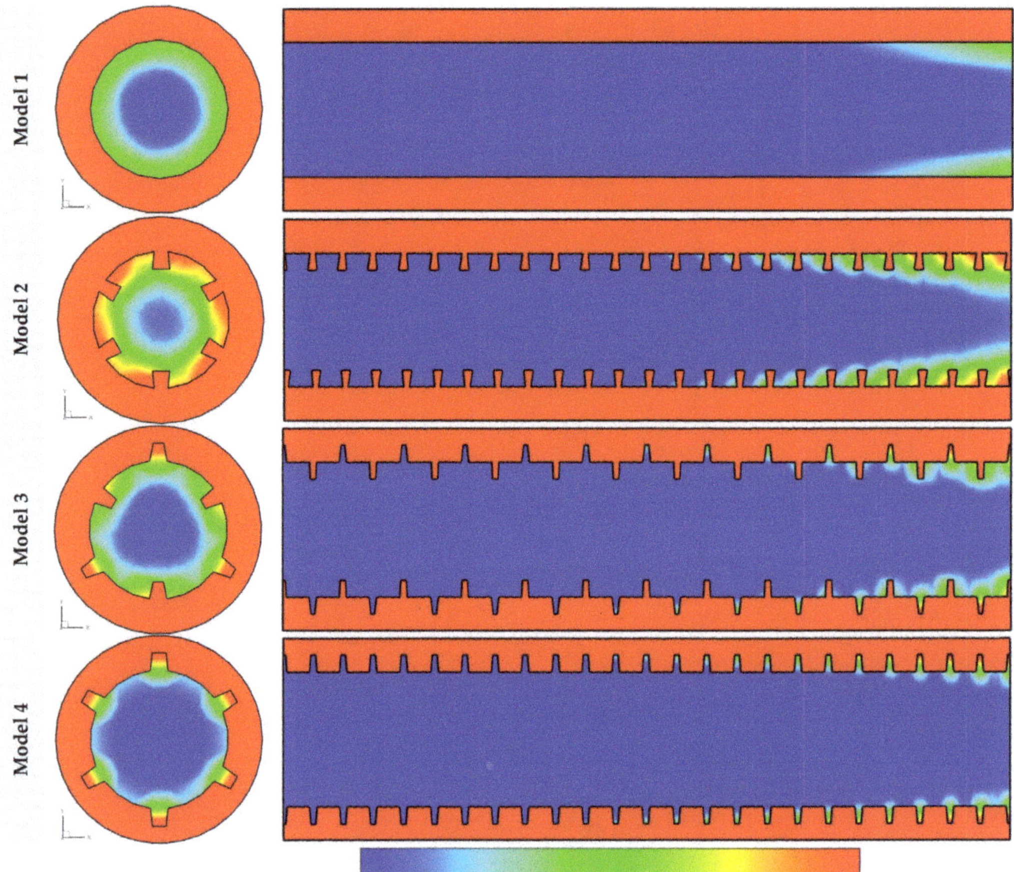

Figure 10. Contours of temperature at outlet and a slice (X = 0) for various models at P = 500 mm and V_{Inlet} = 0.1256 m/s.

The best parameter to comprehensively study the impact of efficient parameters on the performance of the proposed collector is thermal efficiency (η), which is calculated by Equation (6). In this Equation, the index 0 refers to the base collector (without fins). The variation of this parameter versus various inlet velocity for variant models are presented in Figure 11. Hence, it can be firstly concluded that all the models illustrate higher thermal performance than Model 1 (simple collector without fins), which means that the values of thermal performance are higher than unity. Furthermore, the highest and lowest thermal performance improvements (compared to Model 1) belong to the Model 3 by 23.1% (at $V_{inlet} = 0.2512$ m/s) and Model 3 by 20.7% (at $V_{inlet} = 0.1884$ m/s), respectively. Contours of velocity magnitude at surface X = 0 for various models at P = 500 mm and $V_{Inlet} = 0.1256$ m/s are illustrated in Figure 12. It can be realized that among the considered models in this section, Model 2 depicts higher velocity magnitude and consequently more swirl flows in the proposed collector, which leads to a higher heat transfer rate.

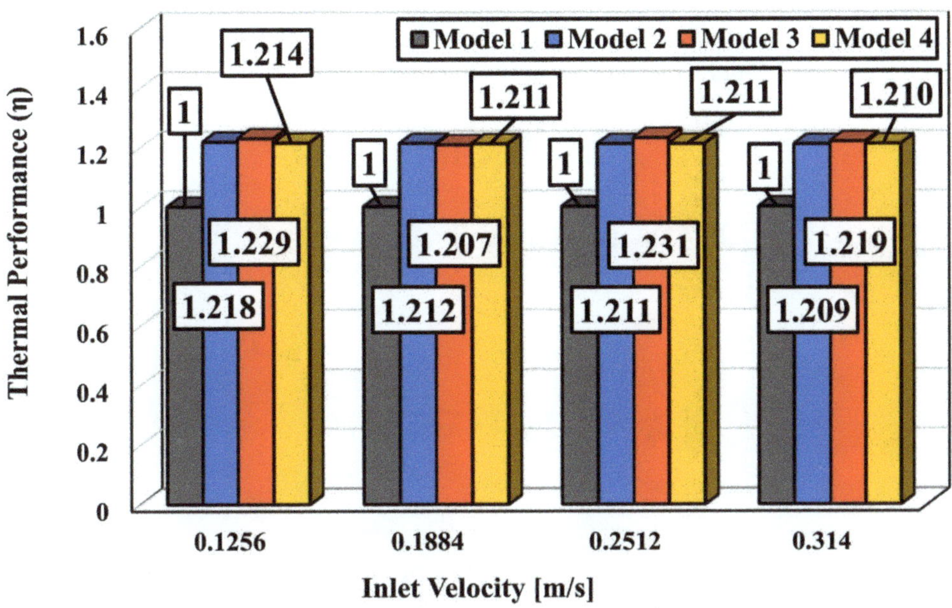

Figure 11. The thermal performance (η) versus inlet velocity for various models.

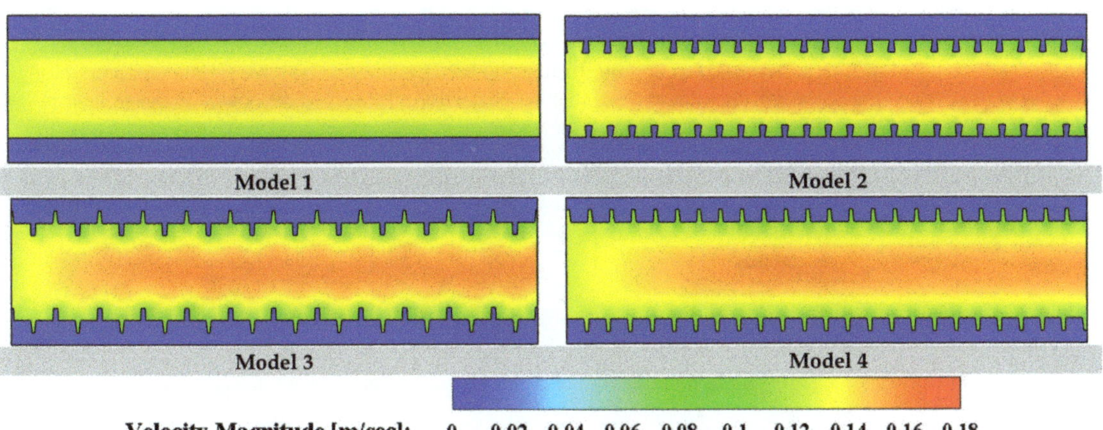

Figure 12. Contours of velocity magnitude at surface X = 0 for various models at P = 500 mm and $V_{Inlet} = 0.1256$ m/s.

4.4. The Impact of the Inner Helical Fins Pitch (P)

In the second section, according to the obtained numerical results from the last section (Section 4.3), Model 2 (see Figure 2) is considered as the selected model here. Results from the last section showed that Model 2 has the maximum heat transfer coefficient with the highest pressure drop. Therefore, to study the impact of the inner helical fins pitch, this model is used. In this section, the impact of the inner helical fins pitch on the thermal efficiency of the system is evaluated numerically. Four various pitches of the inner helical fins including 250, 500, 750 and 1000 mm are considered, and the obtained outcomes are compared with the simple collector (without fins). The considered models here are shown in Figure 3. The variation of heat transfer coefficient versus different inlet velocity for various models is illustrated in Figure 13.

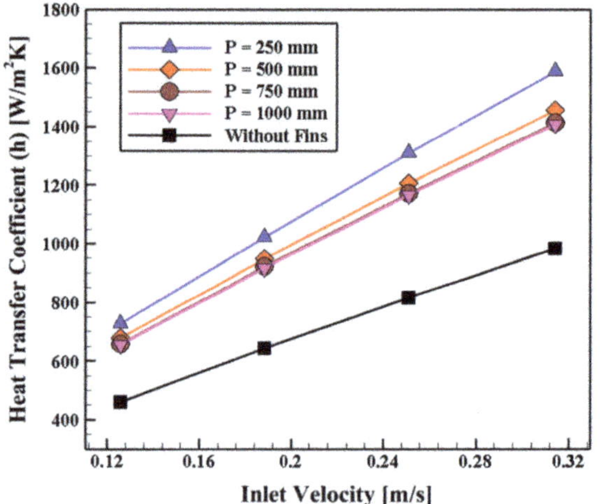

Figure 13. Heat transfer coefficient variation with the inlet velocity for various pitches of inner helical axial fins.

According to Figure 13, it can be concluded that firstly, all the proposed models show higher heat transfer coefficient more than the simple collector (without fins). Secondly, it is depicted that among the studied models, the unmost heat transfer coefficient belongs to case with P = 250 mm in all studied inlet velocities. Additionally, the least heat transfer coefficient belongs to the case with P = 1000 mm in all considered inlet velocities. To realize better the impact of the pitch of inner helical fins on the thermal performance of the proposed collector, the contour of temperature at outlet and a slice in the middle of the computational domain (X = 0) for all cases are shown in Figure 14. It can be obviously seen that higher temperature in the proposed collector is achieved at lower pitch of the inner helical fins.

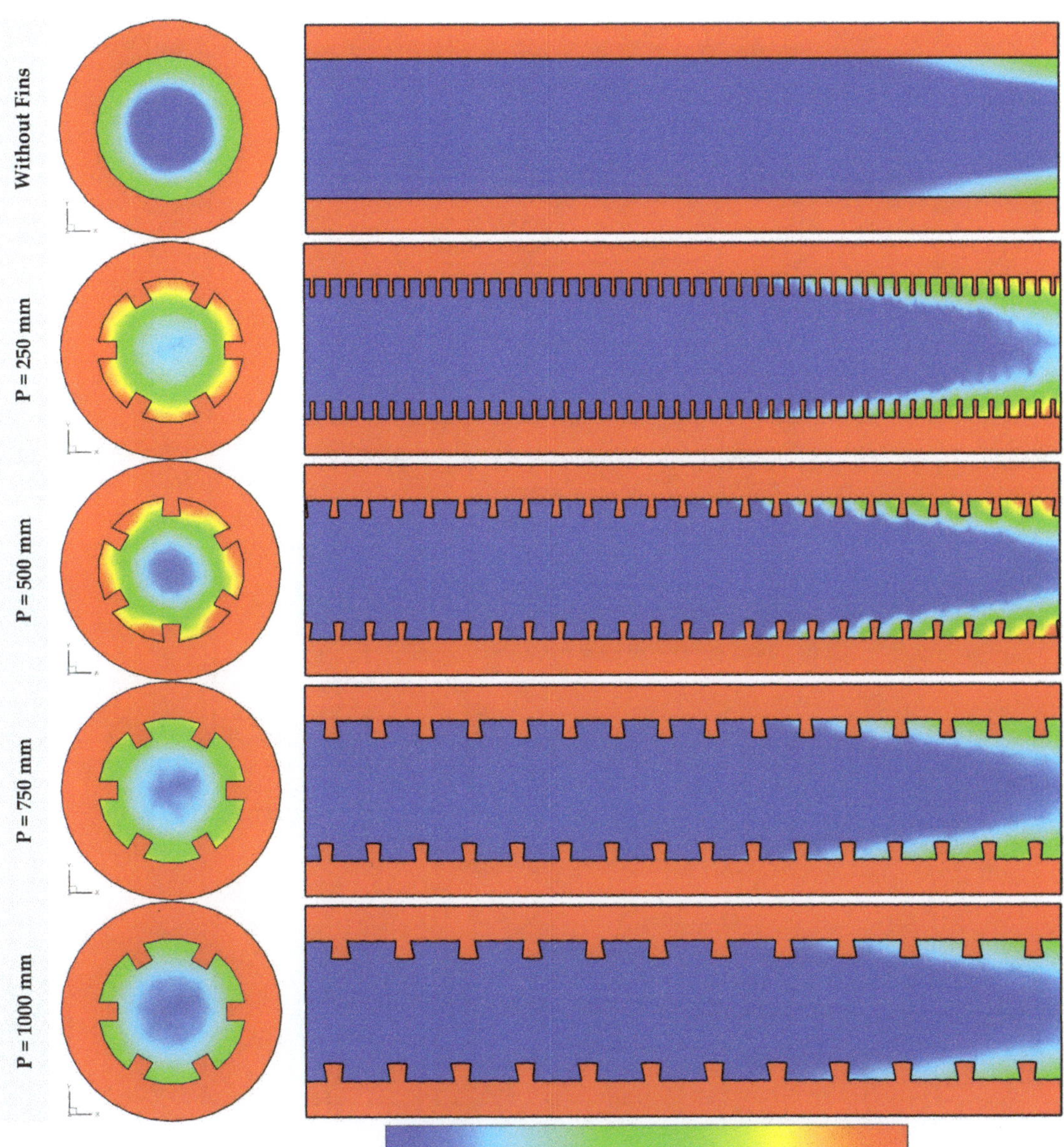

Figure 14. Contours of temperature at outlet and a slice (X = 0) for various models at V_{Inlet} = 0.1256 m/s.

The variations of pressure drop and friction factor (f) versus various inlet velocities for various models are illustrated in Figure 15a,b, respectively. Figure 15a depicts that all models have higher pressure drop compared to the case without helical fins obviously because of the presence of swirl flows. The greatest pressure drop belongs to the case with P = 250 mm, and the case with P = 1000 mm shows the lowest pressure drop among the investigated models. Additionally, as the inlet velocities rise, the pressure drop increases due to the effect of forced convection in the channel. This trend is the same for all models. Moreover, by increasing the inlet velocities, the differences between the studied models rise. This point could be important considering the pressure drop factor, as the differences between the cases with P = 500, 750 and 1000 mm are very low.

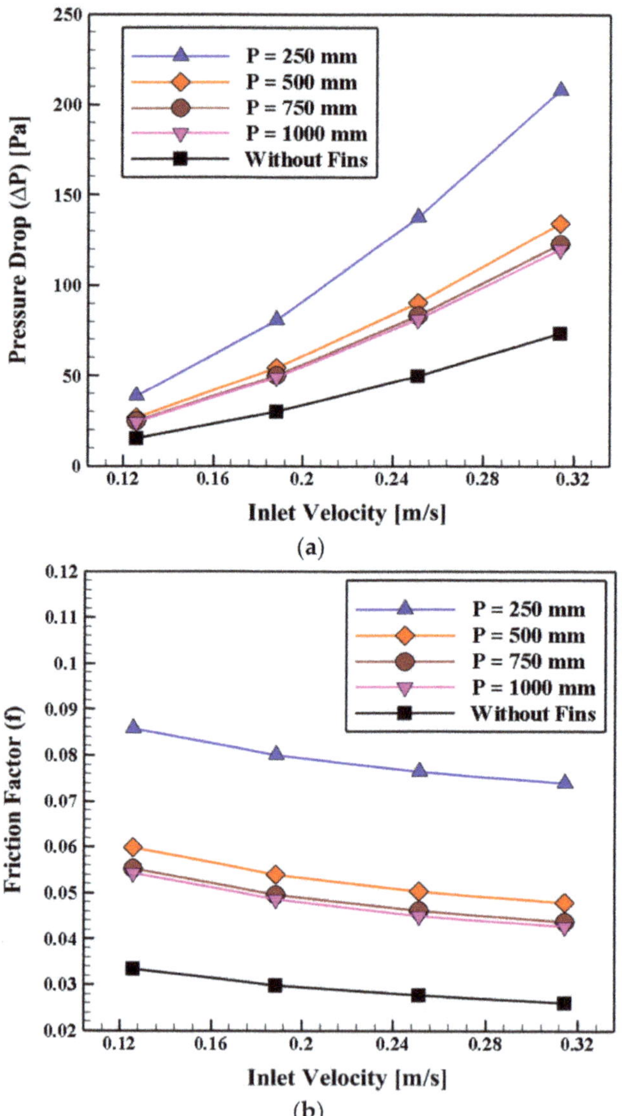

Figure 15. (**a**) Pressure drop and (**b**) friction factor (f) variation with the inlet velocity for various models at P = 500 mm.

Figure 15b shows that the trend of variation of fraction factor among the models is the same with pressure drop (Figure 15a). On the other hand, as same as Figure 15a, the highest and lowest friction factor belong to the cases with P = 250 mm and P = 1000 mm, respectively. However, it should be focused that the variation of friction factor with inlet velocities is not the same with pressure drop and they are exactly the opposite. By increasing the inlet velocity, the friction factor declines. According to Equation (5), the friction factor is directly related to pressure drop; however, the squared velocity is placed at the denominator of the equation.

The streamline with the contour of temperature for various models are presented in Figure 16. The impact of changing the schematic of the collector on temperature distribution in the outlet port and also the heat transfer rate is presented clearly in this figure. Accordingly, all models show better temperature distribution than the case without fins.

Figure 16. *Cont.*

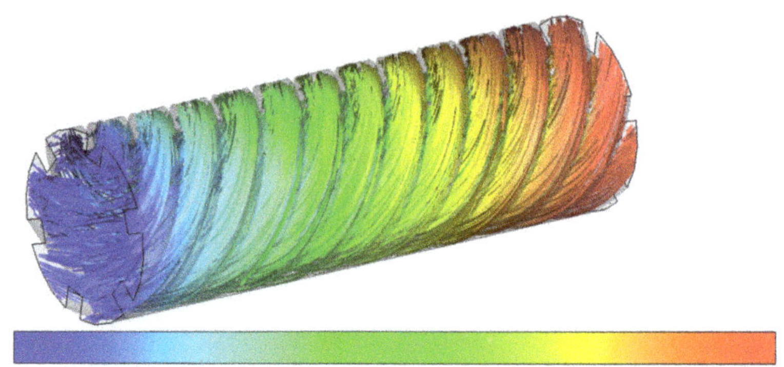

Figure 16. The streamline with contours of temperature for various models.

The best parameter to study comprehensively the impact of efficient parameters on the performance of the proposed collector is thermal efficiency (η), which is calculated by Equation (6). The variation of this parameter is presented in Figure 17. Therefore, it can be concluded firstly that all the models illustrate higher thermal performance than the case without fins, which means that the values of thermal performance are higher than unity. Furthermore, the greatest and smallest thermal performance improvement (compare to the case without fins) belong to the case with P = 1000 mm by 21.53% and the case with P = 250 mm by 14.1% for the latest velocity, respectively.

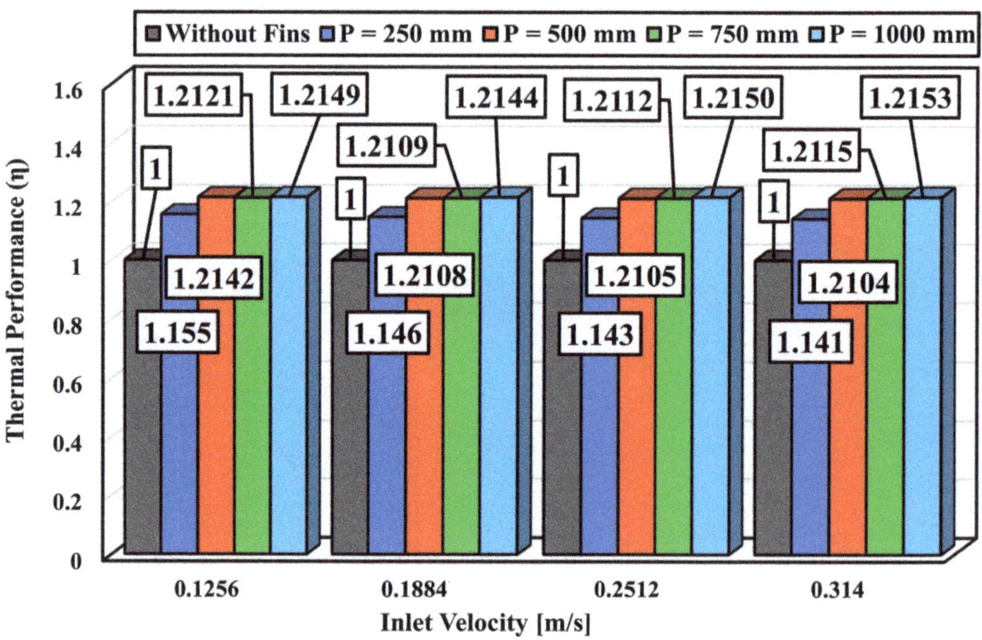

Figure 17. The thermal performance (η) versus various inlet velocities for different models.

5. Conclusions and Future Scope

In the present work, a PTC collector with inner helical axial fins as a swirl generator or turbulator is considered and analyzed. This work consists of two parts. In the first part, four various geometries of the collector are appraised. In the second part, the selected schematic of the collector from the first sector is selected and four various pitches of inner helical fins including 250, 500, 750 and 1000 mm are studied. Results presented that the thermal performance improvement by 23.1% could be achieved by using one of the proposed innovative parabolic trough solar collector compare to the simple one. Additionally, the utmost and least thermal performance improvement (compare to the case without fins) belong to the case with P = 1000 mm by 21.53% (at V_{inlet} = 0.314 m/s) and the case with P = 250 mm by 14.1% (at V_{inlet} = 0.314 m/s), respectively.

Although present work has presented new insight into the flow and thermal characteristics of a parabolic trough solar collector combining innovative inner helical axial fins as a swirl generator, additional research is required to determine the capability of this design in more realistic engineering situations by the development of an experimental prototype.

Author Contributions: Investigation, Software, Validation, Writing—original draft, M.Z.; Formal analysis, Methodology, Validation, Writing—review & editing, S.S.M.A.; Supervision, Writing—review & editing, S.S.; Funding acquisition, Writing—review & editing, M.S.P. All authors have read and agreed to the published version of the manuscript.

Funding: This research received no external funding.

Institutional Review Board Statement: Not applicable.

Informed Consent Statement: Not applicable.

Data Availability Statement: Data available on request due to restrictions eg privacy or ethical.

Conflicts of Interest: The authors declare no conflict of interest.

Expects Data: Data available on request from the authors: The data that support the findings of this study are available from the corresponding author upon reasonable request.

Nomenclature

A	Area, m^2	*Greek and Symbols*	
C_P	Specific heat capacity, [kJ/(kg. K)]	ρ	Density, kg/m^3
D_1	Diameter of inner the tube, [m]	μ	Viscosity, kg/m·s
D_2	Diameter of outer the tube, [m]	$\overline{\overline{\tau}}$	Stress tensor
d	Diameter, [m]	σ	Turbulent Prandtl number
f	Darcy friction factor, [nd]	η	Thermal performance, [nd]
G_b	Generation of turbulence kinetic energy due to buoyancy, [J/kg]	α	Helical angle of the Fins, [degree]
G_k	Generation of turbulence kinetic energy due to the mean velocity gradients, [J/kg]	k	turbulent kinetic energy per unit mass, [J/kg]
g	Gravity, $[m/s^2]$	ε	energy dissipation rate per unit mass, [W/kg]
H	Height of the Fins, [m]	*Dimensionless Parameter*	
h	Heat transfer coefficient, [W/(m^2·K)]	Pe	Péclet number, Pe = Re. Pr, [nd]
k	Thermal conductivity, [W/(m·K)]	Pr	Prandtl number, $Pr = \frac{c_p \mu}{k}$, [nd]
L	Length of the tube, [m]	Nu	Nusselt number, $Nu = \frac{hd}{k}$, [nd]
M_t	Turbulent Mach number, [nd]	Re	Reynolds number, $Re = \frac{\rho u d}{\mu}$, [nd]

P	Helix pitch of the Fins, [m]	*Index*		
p	Pressure, [Pa]	0	Reference	
S_h	Heat generation, [J/m^3]	eff	Effective	
S_m	Mass generation, [kg/m^3]	h	Hydraulic	
T	Temperature, [°C]	m	Average	
t	Time, [s]	PR	Performance Ratio	
th	Thickness of the Fins, [m]	ref	reference	
v	Velocity, [m·s^{-1}]	t	turbulence	

Abbreviation

PTCs   Parabolic Trough Collectors
Exp   Experimental
Num   Numerical

References

1. Berger, S.A.; Talbot, L.; Yao, L.S. Flow in curved pipes. *Annu. Rev. Fluid Mech.* **1983**, *15*, 461–512. [CrossRef]
2. Sheikholeslami, M.; Gorji-Bandpy, M.; Ganji, D.D. Review of heat transfer enhancement methods: Focus on passive methods using swirl flow devices. *Renew. Sustain. Energy Rev.* **2015**, *49*, 444–469. [CrossRef]
3. Esfe, M.H.; Saedodin, S.; Wongwises, S.; Toghraie, D. An experimental study on the effect of diameter on thermal conductivity and dynamic viscosity of Fe/water nanofluids. *J. Therm. Anal. Calorim.* **2015**, *119*, 1817–1824. [CrossRef]
4. Zaboli, M.; Ajarostaghi, S.S.; Noorbakhsh, M.; Delavar, M.A. Effects of geometrical and operational parameters on heat transfer and fluid flow of three various water based nanofluids in a shell and coil tube heat exchanger. *SN Appl. Sci.* **2019**, *1*, 1387. [CrossRef]
5. Kim, K.; Lee, K.S. Frosting and defrosting characteristics of surface-treated louvered-fin heat exchangers: Effects of fin pitch and experimental conditions. *Int. J. Heat Mass Transf.* **2013**, *60*, 505–511. [CrossRef]
6. Gao, D.; Wang, Y. Generating high-purity orbital angular momentum vortex waves from Cassegrain meta-mirrors. *Eur. Phys. J. Plus* **2020**, *135*. [CrossRef]
7. Hobbi, A.; Siddiqui, K. Experimental study on the effect of heat transfer enhancement devices in flat-plate solar collectors. *Int. J. Heat Mass Transf.* **2009**, *52*, 4650–4658. [CrossRef]
8. Nuntaphan, A.; Vithayasai, S.; Vorayos, N.; Vorayos, N.; Kiatsiriroat, T. Use of oscillating heat pipe technique as extended surface in wire-on-tube heat exchanger for heat transfer enhancement. *Int. Commun. Heat Mass Transf.* **2010**, *37*, 287–292. [CrossRef]
9. Dizaji, H.S.; Jafarmadar, S.; Asaadi, S. Experimental exergy analysis for shell and tube heat exchanger made of corrugated shell and corrugated tube. *Exp. Therm. Fluid Sci.* **2017**, *81*, 475–481. [CrossRef]
10. Tang, L.H.; Zeng, M.; Wang, Q.W. Experimental and numerical investigation on air-side performance of fin-and-tube heat exchangers with various fin patterns. *Exp. Therm. Fluid Sci.* **2009**, *33*, 818–827. [CrossRef]
11. Azari, M.; Sadeghi, A.; Chakraborty, S. Electroosmotic flow and heat transfer in a heterogeneous circular microchannel. *Appl. Math. Model.* **2020**, *87*, 640–654. [CrossRef]
12. Darzi, A.A.R.; Farhadi, M.; Sedighi, K. Experimental investigation of convective heat transfer and friction factor of Al$_2$O$_3$/water nanofluid in helically corrugated tube. *Exp. Therm. Fluid Sci.* **2014**, *57*, 188–199. [CrossRef]
13. Du, T.; Chen, Q.; Du, W.; Cheng, L. Performance of continuous helical baffled heat exchanger with varying elliptical tube layouts. *Int. J. Heat Mass Transf.* **2019**, *133*, 1165–1175. [CrossRef]
14. Bahiraei, M.; Mazaheri, N.; Rizehvandi, A. Application of a hybrid nanofluid containing graphene nanoplatelet-platinum composite powder in a triple-tube heat exchanger equipped with inserted ribs. *Appl. Therm. Eng.* **2019**, *149*, 588–601. [CrossRef]
15. Abolarin, S.M.; Everts, M.; Meyer, J.P. Heat transfer and pressure drop characteristics of alternating clockwise and counter clockwise twisted tape inserts in the transitional flow regime. *Int. J. Heat Mass Transf.* **2019**, *133*, 203–217. [CrossRef]
16. Zhang, Y.; Zhou, F.; Kang, J. Flow and heat transfer in drag-reducing polymer solution flow through the corrugated tube and circular tube. *Appl. Therm. Eng.* **2020**, 115185. [CrossRef]
17. Rashidi, S.; Hormozi, F.; Sundén, B.; Mahian, O. Energy saving in thermal energy systems using dimpled surface technology—A review on mechanisms and applications. *Appl. Energy* **2019**, *250*, 1491–1547. [CrossRef]
18. Kareem, Z.S.; Abdullah, S.; Lazim, T.M.; Jaafar, M.M.; Wahid, A.F. Heat transfer enhancement in three-start spirally corrugated tube: Experimental and numerical study. *Chem. Eng. Sci.* **2015**, *134*, 746–757. [CrossRef]
19. Lu, G.; Zhou, G. Numerical simulation on performances of plane and curved winglet type vortex generator pairs with punched holes. *Int. J. Heat Mass Transf.* **2016**, *102*, 679–690. [CrossRef]
20. Mashoofi, N.; Pesteei, S.M.; Moosavi, A.; Dizaji, H.S. Fabrication method and thermal-frictional behavior of a tube-in-tube helically coiled heat exchanger which contains turbulator. *Appl. Therm. Eng.* **2017**, *111*, 1008–1015. [CrossRef]
21. Aghaei, A.; Sheikhzadeh, G.A.; Goodarzi, M.; Hasani, H.; Damirchi, H.; Afrand, M. Effect of horizontal and vertical elliptic baffles inside an enclosure on the mixed convection of a MWCNTs-water nanofluid and its entropy generation. *Eur. Phys. J. Plus* **2018**, *133*, 486. [CrossRef]
22. Noorbakhsh, M.; Zaboli, M.; Ajarostaghi, S.S. Numerical evaluation of the effect of using twisted tapes as turbulator with various geometries in both sides of a double-pipe heat exchanger. *J. Therm. Anal. Calorim.* **2020**, *140*, 1341–1353. [CrossRef]

23. Kwon, B.; Liebenberg, L.; Jacobi, A.M.; King, W.P. Heat transfer enhancement of internal laminar flows using additively manufactured static mixers. *Int. J. Heat Mass Transf.* **2019**, *137*, 292–300. [CrossRef]
24. Ho, C.J.; Hsieh, Y.J.; Rashidi, S.; Orooji, Y.; Yan, W.M. Thermal-hydraulic analysis for alumina/water nanofluid inside a mini-channel heat sink with latent heat cooling ceiling—An experimental study. *Int. Commun. Heat Mass Transf.* **2020**, *112*, 104477. [CrossRef]
25. Liu, H.L.; Qi, D.H.; Shao, X.D.; Wang, W.D. An experimental and numerical investigation of heat transfer enhancement in annular microchannel heat sinks. *Int. J. Therm. Sci.* **2019**, *142*, 106–120. [CrossRef]
26. Ramalingam, S.; Dhairiyasamy, R.; Govindasamy, M. Assessment of heat transfer characteristics and system physiognomies using HYBRID nanofluids in an automotive radiator. *Chem. Eng. Process. Process. Intensif.* **2020**, 107886. [CrossRef]
27. Qi, C.; Fan, F.; Pan, Y.; Liu, M.; Yan, Y. Effects of turbulator with round hole on the thermo-hydraulic performance of nanofluids in a triangle tube. *Int. J. Heat Mass Transf.* **2020**, *146*, 118897. [CrossRef]
28. Nakhchi, M.E.; Esfahani, J.A. Numerical investigation of heat transfer enhancement inside heat exchanger tubes fitted with perforated hollow cylinders. *Int. J. Therm. Sci.* **2020**, *147*, 106153. [CrossRef]
29. Rahbarshahlan, S.; Esmaeilzadeh, E.; Khosroshahi, A.R.; Bakhshayesh, A.G. Numerical simulation of fluid flow and heat transfer in microchannels with patterns of hydrophobic/hydrophilic walls. *Eur. Phys. J. Plus* **2020**, *135*, 157. [CrossRef]
30. Jamesahar, E.; Sabour, M.; Shahabadi, M.; Mehryan, S.A.; Ghalambaz, M. Mixed convection heat transfer by nanofluids in a cavity with two oscillating flexible fins: A fluid-structure interaction approach. *Appl. Math. Model.* **2020**, *82*, 72–90. [CrossRef]
31. Benabderrahmane, A.; Benazza, A.; Hussein, A.K. Heat transfer enhancement analysis of tube receiver for parabolic trough solar collector with central corrugated insert. *J. Heat Transf.* **2020**, *142*. [CrossRef]
32. Akbarzadeh, S.; Valipour, M.S. Experimental study on the heat transfer enhancement in helically corrugated tubes under the non-uniform heat flux. *J. Therm. Anal. Calorim.* **2020**. [CrossRef]
33. Chakraborty, O.; Das, B.; Gupta, R.; Debbarma, S. Heat transfer enhancement analysis of parabolic trough collector with straight and helical absorber tube. *Therm. Sci. Eng. Prog.* **2020**, *20*, 100718. [CrossRef]
34. Amani, K.; Ebrahimpour, M.; Akbarzadeh, S.; Valipour, M.S. The utilization of conical strip inserts in a parabolic trough collector. *J. Therm. Anal. Calorim.* **2020**. [CrossRef]
35. Khan, M.S.; Yan, M.; Ali, H.M.; Amber, K.P.; Bashir, M.A.; Akbar, B.; Javed, S. Comparative performance assessment of different absorber tube geometries for parabolic trough solar collector using nanofluid. *J. Therm. Anal. Calorim.* **2020**, *142*, 2227–2241. [CrossRef]
36. Saedodin, S.; Zaboli, M.; Ajarostaghi, S.S.M. Hydrothermal analysis of heat transfer and thermal performance characteristics in a parabolic trough solar collector with Turbulence-Inducing elements. *Sustain. Energy Technol. Assess.* **2021**, *46*, 101266. [CrossRef]
37. Vasanthi, P.; Reddy, G.J.C. Experimental Investigations on Heat Transfer and Friction Factor of Hybrid Nanofliud Equiped with Angular Twisted Strip Inserts in a Parabolic Trough Solar Collector under Turbulent Flow. *Int. J. Innov. Sci. Eng. Technol.* **2021**, *8*, 315–336.
38. Chakraborty, O.; Roy, S.; Das, B.; Gupta, R. Effects of helical absorber tube on the energy and exergy analysis of parabolic solar trough collector—A computational analysis. *Sustain. Energy Technol. Assess.* **2021**, *44*, 101083. [CrossRef]
39. Akbarzadeh, S.; Valipour, M.S. The thermo-hydraulic performance of a parabolic trough collector with helically corrugated tube. *Sustain. Energy Technol. Assess.* **2021**, *44*, 101013. [CrossRef]
40. Peng, H.; Li, M.; Hu, F.; Feng, S. Performance analysis of absorber tube in parabolic trough solar collector inserted with semi-annular and fin shape metal foam hybrid structure. *Case Stud. Therm. Eng.* **2021**, 101112. [CrossRef]
41. Sheikholeslami, M.; Farshad, S.A. Nanofluid flow inside a solar collector utilizing twisted tape considering exergy and entropy analysis. *Renew. Energy* **2019**, *141*, 246–258. [CrossRef]
42. Moghaddaszadeh, N.; Esfahani, J.A.; Mahian, O. Performance enhancement of heat exchangers using eccentric tape inserts and nanofluids. *J. Therm. Anal. Calorim.* **2019**, *137*, 865–877. [CrossRef]
43. Ghasemi, S.E.; Ranjbar, A.A. Thermal performance analysis of solar parabolic trough collector using nanofluid as working fluid: A CFD modelling study. *J. Mol. Liq.* **2016**, *222*, 159–166. [CrossRef]
44. Reddy, K.S.; Ravi, K.K.; Satyanarayana, G.V. Numerical Investigation of Energy-Efficient Receiver for Solar Parabolic Trough Concentrator. *Heat Transf. Eng.* **2008**, *29*, 961–972. [CrossRef]
45. He, W.; Toghraie, D.; Lotfipour, A.; Pourfattah, F.; Karimipour, A.; Afrand, M. Effect of twisted-tape inserts and nanofluid on flow field and heat transfer characteristics in a tube. *Int. Commun. Heat Mass Transf.* **2020**. [CrossRef]
46. Eiamsa-Ard, S.; Thianpong, C.; Promvonge, P. Experimental investigation of heat transfer and flow friction in a circular tube fitted with regularly spaced twisted tape elements. *Int. Commun. Heat Mass Transf.* **2006**, *33*, 1225–1233. [CrossRef]
47. Fluent, A.N. *18.2, Theory Guide*; ANSYS Inc.: Canonsburg, PA, USA, 2017.
48. Bejan, A. *Convection Heat Transfer*; John Wiley & Sons: Hoboken, NJ, USA, 2013.
49. Zaboli, M.; Nourbakhsh, M.; Ajarostaghi, S.S. Numerical evaluation of the heat transfer and fluid flow in a corrugated coil tube with lobe-shaped cross-section and two types of spiral twisted tape as swirl generator. *J. Therm. Anal. Calorim.* **2020**. [CrossRef]
50. Saedodin, S.; Zaboli, M.; Rostamian, S.H. Effect of twisted turbulator and various metal oxide nanofluids on the thermal performance of a straight tube: Numerical study based on experimental data. *Chem. Eng. Process. Process. Intensif.* **2020**, *158*, 108106. [CrossRef]

51. Baragh, S.; Shokouhmand, H.; Ajarostaghi, S.S.M.; Nikian, M. An experimental investigation on forced convection heat transfer of single-phase flow in a channel with different arrangements of porous media. *Int. J. Therm. Sci.* **2018**, *134*, 370–379. [CrossRef]
52. Baragh, S.; Shokouhmand, H.; Ajarostaghi, S.S.M. Experiments on mist flow and heat transfer in a tube fitted with porous media. *Int. J. Therm. Sci.* **2019**, *137*, 388–398. [CrossRef]
53. Karouei, S.H.H.; Ajarostaghi, S.S.M.; Gorji-Bandpy, M.; Fard, S.R.H. Laminar heat transfer and fluid flow of two various hybrid nanofluids in a helical double-pipe heat exchanger equipped with an innovative curved conical turbulator. *J. Therm. Anal. Calorim.* **2020**. [CrossRef]
54. Hamedani, F.A.; Ajarostaghi, S.S.M.; Hosseini, S.A. Numerical evaluation of the effect of geometrical and operational parameters on thermal performance of nanofluid flow in convergent-divergent tube. *J. Therm. Anal. Calorim.* **2019**. [CrossRef]
55. Moghadam, H.K.; Ajarostaghi, S.S.M.; Poncet, S. Extensive numerical analysis of the thermal performance of a corrugated tube with coiled wire. *J. Therm. Anal. Calorim.* **2019**. [CrossRef]
56. Outokesh, M.; Ajarostaghi, S.S.M.; Bozorgzadeh, A.; Sedighi, K. Numerical evaluation of the effect of utilizing twisted tape with curved profile as a turbulator on heat transfer enhancement in a pipe. *J. Therm. Anal. Calorim.* **2020**, *140*, 1537–1553. [CrossRef]
57. Olfian, H.; Sheshpoli, A.Z.; Ajarostaghi, S.S.M. Numerical evaluation of the thermal performance of a solar air heater equipped with two different types of baffles. *Heat Transf.* **2020**, *49*, 1149–1169. [CrossRef]
58. Ajarostaghi, S.S.M.; Delavar, M.A.; Poncet, S. Thermal mixing, cooling and entropy generation in a micromixer with a porous zone by the lattice Boltzmann method. *J. Therm. Anal. Calorim.* **2020**, *140*, 1321–1339. [CrossRef]
59. Bergman, T.L.; Lavine, A.S.; Incropera, F.P.; DeWitt, D.P. *Introduction to Heat Transfer*; John Wiley & Sons: Hoboken, NJ, USA, 1990.
60. Zou, B.; Dong, J.; Yao, Y.; Jiang, Y. An experimental investigation on a small-sized parabolic trough solar collector for water heating in cold areas. *Appl. Energy* **2016**, *163*, 396–407. [CrossRef]

Article

Experimental and CFD Investigation into Using Inverted U-Tube for Gas Entrainment

Khaled Yousef [1], Ahmed Hegazy [1] and Abraham Engeda [2,*]

[1]   Mechanical Power Engineering Department, Menoufia University, Shebin Elkom 32511, Egypt; khyousef@msu.edu (K.Y.); ahegazy7@yahoo.com (A.H.)

[2]   Turbomachinery Lab, Michigan State University, East Lansing, MI 48824, USA

*   Correspondence: engeda@msu.edu

Received: 3 November 2020; Accepted: 14 December 2020; Published: 18 December 2020

Abstract: An experimental and numerical study is presented in the current work for gas entrainment using an inverted vertical U-tube. Water flows vertically up in an inverted U-tube which creates a low-pressure region in the tube upper portion. This low-pressure region can be used to extract gases by connecting it to a branch pipe. The extracted gases considered in this work are a mixture of air and water vapor. The water vapor from the side branch pipe is mixed with the flowing water under the siphon effect. This results in a progressive water vapor condensation as the mixture proceeds towards the exit due to an increase in vapor partial pressure. The air is drawn by inertia to be released out at the tube lower exit of the inverted U-pipe. The current study deals with these complicated flow behaviors due to the mixing undergoing condensation. A test rig is designed for experimentally studying the behavior of water flow in an inverted U-tube where the air is mixed with the flowing water at the top region of this tube. The CFD computations are accomplished for a side gas mixture with volume fractions up to 0.7 with water vapor mass fractions in this mixture to be 0.1–0.5. The tested water mass flow rates in the main tube are 2, 4, 6, 8 kg/s to account for all possible flow mass ratios. The CFD computations are validated with water and air two phase flow with the measurements of both the experiments of the current research and the literature. The present results reveal that slightly raising the water mass flow rate at a constant side mixture mass ratio produces a reduced generated pressure in the upper tube part. This is attributed to extra water vapor condensation taking place rapidly by increasing the water flow rate in the tube upper part. Furthermore, the turbulence quantities begin to break down at a side mixture volume fraction of 0.55 with water and air mass flow rates of 2 kg/s and 0.002 kg/s, respectively. On the other side, raising the air mass flow rate at the higher values of water vapor and water mass flow rates breaks the generated vacuum pressure and turbulence due to entrainment. Moreover, this proposed framework can produce a lower static pressure, reaching 55.1 kPa, which makes it attractive for gas extraction. This new technique presents innovative usage with less consumable energy for extracting gases in engineering equipment.

Keywords: inverted U-tube; two-phase flow; air/water vapor entrainment; steam condenser

1. Introduction

Few engineering and medical applications need evacuated cavities during certain manufacturing and testing processes. These evacuated cavities may contain a variety of gases and liquids. So, the continuous removal of these undesired fluids and maintenance of the required evacuated cavities at a specific pressure becomes mandatory for these applications. The most used devices for venting these gases are categorized as a steam trap, liquid ring, vacuum pumps, manual valve, automatic valve and steam ejector. All these devices consume a considerable amount of energy to accomplish their duties. The use of an inverted U-tube is an innovative method for entraining such gases, which may lead to the consumption of less energy [1]. When water flows up in an inverted U-tube, it produces a low-pressure region at its upper part. This produces a low-pressure region, which can be exploited to draw gases from the required cavity by connecting the top of the inverted U-tube to the cavity through a side pipe. In the present paper, two-phase flow (water and gas) and three components (water liquid, air and water vapor) are considered in the proposed CFD computations. One of the most important pieces of equipment working under vacuum is the steam condenser of the steam power plant. The main function of the steam condenser, besides steam condensation, is to preserve the generated operational vacuum by evacuating the gases in the cycle. Currently, steam power plants share the largest percentage of electricity generation worldwide [2,3].

As commonly known in a condenser, the volume fractions of air, vapor, and water change rapidly, and this complicated behavior has a significant impact on plant efficiency. Water vapor can be taken out of the condenser cavity through condensation and it is pumped back to the water feeding system of the plant boiler. The non-condensable gas is mainly air that has leaked into the plant turbine and condenser, which are working below the atmospheric pressure or it comprises air and non-condensable gases released from the plant working fluid. Furthermore, these non-condensable gases may also be formed by water decomposition. For efficient plant operation, these gases are extracted out of the condenser cavity to avoid increasing the condenser operating pressure. Raising the condenser operating pressure diminishes the steam turbine output power. Additionally, these gases severely reduce the heat transfer rates in the condenser by the covering wrap of its tubes. Furthermore, the oxygen content raises the tube corrosiveness in the condenser, which lowers the condenser's operational life. So, these gases should be evacuated continuously for better operation and longer life of the steam condenser [4–6].

Few experiments consider the two-phase flow of steam condensation with the presence of gas in vertical tubes [7–9]. The main finding of these experiments was that raising both the gas mass fraction and the inlet properties reduced the heat transfer rate. Additionally, the local heat transfer was measured in the case of steam only and steam mixed with helium and air. The obtained temperature profiles for the considered cases provided a reasonable way of validating the proposed numerical model. The CFD computations of condensation in the existence of gases encountered some implementing difficulties but the resulted CFD flow profiles helped to investigate the physical mechanisms due to condensation. The effect of non-condensing gases on the steam condenser performance is presented theoretically by Strušnik et al. [10]. In their study, the steam ejector pump system is modeled by monitoring the gas extraction. Optimizing the steam ejector pump geometry was recommended for enhancing plant efficiency [10]. Strušnik et al. [11] made a comparative study for the power consumptions in the case of using the ejector and electric vacuum pump. They presented an economic analysis and operating costs guidelines for the appropriate selection of the condenser evacuating system in the steam power plant. It was found that the considered ejector is more consuming for the plant energy, as compared to the liquid ring vacuum pump. The condensation of steam with the existence of non-condensing gas was studied in [12,13]. They explored that the existence of the non-condensing gases severely reduced the heat transfer coefficient in the steam condenser. A hybrid system for gas extraction in the plant condenser was studied by Kapooria et al. [4]. In this hybrid system, a steam ejector integrated with a liquid ring vacuum pump was utilized. This proposed hybrid system was able to generate more vacuum in the steam condenser, while an observed increase in energy consumption was recorded.

Significant attention has been directed toward the complicated air and water flow patterns and their application in the steam condenser. In this regard, Yao et al. [14] studied numerically the two-phase flow of air/water in a small horizontal pipe. The obtained CFD results had an acceptable agreement with the considered experimental data and the CFD simulation accurately predicted the flow structures. The noticed small discrepancies between the experimental data and the CFD results may be attributed to the requirement for the full three-dimensional CFD simulations. In the same way, Juggurnath et al. [15] numerically presented the various two-phase flow profiles created in an inclined tube. They considered the effect of both surface tension and gravitational forces on the generated flow patterns. The air/water flow in a horizontal pipe was investigated numerically by Vásquez et al. [16].

Different multiphase flow models were used for obtaining the flow patterns and the proposed CFD model was validated. The validation showed a reasonable agreement with slight deviations. Seven experimental facilities for the air and water co-current flow in sloping pipes were presented by Pothof and Clemens [17]. A CFD geometrical and operational analysis for preventing the air accumulation in the tube was proposed. Panella [18] numerically and experimentally predicted the air and water flow in pipes. In this study, a framework for mixing air and water in the tube is fabricated. A circulating pump was utilized for controlling the water mass flow rate. This study revealed good validation in a wide range of data. Air and water two-phase flow in a tube with gas separator was presented by Afolabi and Lee [19]. This study ensured the ability of CFD models to accurately predict the behavior of the two-phase flow. Besides, the computational results had a good validation with the experiments.

The objective of the present work is to numerically capture the two-phase flow patterns in the inverted U-tube when mixing water with a side mixture of air and water vapor. Furthermore, the required flow conditions and operational parameters for using the inverted U-tube for the side entrainment of the air-water vapor mixture will be determined. Additionally, a reduced scale experimental test rig is fabricated to validate the present CFD results.

2. Experimental Setup

An experimental setup is fabricated to disclose the siphon flow characteristics with side air entrainment in the inverted U-tube. The experimental test rig with half scaled dimensions, compared to the CFD model dimensions, is fabricated. The experimental set up comprises two tanks with a water elevation difference of ΔZ—see Figure 1. An inverted U-tube of 2.54 cm inner diameter, with its inlets immersed into the two water tanks, is utilized. The water levels in the tanks are changeable by lowering/lifting the right-side tank. The main filling port (2) in the left side tank and a water exit port in the right-side tank (12) is used to maintain the water levels between the two tanks constant during measurements. Bourdon pressure gauges (4) are utilized to measure the flow pressure around the system. The flowing water through the inverted U-tube is measured by using a calibrated tank and stopwatch while the side air is measured by using a variable area rotameter (7). The rotameter has a full-scale accuracy/repeatability of 6/2%. The rotameter has a variable flow area valve (8) to control the inlet air flow rate (9). The siphon flow generation is one of the tricky points before measurements. The pre-filling port (5) is used to fill the whole system with water while the air venting port (6) is used to release the air from the system cavity. The inverted U-tube inlets in the right- and left-side tanks are blocked during this system pre-filling by using manual valves.

Once the whole system is full of water and free of air, the inverted U-tube inlets in the right- and left-side tanks are unblocked while maintaining the water pre-filling port (5) opened to help initiate the siphon flow for a while. Reaching a steady siphon flow is the main task that should be performed first. Then, the desired water elevation difference between the left and right tanks is obtained and kept constant. The water flow rate is measured before and after side air enters the system at the same ΔZ. The air flow rate is changeable by using the variable area valve (8), which is also used to close the side pipe during the siphon flow initiation. The water level difference between the two tanks can be changed by adjusting the side air entrainment and the experiments are repeated for different flow conditions.

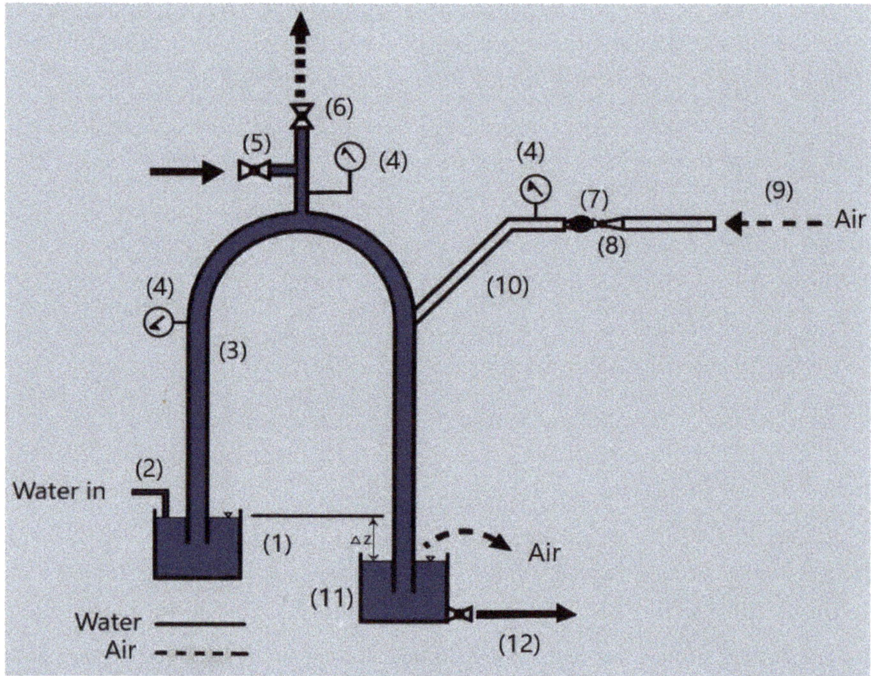

(1)	Left side water tank		(7)	Globe valve
(2)	Main water filling port		(8)	Rotameter
(3)	Inverted U-tube		(9)	Side air inlet
(4)	Pressure gauges		(10)	Side-pipe
(5)	water pre-filling port		(11)	Right side water tank
(6)	Air pre-venting port		(12)	Water outlet

Figure 1. Experimental test rig and its components.

3. Computational Model

The two-phase model is a set of equations and mathematical relations that are used to describe the phase's motion and interactions. The most commonly utilized two-phase models in the literature are homogeneous, drift flux and separated flow models [20–25]. The homogeneous model is less complicated and can be applied to many complex two-phase flows with good accuracy. The homogeneous model is represented by the mixture two-phase model in CFD computations, as will be discussed later in this paper. In homogenous models, the average properties for the flow variables are considered in the calculations and the relative motion between phases can be accurately calculated [20,26]. In addition, the homogenous model is easy to implement as its governing equations are close to the single-phase flow equations. In the present study, the mixture model for calculating the two-phase flow variables in the inverted U-tube is utilized.

Figure 2 shows the CFD model with its boundary conditions. In the proposed work, the siphon flow generated by water flowing up in an inverted U-tube creates a vacuum in the tube upper region. This vacuum region can be connected by a side pipe to extract a mixture of air and water vapor from the equipment cavity, the equipment cavity is not shown in this figure. The equipment cavity is considered the steam condenser in the present case. This results in mixing the water with the induced side mixture of air and water vapor. This mixing will cause the water vapor condensation while all the flow mixture will continue flowing down due to flow inertia and gravity. At the lower outlet of the inverted U-tube, the air is released to the environment while water is re-circulated through the inverted U-tube. Furthermore, a variable speed circulating pump is used to transfer the water between the two tanks at the inlet and outlet of the inverted U-tube (the pump is not shown in the figure). Additionally, the hydrostatic head difference between the water surfaces in both tanks can be altered to control the water mass flow rate. The considered tube has low maintenance and operational cost, which makes it more favorable compared to the other energy consuming systems, i.e., water ring pump and steam ejector [27].

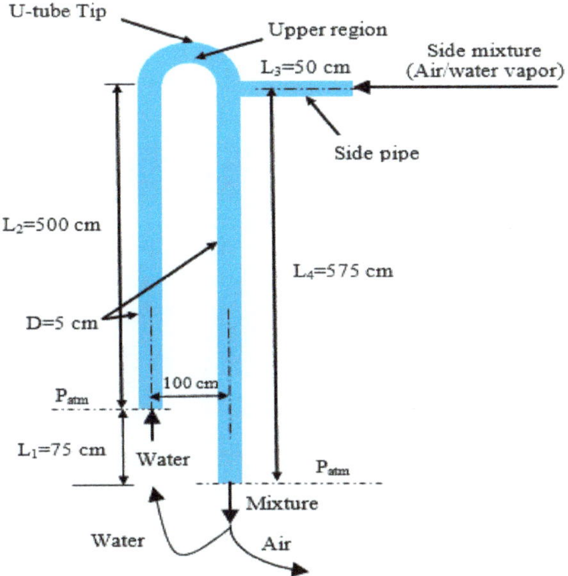

Figure 2. CFD model for the inverted U-tube with dimensions.

3.1. Governing Equations

Steady-state three dimensional CFD computations were performed for the considered geometry shown in Figure 2. Differential equations governing the fluid flow are given as [28,29].

$$\Delta\,(\rho u) = 0 \tag{1}$$

$$\Delta\,(\rho u\,u) = -\frac{\partial p}{\partial x} + \Delta\,(\mu\,\text{grad}\,u) + \left[-\frac{\partial \overline{\rho u'^2}}{\partial x} - \frac{\partial \overline{\rho u'\,v'}}{\partial y} - \frac{\partial \overline{\rho u'\,w'}}{\partial z} \right] \tag{2}$$

$$\Delta\,(\rho u\,v) = -\frac{\partial p}{\partial y} + \Delta\,(\mu\,\text{grad}\,v) + \left[-\frac{\partial \overline{\rho\,u'\,v'}}{\partial x} - \frac{\partial \overline{\rho\,v'^2}}{\partial y} - \frac{\partial \overline{\rho\,v'\,w'}}{\partial z} \right] \tag{3}$$

$$\Delta\,(\rho u\,w) = -\frac{\partial p}{\partial z} + \Delta\,(\mu\,\text{grad}\,w) + \left[-\frac{\partial \overline{\rho u'\,w'}}{\partial x} - \frac{\partial \overline{\rho v'\,w'}}{\partial y} - \frac{\partial \overline{\rho\,w'^2}}{\partial z} \right] \tag{4}$$

The energy equation can be found as follows to consider the heat transfer, as in the current study.

$$\Delta\,(\rho I u) = \Delta\,(k\,\text{grad}\,T) + \Phi + \left[-\frac{\partial \overline{\rho\,u'\,I'}}{\partial x} - \frac{\partial \overline{\rho\,v'\,I'}}{\partial y} - \frac{\partial \overline{\rho\,w'\,I'}}{\partial z} \right] \tag{5}$$

I is the total energy, $I = h - \frac{P}{\rho} + \frac{(u^2 + v^2)}{2}$, while Φ is the function of dissipation. h is taken to be the sensible enthalpy where $h = \sum_i Y_i h_i$ and $h_i = \int_{T_{ref}}^{T} C_{pi} dT$. T_{ref} is the reference temperature and it is equal to 298.2 K.

3.2. Realizable $k - \varepsilon$ Model

In the present study, the realizable $k - \varepsilon$ model in the mixture form is utilized for computing the turbulence fields. The choice of turbulence model is very critical in any CFD simulation. The realizable $k - \varepsilon$ model is can provide outstanding performance in many complicated flows. Additionally, this turbulent modeling is characterized over the other turbulence models as it has a modified equation for calculating turbulent viscosity and a new transport equation for the dissipation rate ε [28,29].

$$\frac{\partial}{\partial x_i}(\rho k u_i) = \frac{\partial}{\partial x_i}\left[\left(\mu + \frac{\mu_t}{\sigma_k}\right)\frac{\partial k}{\partial x_i}\right] + G_b + G_k - \rho\varepsilon - Y_M + S_k \tag{6}$$

$$\frac{\partial}{\partial x_i}(\rho\varepsilon u_i) = \frac{\partial}{\partial x_i}\left[\left(\mu + \frac{\mu_t}{\sigma_\varepsilon}\right)\frac{\partial \varepsilon}{\partial x_i}\right] + \rho C_1 S\varepsilon - \rho C_2\frac{\varepsilon^2}{k + \sqrt{\upsilon\varepsilon}} + C_{1\varepsilon}\frac{\varepsilon}{k}C_{3\varepsilon}G_b + S_\varepsilon \tag{7}$$

In the mixture turbulence model, the same turbulence field is given for all phases. This means that all phase properties are replaced with mixture properties in the equations. G_k is the turbulent kinetic energy generation, due to the mean velocity gradients, while G_b is the turbulence kinetic energy generation due to buoyancy. S_k and S_ε are user-defined source terms. The constants in these equations are [28,29].

$$C_1 = \max\left[0.43, \frac{\eta}{\eta + 5}\right], \ \eta = S\frac{\varepsilon}{k}, \ S\sqrt{2S_{ij}S_{ij}} \tag{8}$$

In turbulence modeling, $\mu_t = \rho\,C_\mu\frac{k^2}{\varepsilon}$, $\mu_{eff} = \mu + \mu_t$.

The favorable features of the considered turbulence model, as discussed earlier, is C_μ, which does not have a fixed value and can be correlated as [28–30].

$$C_\mu = \frac{1}{A_0 + A_s\frac{kU^*}{\varepsilon}}, \ U^* = \sqrt{S_{ij}S_{ij} + \hat{\Omega}_{ij}\,\hat{\Omega}_{ij}}, \ \hat{\Omega}_{ij} = \overline{\Omega}_{ij} - \varepsilon_{ijk}\omega_k \tag{9}$$

Here, $\overline{\Omega}_{ij}$ is the rotation tensor rate with angular velocity ω_k and $A_0 = 4.04$, $A_s = \sqrt{6}\cos\varphi$, $\varphi = \frac{1}{3}\cos^{-1}\left(\sqrt{6}W\right)$, $W = \frac{S_{ij}S_{jk}S_{ki}}{\tilde{S}^3}$, $\tilde{S} = \sqrt{S_{ij}S_{ij}}$, $S_{ij} = \frac{1}{2}\left(\frac{\partial U_i}{\partial x_j} + \frac{\partial U_j}{\partial x_i}\right)$. In the current study, $C_{1\varepsilon} = 1.44$, $C_2 = 1.9$, $\sigma_k = 1.0$, and $\sigma_\varepsilon = 1.2$.

3.3. Two-Phase Model

The mixture two-phase model in Ansys Fluent is used for the present computations [31]. This model is the best representation of the homogenous two-phase flow for water, air and water vapor flow in the U-tube. In the mixture model, all phases share the same cell volume and can interchange energy and momentum. The cell volume fraction for the phase is α, and all phase volume fraction summation is equal to 1. In the following equations, α_m is the volume ratio of the side mixture in a combination of the mixture (water vapor and air) and water flows. X_w and X_v are the water and water vapor mass flow rate ratios concerning the side mixture mass flow rate, respectively. Y_a is the ratio of air mass flow rate to the total fluid mass flow rate, while $V_{a,v}$ is the volume for the side mixture [29,32,33].

$$\sum_{j=1}^{n} \alpha_j = 1 \tag{10}$$

$$\alpha_m = \frac{V_{a,v}}{(V_w + V_{a,v})}, \ \dot{m}_{a,v} = \dot{m}_a + \dot{m}_v \tag{11}$$

$$X_w = \frac{\dot{m}_w}{\dot{m}_{a,v}}, \; X_v = \frac{\dot{m}_v}{\dot{m}_{a,v}}, \; Y_a = \frac{\dot{m}_a}{\dot{m}_w + \dot{m}_a} \tag{12}$$

For the filled cells with water in two-phase flows, the volume fraction of water, α_w, is 1 while α_v and α_a are taken to be 0 and the same way for the filled water vapor cells. If the cell combines the mixing of air, water, and water vapor, then these cell volume fractions are between 0 and 1. For calculating the secondary and primary phase volume fractions, FLUENT® solves separate continuity equations for secondary phases, while Equation (10) is utilized for computing the primary phase volume fraction. In the condensation process, the water is generated from water vapor due to phase change. Therefore, the water is taken to be the primary phase while water vapor and air are the secondary phases. The continuity equation for calculating the secondary phase volume fraction is given as follows [29].

$$\nabla . \alpha_{sec} \rho_{sec} \vec{v}_m = \sum_{j=1}^{n} \dot{m}_{pri-sec} - \dot{m}_{sec-pri} \tag{13}$$

$\dot{m}_{pri-sec}, \dot{m}_{sec-pri}$ is a mass source term that represents the mass transfer rate from primary phase to secondary phase in the case of evaporation or mass transfer rate from secondary phase to primary phase in the case of condensation, respectively. The average velocity of the mixture is, $\vec{v}_m = \frac{\sum_{j=1}^{n} \alpha_j \rho_j \vec{v}_j}{\rho_m}$ and the mixture density is computed for all phase densities, $\rho_m = \sum_{j=1}^{n} \alpha_j \rho_j$. All the mixture properties are correlated, likewise, with these mixture density calculations.

The source term added in the energy equation and the internal energy is found as

$$S_I = \left(\dot{m}_{pri-sec} - \dot{m}_{sec-pri}\right)h_{fg} \tag{14}$$

$$E_j = h_j - \frac{P}{\rho_j} + \frac{\left(u_j^2 + v_j^2 + w_j^2\right)}{2} \tag{15}$$

In Equation (15), h_j can be correlated as $h_j = h_{sse,j} + \int_{T_{ref}}^{T} c_{p,j} dT$, where $h_{sse,j}$ is defined as the phase standard state enthalpy and h_{fg} is the vaporization latent heat, which is the difference between water vapor and water-sensible heat. The Evaporation–Condensation model is utilized in this work to describe the mass transfer rate between phases. This model relies on the relation of the phase temperature inside the cell to the saturation temperature. Based on this temperature comparison, a mass transfer occurs with releasing or gaining the latent heat in the case of condensation or evaporation, respectively [34]. The mass transfer rate in the case of condensation, as in the present case, is calculated by Equation (16).

$$\dot{m}_{v-w} = X_{cs} \, \alpha_v \, \rho_v \left(\frac{T_{sat} - T_{cell}}{T_{sat}}\right) \tag{16}$$

X_{cs} is defined as the condensation parameter and can be calculated by Equation (17). In addition, it is assumed that the condensation from water vapor to water liquid generates a droplet of spherical shape [29,35].

$$X_{cs} = \frac{6}{d_{drop}} \beta_{cs} \sqrt{\frac{M}{2\pi r T_{sat}}} \left[\frac{\rho_w h_{fg}}{\rho_w - \rho_v}\right] \tag{17}$$

In the above equation, d_{drop} is the diameter of the water droplet and it can be determined by dividing the surface area of the droplet by its volume, as shown in Equation (18) [29,35].

$$A_{w-drop} = \frac{\pi d_{drop}^2}{\frac{\pi d_{drop}^3}{6}} = \frac{6}{d_{drop}} \tag{18}$$

The concentration of interfacial area is calculated by Equation (19) by assuming equal diameters for all generated water droplets due to condensation [35].

$$A_{\text{intf–drop}} = \frac{6\alpha_v}{d_{\text{drop}}} \tag{19}$$

β_{cs} in Equation (17) is the coefficient of accommodation and can be computed by Equation (20) after setting the tunable coefficient of condensation ξ_{cs} [29,35].

$$\beta_{cs} = \frac{2\xi_{cs}}{2 - \xi_{cs}} \tag{20}$$

3.4. Solution Procedure

In the present simulation, the governing equations are solved with the Evaporation–Condensation model in the three-dimensional domain using the commercial package Ansys Fluent [34]. Additionally, a quick discretization and first-order upwind scheme are used for all conservation equations to save computational time. For more accurate results, the third-order MUSCL discretization scheme is used. The coupled scheme is used for pressure-velocity coupling and the pressure is calculated with the presto scheme, as it is recommended for two-phase complex flow. The coupled algorithm solves the momentum and pressure-based continuity equations together which makes the solution robust and more efficient over the segregated solution scheme. The gravitational acceleration is set equal to -9.81 m/s^2 in the vertical direction to account for the buoyancy effect with the density gradient. To accomplish robust convergence, various pseudo time values are tested for the fluid and solid zones of the computational domain. The selected pseudo time scale factors for fluid and solid zones are 0.7 and 1, respectively. The pseudo-transient fashion is recommended for the complex two-phase flow, which has high mass transfer through simulation [29]. A grid independence study is performed to grantee the computational results and the generated mesh is shown in Figure 3. The total number of the used tetrahedral mesh is 534,136. The mesh is refined near the tube wall and y+ is approximately equal to 1. Moreover, enhanced wall treatment is used for a smooth transition between the boundary layers. The enhanced wall treatment is recommended in the case of heat transfer simulation and it needs fine mesh in the viscous sublayer.

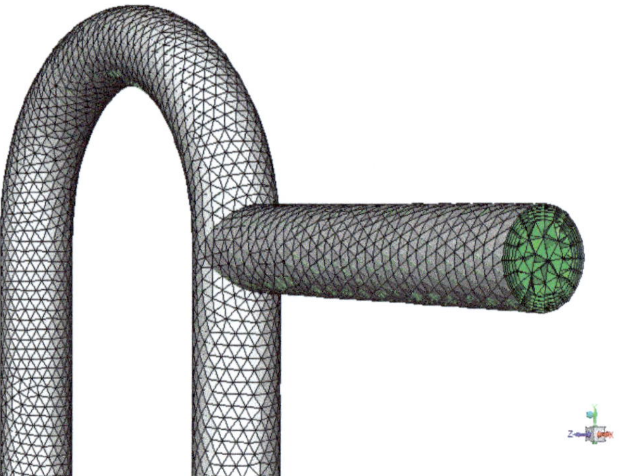

Figure 3. The computational mesh of the inverted U-tube.

Boundary conditions specify the flow and thermal variables on the physical domain boundaries. Figure 2 shows the boundary conditions utilized in the present study. The mass flow inlet applied to each phase and mixture temperature are specified at the flow inlets. The pressure outlet boundary condition is specified as a constant value and is equal to zero-gauge pressure at the pipe outlet. The isolated wall of the duct has been set as hydraulically smooth walls with non-slip boundary conditions.

4. Results and Discussion

In the proposed framework, low energy consumption with a simple construction device for maintaining the low generated pressure for engineering applications is presented. This flow is captured using Ansys Fluent with some model adjustments. The full details for the computational model are shown in Figures 2 and 3. Furthermore, the free predicted water mass flow rate in the U-tube shown in Figure 2 is 3.72 kg/s at a water elevation difference of 75 cm between the two tanks while the mixture inlet at the side pipe is blocked. This free predicted water flow rate is a function of the elevation difference between two water levels. The main objective of the current study is to investigate the two-phase flow interactions when it encounters condensation in the considered U-tube. Besides, the effect of water flow rate ratios and the critical operational conditions are highlighted.

First, the computational results are validated with the results obtained in the experimental part of the present work and with the experimental data, reported in [18] as well, for air and water two-phase flow in pipes, as shown in Figure 4. It is to be noticed here that, in carrying out the computation for validation, the values of the mass flow rate of water siphon flow and air streaming from the side pipe were taken from the present experimental record. Additionally, the air-water two-phase flow in the inverted U-tube is previously studied in [31] and the main findings encouraged the present computations. The air mass fraction in the air/water mixture is represented on the horizontal axis while the air volume fraction appears on the vertical axis. Figure 4 shows an acceptable qualitative validation for the present computational model. It is expected that the water mass flow rate affected by the siphon flow would decrease as the air and water vapor mass flow rate is raised. This is ascribed to the increase in flow resistance. The effect of side air entrainment on the water siphon flow at different water levels between tanks (ΔZ) appears in Figure 5, as obtained by the aid of the experimental data in this work. Side air suction from the branch side tube lowers the water siphon flow rate at different water levels between tanks. This is attributed to the effect of entrained air on the vacuum pressure at the tube upper region which is found weakening the siphon flow. The experimentally recorded vacuum pressure at the inverted U-tube tip is shown in Figure 6. Side air entrained into the inverted U-tube decreases the generated vacuum pressure at its tip.

The computational model is executed for the operational conditions of water mass flow rate of 2, 4, 6 and 8 kg/s with side gas mixture fraction by volume of 0.2–0.7. The water vapor fraction in the side mixture is allowed to be in the range of 0.1–0.5. The air mass flow rate effect on the static pressure in the U-tube upper region is shown in Figure 7 at various water vapor mass ratio, MR. Reducing the U-tube upper part static pressure raises the suction effect for the side mixture, which results in improving the present tube function. Increasing both air and water vapor mass flow rates elevates the static pressure in the upper part of the tube, which is undesirable. Additionally, the lowest generated static pressure can be found at a water flow rate of 2 kg/s at lower than or equal to the water/air mass ratio of 1000. For a water/air mass ratio of more than 1000, raising the water vapor flow rates magnifies the static pressure; this is attributed to increasing the water vapor flow rate at a higher air mass flow rate, which decelerates the main flow and pushes the side flow down in the tube and, hence, the condensation takes place away from the upper region. The gradient of increasing the pressure with air mass flow rate in the upper region is more pronounced at a water mass flow rate of 2 kg/s, as noticed in Figure 7a, compared to the static pressure in Figure 7b at a water mass flow rate of 4 kg/s. This is attributed to raising the water mass flow rate, which magnifies the condensation occurrence around the upper region and hence the static pressure decreases due to phase change. The critical flow rate ratio has the largest value, at which the flow of water caused by the siphon effect withdrawing air and water vapor from the side tube is maintained.

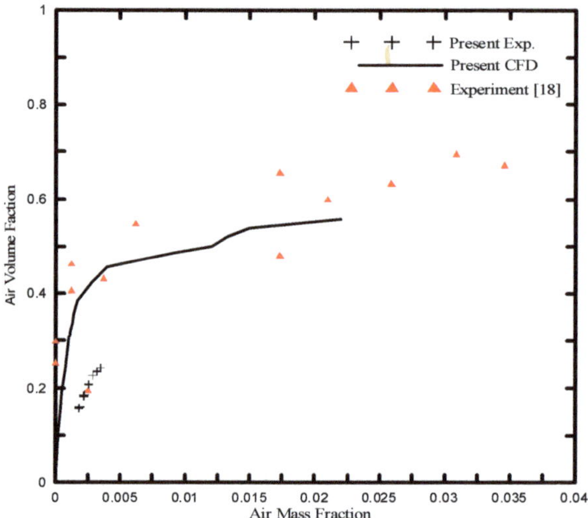

Figure 4. Comparison of the CFD results with those of the present experimental ones and reported in [18].

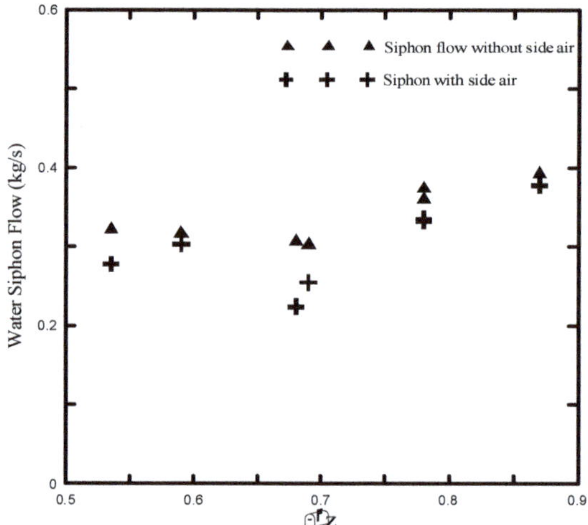

Figure 5. Effect of side air on the water siphon flow.

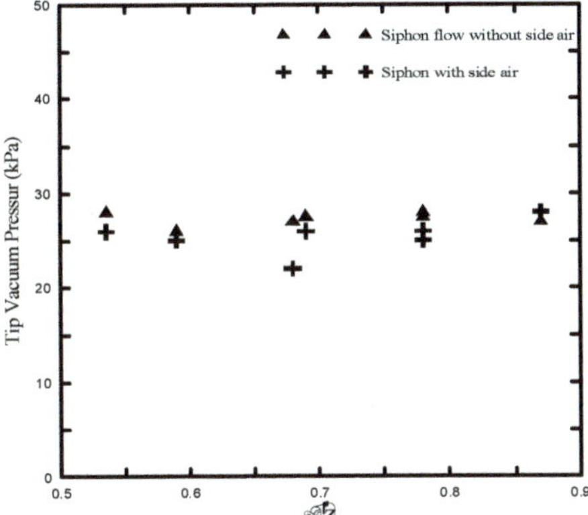

Figure 6. Effect of side air on the vacuum pressure at inverted U-tube tip.

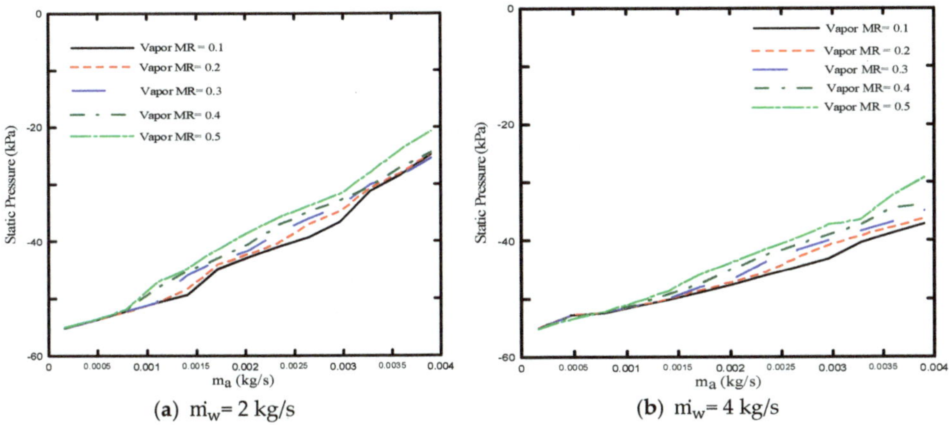

(**a**) $\dot{m}_w$= 2 kg/s   (**b**) $\dot{m}_w$= 4 kg/s

Figure 7. Effect of air and vapor mass flow rates on the static pressure generation in the tube upper region.

The turbulent intensity and kinetic energy at a point measure the degree of interactions and interruption due to fluids mixing. Figures 8 and 9 record the turbulent values due to mixing the water liquid with the side mixture at the tube upper region. The turbulent intensity and kinetic energy increase with air mass flow rate until $\dot{m}_a = 0.002$ kg/s with a water to air mass ratio of 1000 and water mass flow rate of 2 kg/s, as depicted in Figures 8a and 9a. Increasing the water mass flow rate to 4 kg/s has no significant effect on these turbulent values at the tube upper region as, by increasing the water liquid mass flow rate, the side entrained mixture is directed to flow down the tube. So, the generated turbulent values due to the mixing process in Figures 8 and 9 are less noticeable at the tube upper region at water mass flow rates of 4 kg/s compared to 2 kg/s. That means that working at a water siphon mass flow rate of 2 kg/s requires a water/air mass ratio of 1000 or less for avoiding system flow discontinuity. This can be ensured by tracking the flow velocity inside the upper region at the same operational conditions as demonstrated in Figure 10. The flow velocity rises with increasing the air mass flow rate until a mass ratio of 1000 then it recedes, and this inverted tube breaks down water siphon flow at a vapor mass ratio of 0.5, as shown in Figure 10a.

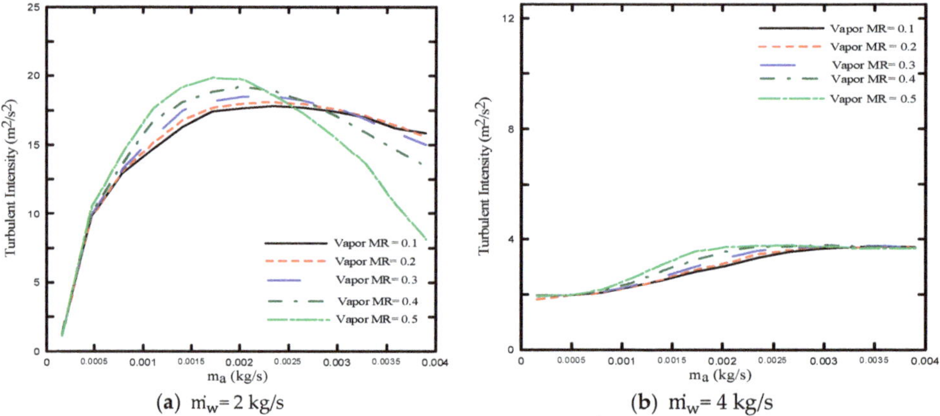

(a) $\dot{m}_w = 2$ kg/s **(b)** $\dot{m}_w = 4$ kg/s

Figure 8. Effect of air and vapor mass flow rates on the turbulent intensity in the tube upper region.

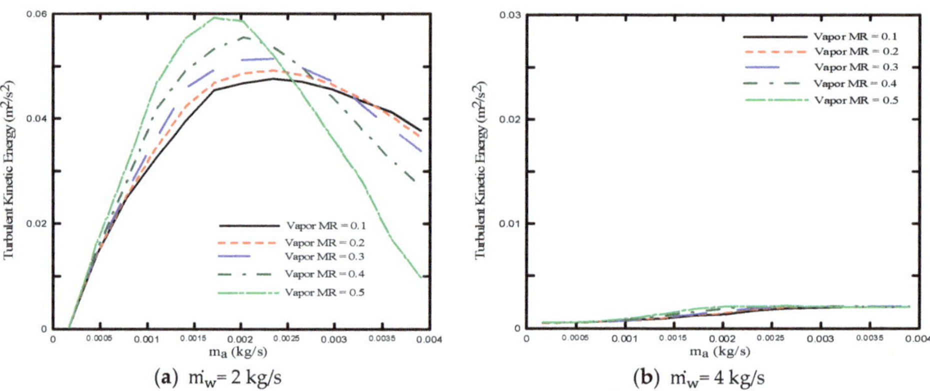

(a) $\dot{m}_w = 2$ kg/s **(b)** $\dot{m}_w = 4$ kg/s

Figure 9. Effect of air and vapor mass flow rates on the turbulent kinetic energy in the tube upper region.

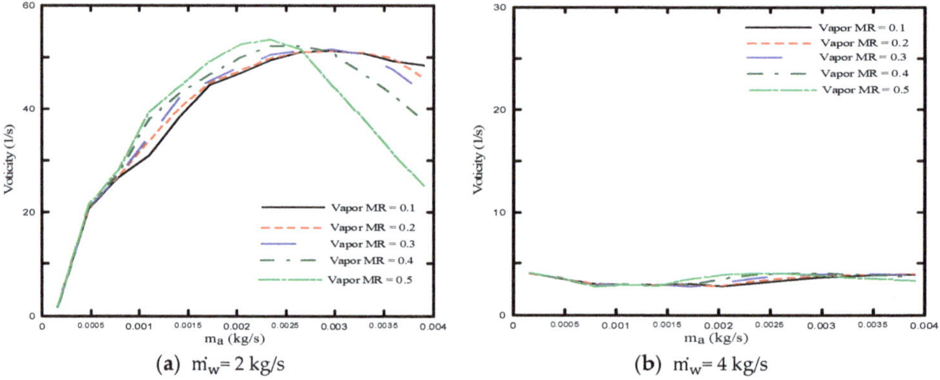

(a) $\dot{m}_w = 2$ kg/s **(b)** $\dot{m}_w = 4$ kg/s

Figure 10. Effect of air and vapor mass flow rates on the flow velocity in the tube upper region.

The effect of flow mass ratios of water, water vapor, and air on the static pressure contours are visualized in Figures 11–14. Low values of air and vapor flow rates satisfy the minimum pressure inside the U-tube, see Figure 11. On the other side, raising the side mixture volume ratio to 0.7 with preserving the vapor mass ratio to be 0.1 increases the static pressure at the top of the U-tube as demonstrated in Figures 11 and 12. Raising the vapor mass ratio from 0.1 to 0.5 with maintaining the side mixture at 0.2 reduces the static pressure generation along the inverted U-tube as depicted in Figures 11 and 13. Meanwhile, increasing both ratios of water and air magnify the static pressure inside the considered tube, but water vapor has the noticed effect on preserving the lowest static pressure as preferred for this tube operation compared to full side mixture effect, as shown in Figures 12 and 14. This means that more condensation, due to increasing the vapor flow rate, enhances the present desired operations of this U-tube and it magnifies the required induced vacuum.

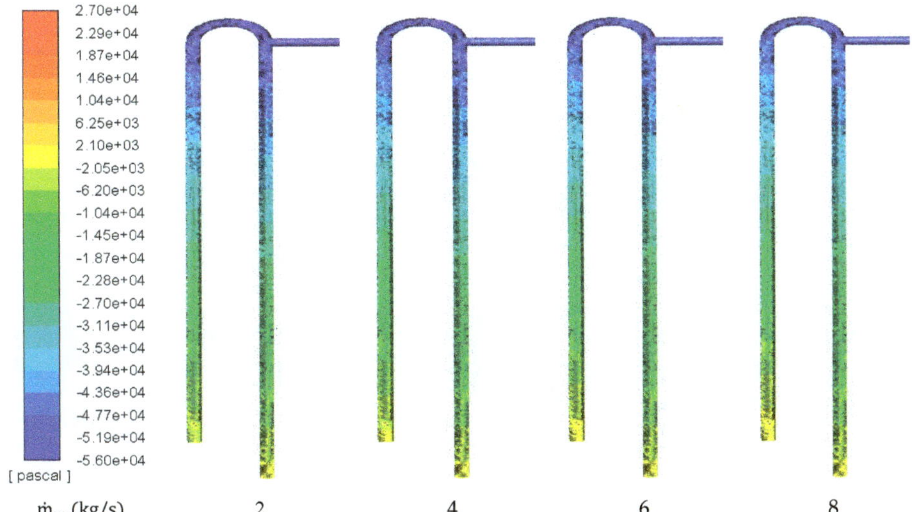

Figure 11. Pressure contours at different flow ratios and $X_w = 4306.59$, $\alpha_m = 0.2$ and $X_v = 0.1$.

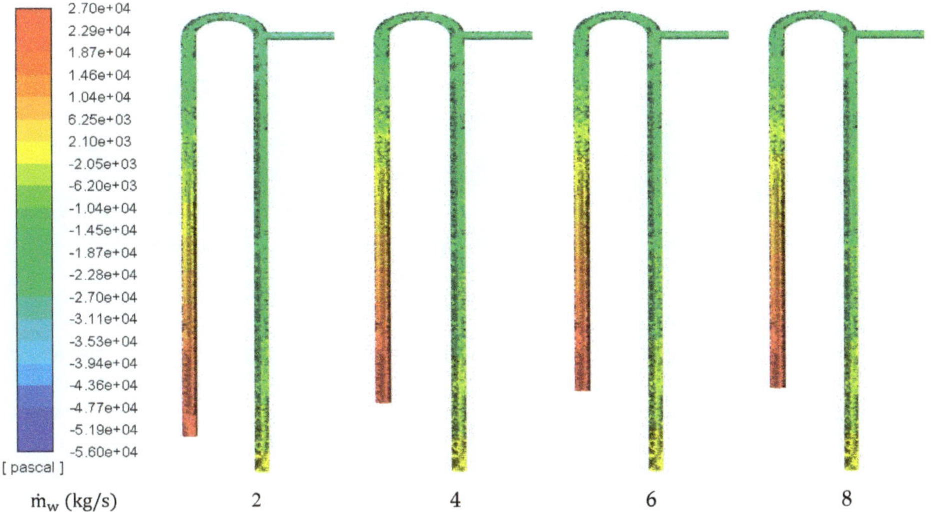

Figure 12. Pressure contours at different flow ratios and $X_w = 461.42$, $\alpha_m = 0.7$ and $X_v = 0.1$.

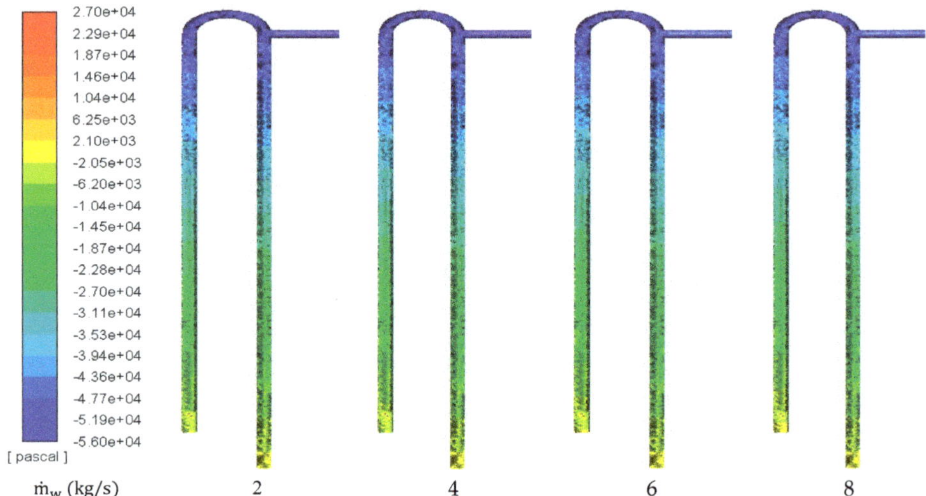

Figure 13. Pressure contours at different flow ratios and $X_w = 6187.64$, $\alpha_m = 0.2$ and $X_v = 0.5$.

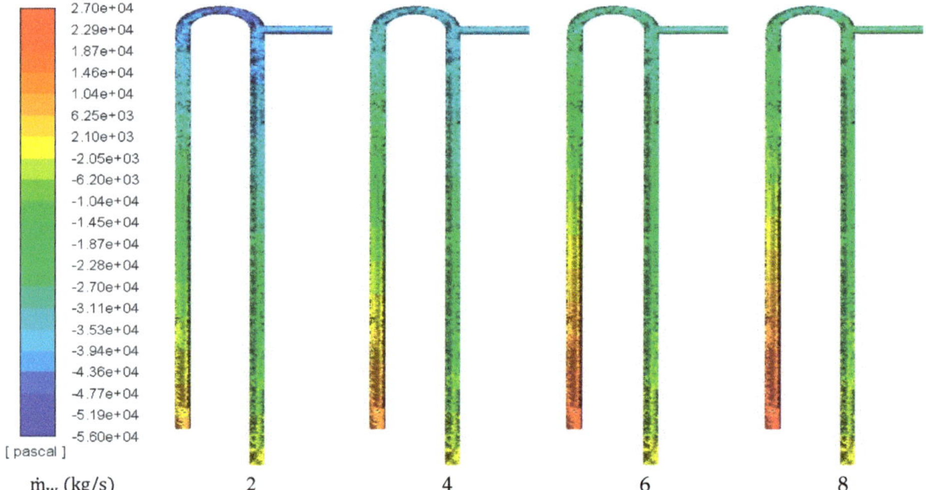

Figure 14. Pressure contours at different flow ratios and $X_w = 662.96$, $\alpha_m = 0.7$ and $X_v = 0.5$.

The effect of flow mass ratios in the tube on the side air distribution is shown in Figures 15–18. Increasing the main water flow rate moves the side entrained air to flow downward, not to the suction region in the upper part of the U-tube. This is attributed to force balancing among the inertia of water flow and gravity with the force due to flow suction in the upper portion of the tube. So, reducing the force due to flow suction by magnifying the resulted force due to flow inertia weakens the entrainment operation in the inverted U-tube, as expected above in Figure 9. The flow inertia is a function of the flow velocity, which is determined by the water flow rate. This finding ensures that the relation between the flow rates of water in the tube and water vapor in the side mixture controls the entrainment process of the tube. Moreover, raising the side air flow rate converts the flow in the right branch of the U-tube to bubbly flow which is found to break down the entrainment process and siphon the flow of the U-tube by raising the static pressure in the tube upper part. The water vapor contours are seen in Figures 19–22. Lower values of water flow rates even at higher values of the side water vapor flow decrease the pressures in the U-tube upper part. This is attributed to the instantaneous local condensation of water vapor in the upper part which enhances the continuous entrainment of the side mixture. Additionally, the water vapor can be seen at the right branch of the U-tube exit without condensation at higher values of water and side mixture flow rates. This process with time breaks the siphon flow in the proposed system.

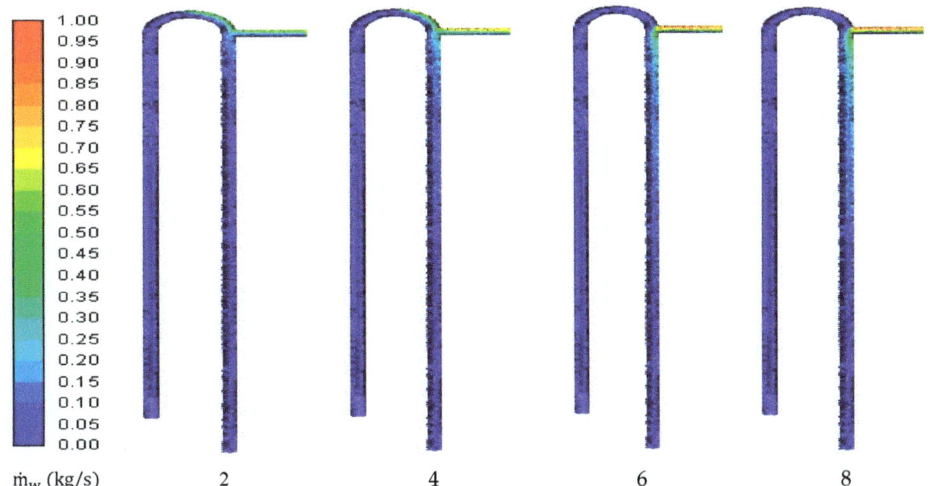

Figure 15. Air volume fraction contours at different flow ratios and $X_w = 4306.59$, $\alpha_m = 0.2$ and $X_v = 0.1$.

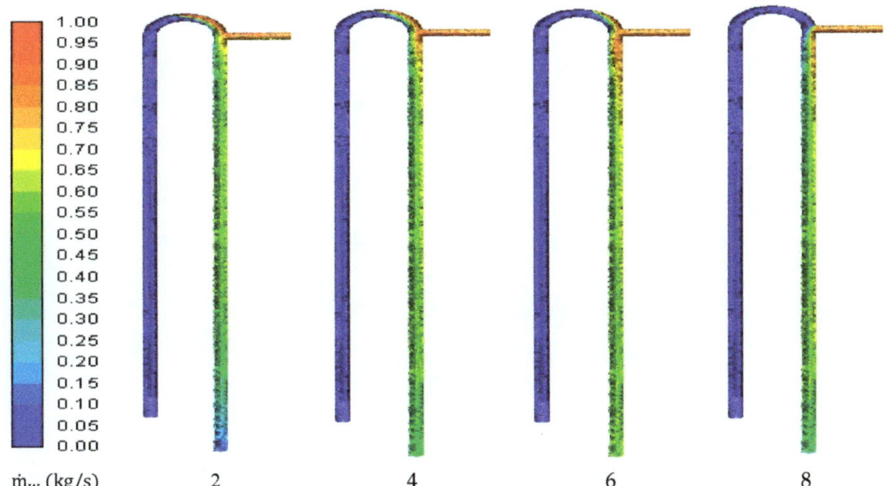

Figure 16. Air volume fraction contours at different flow ratios and $X_w = 461.42$, $\alpha_m = 0.7$ and $X_v = 0.1$.

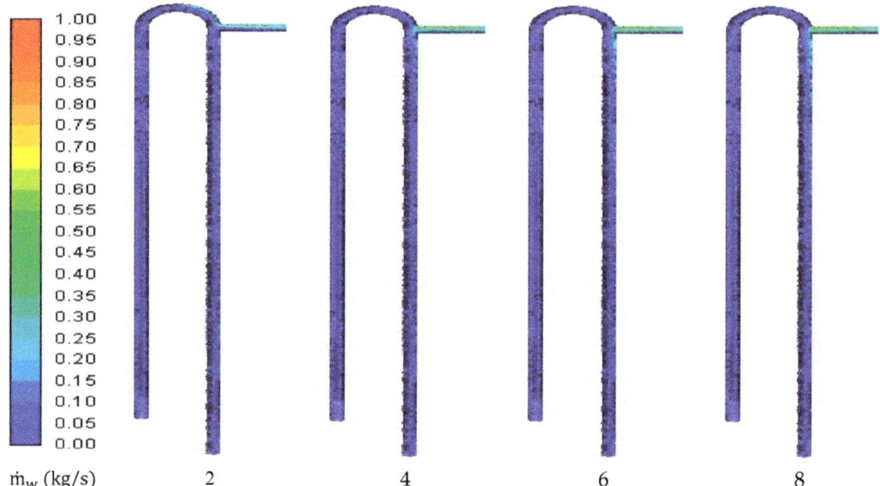

Figure 17. Air volume fraction contours at different flow ratios and $X_w = 6187.64$, $\alpha_m = 0.2$ and $X_v = 0.5$.

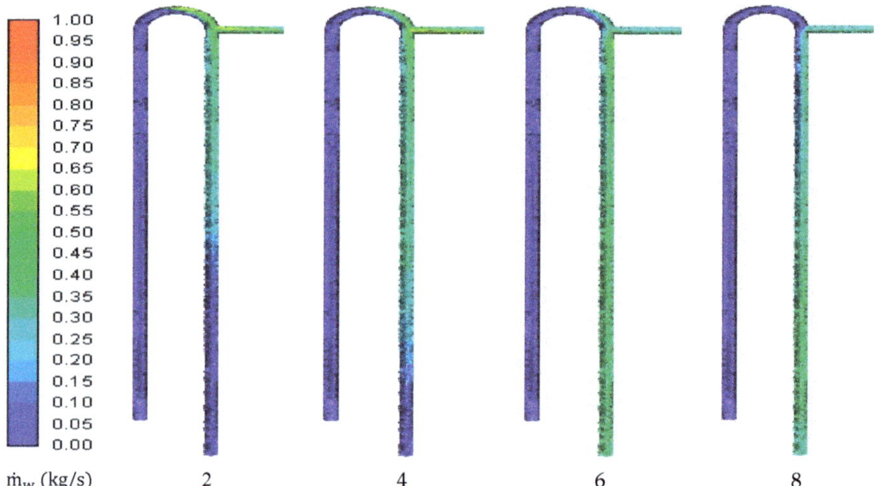

Figure 18. Air volume fraction contours at different flow ratios and $X_w = 662.96$, $\alpha_m = 0.7$ and $X_v = 0.5$.

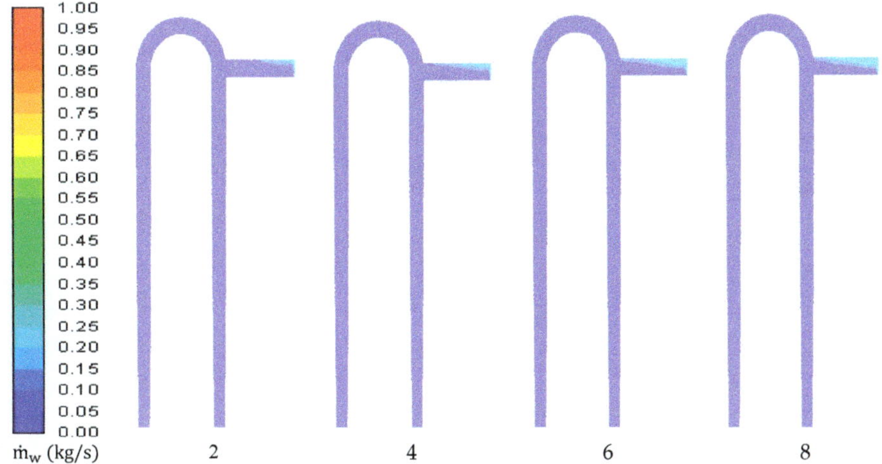

Figure 19. Vapor volume fraction contours at different flow ratios and $X_w = 4306.59$, $\alpha_m = 0.2$ and $X_v = 0.1$.

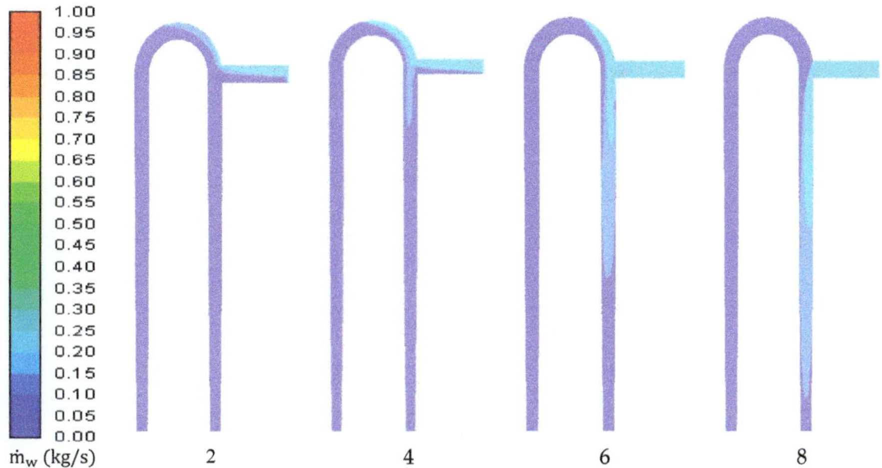

Figure 20. Vapor volume fraction contours at different flow ratios and $X_w = 462.42$, $\alpha_m = 0.7$ and $X_v = 0.1$.

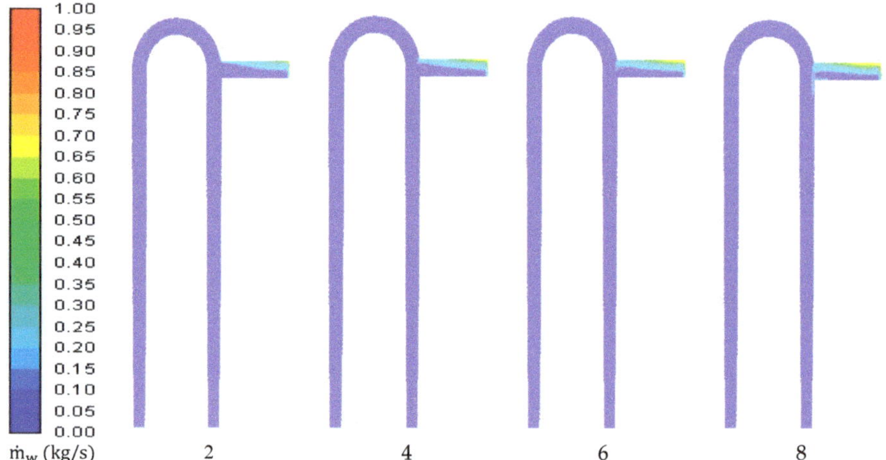

Figure 21. Vapor volume fraction contours at different flow ratios and $X_w = 6187.64$, $\alpha_m = 0.2$ and $X_v = 0.5$.

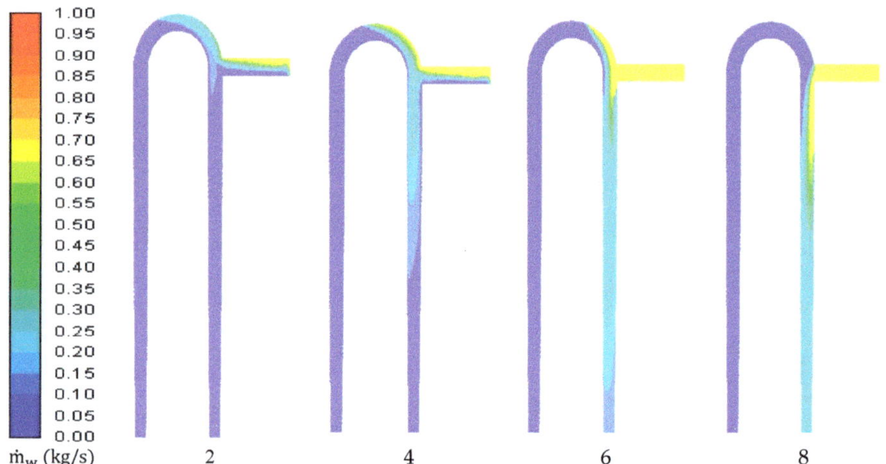

Figure 22. Vapor volume fraction contours at different flow ratios and $X_w = 662.96$, $\alpha_m = 0.7$ and $X_v = 0.5$.

Generally, in mixing two different flows, the turbulent Reynolds number is computed to capture the degree of interaction and turbulence due to mixing. This number is a relation between the turbulent kinetic energies to the dissipation rates and is defined as ($Re_t = k^2/(\varepsilon \vartheta)$). Moreover, the turbulent Reynolds number clarifies the mixing shear layers of the flow. The turbulent Reynolds number contours at different values of flow mass ratios are depicted in Figures 23–26. It is noticed that at a water flow rate of 2 and 4 kg/s and side mixture volume ratio of 0.7, higher values of turbulent Reynolds number and flow eddies occur. This ensures, better entrainment, and mixing can be satisfied with the flow conditions of 2–4 kg/s with a mass ratio in the range of 461–662 or less. Acceptable entrainment can work until a mass ratio of 1000 at $\dot{m}_w = 2$ kg/s, as predicted previously in Figure 10.

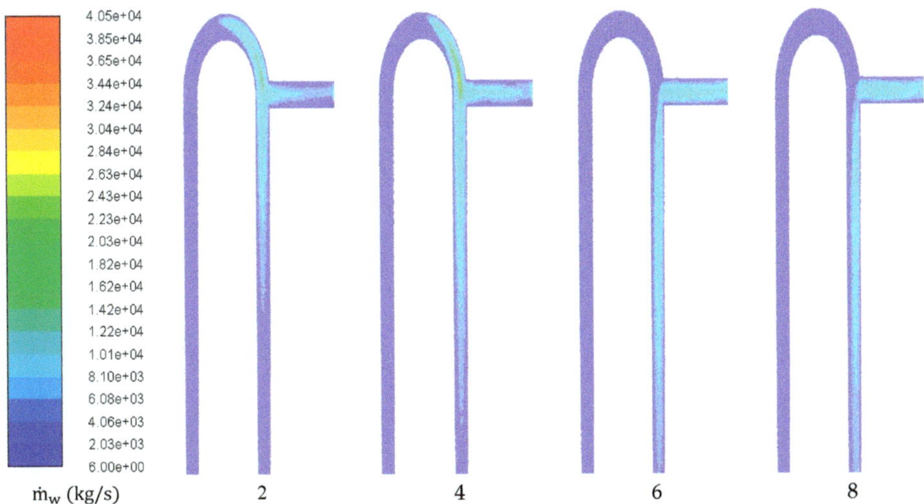

Figure 23. Turbulent Reynolds number contours at different flow ratios and $X_w = 4306.59$, $\alpha_m = 0.2$ and $X_v = 0.1$.

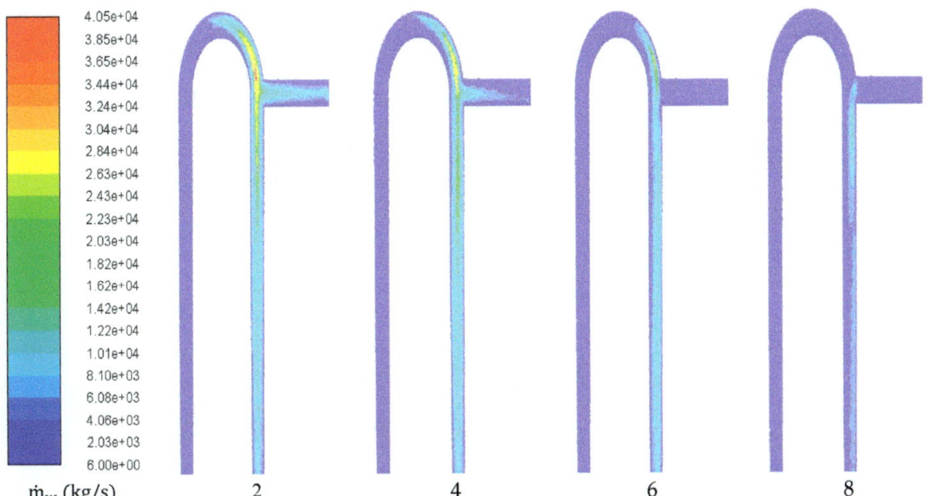

Figure 24. Turbulent Reynolds number contours at different flow ratios and $X_w = 461.42$, $\alpha_m = 0.7$ and $X_v = 0.1$.

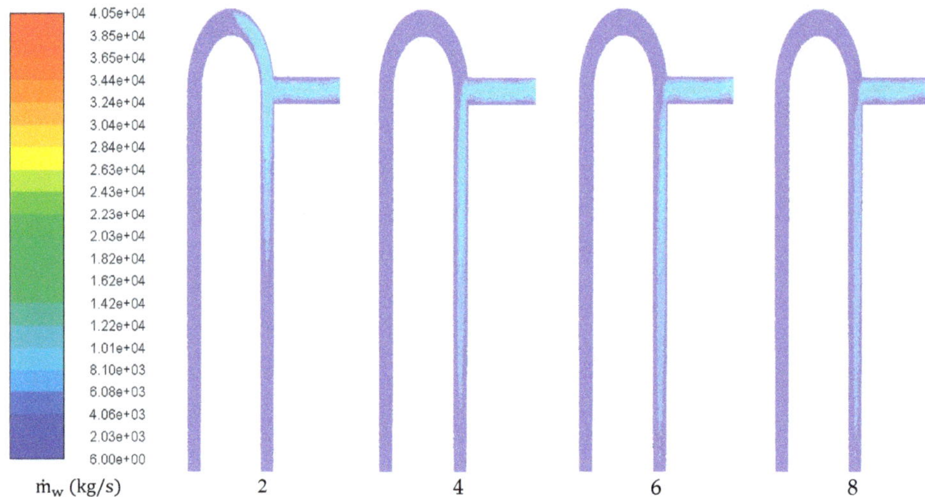

Figure 25. Turbulent Reynolds number contours at different flow ratios and $X_w = 6187.64$, $\alpha_m = 0.2$ and $X_v = 0.5$.

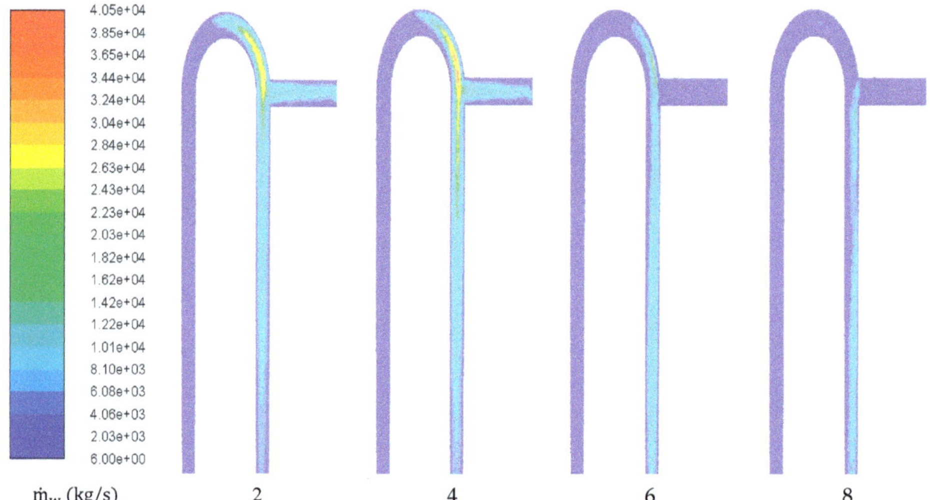

Figure 26. Turbulent Reynolds number contours at different flow ratios and $X_w = 662.96$, $\alpha_m = 0.7$ and $X_v = 0.5$.

The radial distribution of air and vapor volume fraction contours at different mass ratios and positions (L/d = 10, 25, 50, and 100) are shown in Figures 27 and 28. In these Figures, L represents the measured downward distance on the tube from the side pipe centerline, while d is the tube diameter. Air is flowing near the side mixture inlet at closer positions of the side pipe due to the water steam inertia. On the other side, the vapor is instantaneously condensed after mixing with the tube water stream at lower values of water mass flow rate. Increasing the tube water flow rate makes the vapor continuing downward in the tube without condensation and in some extreme cases, the vapor may exit the inverted U-tube in the lower water tank without condensation. In addition, the vapor is attracted to flow in the tube center cavity compared to air, which is found to flow closer to the tube walls.

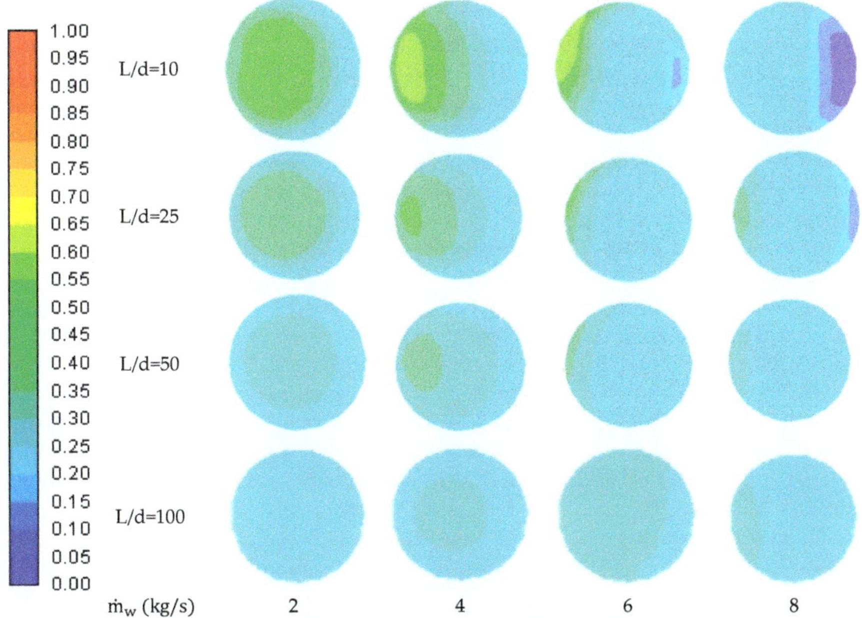

Figure 27. Radial air volume fraction contours at different downward distances from the side pipe centerline at various flow ratios and $X_w = 662.96$, $\alpha_m = 0.7$ and $X_v = 0.5$.

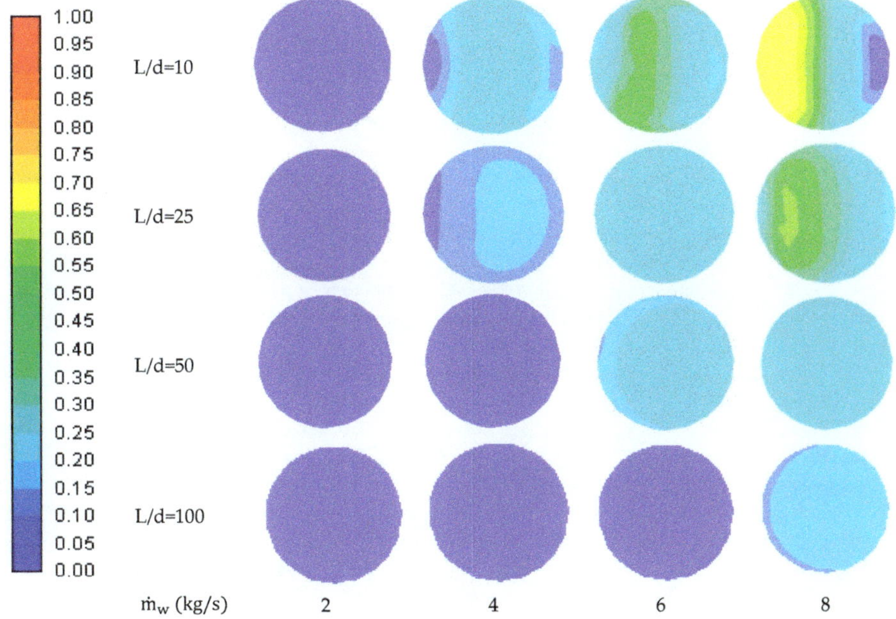

Figure 28. Radial vapor volume fraction contours at different downward distances from the side pipe centerline at various flow ratios and $X_w = 662.96$, $\alpha_m = 0.7$ and $X_v = 0.5$.

5. Conclusions

Using the inverted U-tube for side gas entrainment is experimentally and numerically investigated in this paper. Various gases can be entrained from the required cavities, but the air is suggested in this paper as it is the dominant gas in most applications. Side entrainment is considered a mixture of air and water vapor with different mass ratios. The CFD model is validated first with experimental data and an acceptable comparison is obtained. The present experiment shows the effect of mixing side-air with water flowing in the inverted U-tube on the created water siphon flow at various water levels between water inlet tanks due to altering the generated vacuum at the tube tip. Additionally, the siphon water mass flow rate is found to be reduced with increasing side-air mass flow rate. Then, different mass ratios for the main water stream and side mixtures in the inverted U-tube are examined. The proposed U-tube is capable to bring down the absolute static pressure to.55.1 kPa in the tube upper part. This low-pressure region motives the suction of the side mixture to the tube cavity. This study proves that, at water streams of 2–4 kg/s, better side entrainment can be found at the water/side mixture (air and water vapor) mass ratio of 461–662. In addition, the side entrainment can still exist at a mass ratio of 1000 for water flowing at 2 kg/s. Moreover, the condensation can magnify the generated vacuum at the tube upper portion at a higher water mass flow rate. This means that circulating flow in the range of 2–4 kg/s in the U-tube enhances the function of the tube in entraining the side mixture with maintaining the siphon flow and undergoing vapor condensation. Increasing the water mass flow rate weakens the side entrainment of gases to the tube upper part and hence reduces the tube function at certain mass ratios. These findings can be applied to a variety of engineering applications for evacuating systems.

Author Contributions: Conceptualization A.H., A.E. and K.Y. formal analysis, K.Y., A.E. and A.H.; funding acquisition A.E.; investigation, A.H., K.Y. and A.E.; methodology, K.Y., A.H. and A.E.; project administration, A.E.; Supervision, A.H. and A.E.; writing—original draft, K.Y.; writing—review and editing, A.H., K.Y. and A.E.; All authors have read and agreed to the published version of the manuscript.

Funding: This research received no external funding.

Conflicts of Interest: The authors declare no conflict of interest.

References

1. Engeda, A.; Hegazy, A.S.; Yousef, K. A New Vacuum System for Steam Plant Condenser. *J. Sci. Eng. Appl.* **2020**, *12*, 1–23. [CrossRef]
2. Li, Y.; Chang, S.; Fu, L.; Zhang, S. A technology review on recovering waste heat from the condensers of large turbine units in China. *Renew. Sustain. Energy Rev.* **2016**, *58*, 287–296. [CrossRef]
3. Dennis, M.; Cochrane, T.; Marina, A. A prescription for primary nozzle diameters for solar driven ejectors. *Sol. Energy* **2015**, *115*, 405–412. [CrossRef]
4. Kapooria, R.; Kumar, S.; Kasana, K. Technological investigations and efficiency analysis of a steam heat exchange condenser: Conceptual design of a hybrid steam condenser. *J. Energy S. Afr.* **2017**, *19*, 35–45. [CrossRef]
5. Wölk, G.; Dreyer, M.; Rath, H. Flow patterns in small diameter vertical non-circular channels. *Int. J. Multiph. Flow* **2000**, *26*, 1037–1061. [CrossRef]
6. Zimmerman, R.W.; Gurévich, M.; Mosyak, A.; Rozenblit, R.; Hetsroni, G. Heat transfer to air–water annular flow in a horizontal pipe. *Int. J. Multiph. Flow* **2006**, *32*, 1–19. [CrossRef]
7. Siddique, M. The Effects of Noncondensable Gases on Steam Condensation under Forced Convection Conditions. Ph.D. Thesis, Massachusetts Institute of Technology University, Cambridge, MA, USA, 1992.
8. Kuhn, S.Z. Investigation of Heat Transfer from Condensing Steam-Gas Mixtures and Turbulent Films Flowing Downward inside a Vertical Tube. Ph.D. Thesis, University of California, Berkeley, CA, USA, 1995.
9. Kuhn, S.Z.; Schrock, V.E.; Peterson, P.F. An investigation of condensation from steam-gas mixtures flowing downward inside a vertical tube. *Nucl. Eng. Des.* **1997**, *177*, 53–69. [CrossRef]
10. Strušnik, D.; Golob, M.; Avsec, J. Effect of non-condensable gas on heat transfer in steam turbine condenser and modelling of ejector pump system by controlling the gas extraction rate through extraction tubes. *Energy Convers. Manag.* **2016**, *126*, 228–246. [CrossRef]

11. Strušnik, D.; Marcic, M.; Golob, M.; Hribernik, A.; Zivic, M.; Avsec, J. Energy efficiency analysis of steam ejector and electric vacuum pump for a turbine condenser air extraction system based on supervised machine learning modelling. *Appl. Energy* **2016**, *173*, 386–405. [CrossRef]

12. Havlík, J.; Dlouhý, T. Condensation of the air-steam mixture in a vertical tube condenser. In Proceedings of the 10th Anniversary International Conference on Experimental Fluid Mechanics 2015, EPJ Web Conferences, Prague, Czech Republic, 17–20 November 2015; Volume 114, p. 02037.

13. Zhang, S.; Shen, F.; Cheng, X.; Men, g.X.; He, D. Experimental investigation on condensation of vapor in the presence of non-condensable gas on a vertical tube. *Preprints* **2018**. [CrossRef]

14. Yao, J.; Yao, Y.; Arini, A.; Mchiiwain, S.; Gordon, T. Modelling air and water two-phase annular flow in a small horizontal pipe. *Int. J. Mod. Phys. Conf. Ser.* **2016**, *42*, 1–12. [CrossRef]

15. Juggurnath, D.; Dauhoo, M.Z.; Elahee, M.K.; Khoodaruth, A.; Osowade, A.E.; Olakoyejo, O.T.; Obayopo, S.O.; Adelaja, A.O. Simulations of air-water two-phase flow in an inclined pipe. In Proceedings of the 13th International Conference on Heat Transfer, Fluid Mechanics and Thermodynamics (HEFAT2017), Portoroz, Slovenia, 17–19 July 2017; pp. 77–84.

16. Vásquez, F.; Stanko, M.; Vásquez, A.; De Andrade, J.; Asuaje, M. Air–water: Two phase flow behavior in a horizontal pipe using computational fluids dynamics (CFD). In *Advances in Fluid Mechanics IX*; WIT Transactions on Engineering Sciences; WIT Press: Southampton, UK, 2012; Volume 74, pp. 381–392.

17. Pothof, I.W.M.; Clemens, F.H.L.R. Experimental study of air–water flow in downward sloping pipes. *Int. J. Multiph. Flow* **2011**, *37*, 278–292. [CrossRef]

18. Panella, B.A. Theoretical and Experimental Study of Horizontal Air-Water Two-Phase Flow with a Spool Piece. Master's Thesis, The Polytechnic University of Turin, Turin, Italy, 2011.

19. Afolabi, E.A.; Lee, J.G.M. An eulerian-eulerian CFD simulation of air-water flow in a pipe separator. *J. Comput. Multiph. Flows* **2014**, *6*, 133–149. [CrossRef]

20. Chaudhry, M.H.; Bhallamudi, S.M.; Martin, C.S.; Naghash, M. Analysis of transient pressures in bubbly, homogeneous, gas-liquid mixtures. *ASME J. Fluids Eng.* **1990**, *112*, 225–231. [CrossRef]

21. Faille, I.; Heintze, E. A rough finite volume scheme for modeling two-phase flow in a pipeline. *Comput. Fluids* **1999**, *28*, 213–241. [CrossRef]

22. Hibiki, T.; Ishii, M. One-dimensional drift-flux model and constitutive equations for relative motion between phases in various two-phase flow regime. *Int. J. Heat Mass Trans.* **2003**, *46*, 4935–4948. [CrossRef]

23. Ishii, M. Thermo-fluid dynamic theory of two-phase flow. In *Collection de la Direction des Etudes et Recherches d'Electricit´e de France 22*; Eyrolles: Paris, France, 1975.

24. Tiselj, I.; Petelin, S. Modelling of two-phase flow with second-order accurate scheme. *J. Comput. Phys.* **1997**, *136*, 503–521. [CrossRef]

25. Cerne, S.; Petelin, S.; Tiselj, I. Coupling of the interface tracking and the two-fluid models for the simulation of incompressible two-phase flow. *J. Comput. Phys.* **2001**, *171*, 776–804. [CrossRef]

26. Wylie, E.B.; Streeter, V.L. *Fluid Transients in Systems*; Prentice Hall: Englewood Cliffs, NJ, USA, 1993.

27. Kiameh, P. *Power Generation Handbook, Selection, Applications, Operations, and Maintenance*, 1st ed.; McGraw-Hill Professional: New York, NY, USA, 2002.

28. Versteeg, H.K.; Malalasesekera, W. *An Introduction to Computational Fluid Dynamics, the Finite Volume Method*; Pearson Education Limited: Harlow, UK, 2007; Volume 2.

29. Ansys Fluent 18.1 Theory's Guide. 2018. Available online: http://www.pmt.usp.br/ACADEMIC/martoran/NotasModelosGrad/ANSYS%20Fluent%20Theory%20Guide%2015.pdf (accessed on 10 November 2020).

30. Li, J.D. CFD simulation of water vapor condensation in the presence of non-condensable gas in vertical cylindrical condensers. *Int. J. Heat Mass Transf.* **2013**, *57*, 708–721. [CrossRef]

31. Yousef, K.; Hegazy, A.; Engeda, A. Mixing of dry air with water-liquid flowing through an inverted U-tube for power plant condenser applications. In Proceedings of the ASME-JSME-KSME 2019 Joint Fluids Engineering Conference AJKFLUIDS2019, San Francisco, CA, USA, 28 July–1 August 2019.

32. Yousef, K. Use of Vapor Compression Refrigeration System for Cooling Steam Power Plant Condenser. Ph.D. Thesis, Menoufia University, Shibin el-Kom, Egypt, 2015.

33. Chen, S.; Yang, Z.; Duan, Y.; Chen, Y.; Wu, D. Simulation of condensation flow in a rectangular microchannel. *Chem. Eng. Process.* **2014**, *76*, 60–69. [CrossRef]

34. Lee, W.H. *A Pressure Iteration Scheme for Two-Phase Modeling*; Technical Report LA-UR 79-975; Los Alamos Scientific Laboratory: Los Alamos, NM, USA, 1979.
35. Liu, Z.; Sunden, B.; Yuan, J. VOF modeling and analysis of filmwise condensation between parallel vertical plates. *Heat Transf. Res.* **2012**, *43*, 47–68. [CrossRef]

Publisher's Note: MDPI stays neutral with regard to jurisdictional claims in published maps and institutional affiliations.

 © 2020 by the authors. Licensee MDPI, Basel, Switzerland. This article is an open access article distributed under the terms and conditions of the Creative Commons Attribution (CC BY) license (http://creativecommons.org/licenses/by/4.0/).

Article

In-Vitro Experimental Modeling of Oscillatory Respiratory Flow in a CT-Scanned OSAHS Tract

Zhenshan Zhu [1], Yaping Ju [1] and Chuhua Zhang [1,2,*]

[1]   Department of Fluid Machinery and Engineering, School of Energy and Power Engineering, Xi'an Jiaotong University, Xi'an 710049, China; zzs2010@stu.xjtu.edu.cn (Z.Z.); yapingju@mail.xjtu.edu.cn (Y.J.)

[2]   State Key Laboratory for Strength and Vibration of Mechanical Structures, Xi'an Jiaotong University, Xi'an 710049, China

*   Correspondence: chzhang@mail.xjtu.edu.cn; Tel.: +86-029-866-8723

Received: 19 October 2020; Accepted: 9 November 2020; Published: 10 November 2020

Abstract: Obstructive sleep apnea-hypopnea syndrome (OSAHS) is a highly prevalent respiratory disorder. The knowledge of respiratory flow is an essential prerequisite for the establishment and development of OSAHS physiology, pathology, and clinical medicine. We made the first in-vitro experimental attempt to measure the oscillatory flow velocity in a computed tomography (CT) scanned extra-thoracic airway (ETA) model with OSAHS by using the particle image velocimetry (PIV) technique. In order to mimic respiration flow, three techniques were adopted to address difficulties in in-vitro experimental modeling: (1) fabricating the obstructive ETA measurement section with the CT-scanned data of an OSAHS patient airway; (2) maintaining the measurement accuracy by using the optical index-matching technique; (3) reproducing the oscillatory respiratory flow rates with the compiled clinical data of transient tidal volumes. The in-vitro measurements of oscillatory respiratory flow velocity manifested the time evolution of the complex OSAHS flow patterns, and the potential wall collapse of the ETA model with OSAHS.

Keywords: in-vitro experimental modeling; CT-scanned OSAHS model; oscillatory respiratory flow rates; particle image velocimetry technique; optical index matching

1. Introduction

The human airway is a key organ for the body to exchange air with the external environment. However, this passageway is susceptible to inhaled particulate matters, extreme atmospheric conditions, bad individual habits, and natural aging, resulting in a wide spectrum of respiratory diseases ranging from the simple common cold, snoring, nasal airway blockage and obstructive sleep apnea-hypopnea syndrome (OSAHS) [1]. In the case of such disorders, OSAHS is highly prevalent and poses a high risk of damaging human health and life [2–5]. Its severity is highly correlated to the flow characteristics of the upper airway [6]. Thus, the knowledge of OSAHS respiratory flow in the upper airway is of scientific significance for the establishment and development of OSAHS physiology and pathology, and of clinical medicine references for the diagnosis and therapy of OSAHS diseases.

Compared with the in-vivo measurement of respiratory flow in a realistic human airway, the in-vitro measurement of respiratory flow can effectively provide more detailed flow fields, allowing the respiratory physiologist to deepen the understanding of OSAHS pathology and the respiratory physician to improve diagnosis and therapy for OSAHS diseases. Nonetheless, OSAHS physiological studies of airway anatomy has earned much less attention than healthy airways. Only a small number of studies have focused on respiratory flow in OSAHS-like tract, although much experimental research on the respiratory flow has been carried out in simplified smooth multiple-generation airway models, which were briefly reviewed in Zhu et al. [7]. Following a series of numerical works on steady inspiratory or expiratory flow in smooth symmetrical and asymmetrical multiple-generation bifurcating airway models [8–10], Liu and his collaborators extended their numerical methods to unsteady respiratory flow in a human upper airway (HUA) model before and after OSAHS surgery [11]. Their numerical results confirmed that after surgery, the HUA flow resistance decreases with even more streamlined flow patterns, demonstrating the potential application of OSAHS respiratory flow in OSAHS surgery. Later, they numerically simulated the steady and unsteady flows in a HUA model with OSAHS resulting from nasopharyngeal obstruction [12]. Their numerical results indicated that the respiratory flow with OSAHS features strong flow injection caused by narrowing of the pharynx at both inspiration and expiration stages. Recently, Song et al. [13] numerically simulated steady inspiration flows in a nasopharyngeal obstruction model with OSAHS. Their model is quite similar to that used in Liu et al. [12]. The numerical results found that inspiratory flow velocity reached a maximum at the obstructed area with a strong negative pressure gradient. Very recently, Cui et al. [14] used the large eddy simulation (LES) method to investigate oscillatory respiration flows in an idealized extra-thoracic (ETA) airway. The simulated results indicated that the unsteady airflow structures are highly impacted by inspiration and expiration phases, and the techniques to model the breathing process.

As for the experimental measurements of respiratory flow, Kim and Chung [15] used tomographic particle image velocimetry (PIV) techniques to measure the airflow in disordered and corrected nasal cavity models. Later, Kim and his collaborators evaluated the feasibility of 3D steady tomographic PIV measurements in a normal nasal cavity [16]. More recently, Wu et al. [17] implemented a smoke-wire visualization of steady flow at the mid-sagittal plane of a HUA model with OSAHS by using a high-speed camera. Their experimental visualization, together with numerical simulation, suggested that for OSAHS patients, HUA narrowing is a symptom of OSAHS development and deterioration.

All of the above in-vitro experimental models for OSAHS respiratory flow were limited to steady flow cases. The recent numerical observations [18,19] of the unsteady flow patterns differ significantly from the steady ones in the ETA models. In-vitro measurement of unsteady respiratory flow in human airway models poses a great challenge for fluid experimentalists due to related geometrical complexity, limited optical accessibility and oscillatory flow rate [20]. Specifically, three difficulties need to be overcome if an accurate and reliable mimicking of respiratory flow in in-vitro environment was required, i.e., a realistic respiratory tract model with OSAHS, good optical index matching, and reliable control over the oscillatory respiratory flow rate. It is worth noting that most investigations on oscillatory respiratory flow were based on the simplified sin curve flow rate as the oscillatory boundary condition [18,19]. In fact, the realistic oscillatory respiratory boundary condition is quite different from the sin function [21].

The goal of the present research is twofold. The primary goal is to tackle the above mentioned three critical difficulties arising in the in-vitro experimental modeling of OSAHS flows. The secondary goal is to employ the experimental rig to measure the oscillatory respiratory flows in an ETA model with OSAHS. The ETA model with OSAHS is composed of an oral cavity, pharynx, larynx and trachea with total closure of the nasal passageway.

The remainder of this paper is organized as follows. In Section 2, an in-vitro experimental model will be described, focusing on techniques to overcome the above three difficulties. Section 3 will present PIV techniques tailored to the experimental rig. Section 4 will present the PIV measurement results of oscillatory respiratory flow in the ETA model with OSAHS and their clinical implications. Concluding remarks will be made in Section 5.

2. In-Vitro Experimental OSAHS Flow Model

2.1. ETA Model with OSAHS

Creating an anatomically correct transparent experimental respiratory flow model is essential to measure respiration flow inside a human airway by using the PIV technique. The present ETA model was based an anatomical data, which was obtained from a scanned airway of a 48-year-old Chinese male

OSAHS patient with a 128-slice spiral CT, under the rest breathing condition. The apnea-hypopnea index (AHI) and body mass index (BMI) were 52 and 26.8, respectively. In order to obtain detailed data, CT scanning was performed on the patient along the coronal, sagittal, and cross-sectional directions. The thickness between the CT-scanned slice images was 1 mm. The OSAHS condition of the patient was so severe that the passageway through the nasal cavity was totally blocked in the retro uvula region (RUL) as shown in Figure 1. The ETA model composed of the oral cavity, pharynx, larynx and trachea was adopted to mimic the patient's breathing through the oral cavity. The ETA model with OSAHS was then constructed with the image processing software Mimics [15,16,22]. The digital ETA model is illustrated in Figure 2, from which the obstruction resulting from the retro uvula hypertrophy can be clearly observed in Figure 2a. Two representative measurement sections were selected, i.e., the global A-A section viewed from the lateral side and the local B-B section near the obstructed region, as shown in Figure 2a,b.

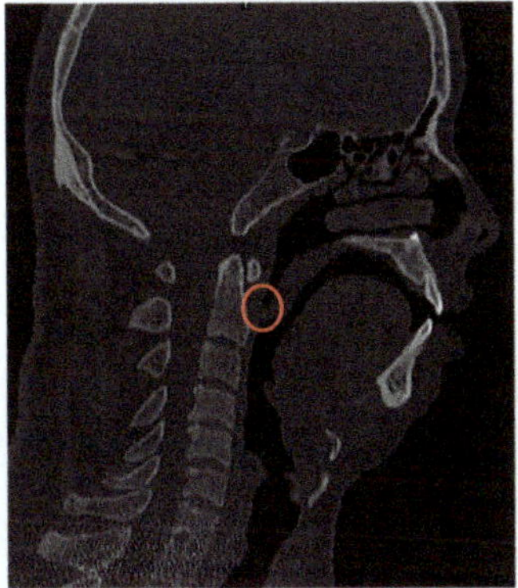

Figure 1. Sagittal plane CT image of OSAHS patient.

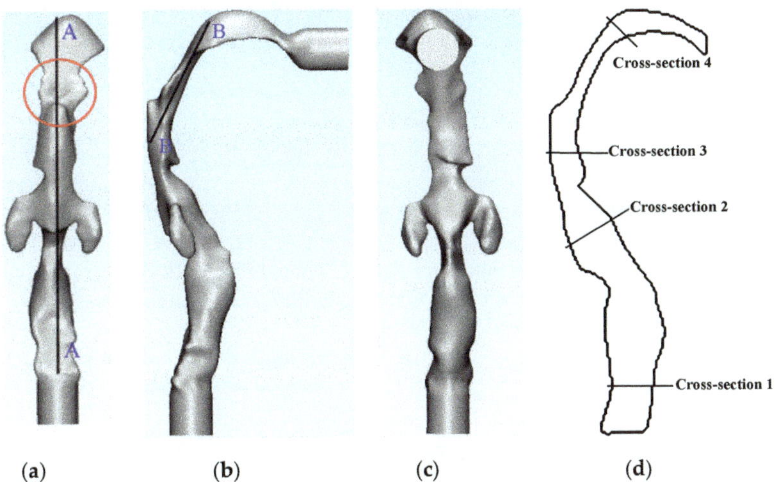

Figure 2. Reconstructed ETA model with OSAHS: (**a**) dorsal view; (**b**) lateral view; (**c**) front view; (**d**) four cross-sections.

To facilitate an accurate experimental measurement, the experimental model was finely fabricated as follows. First, a negative ETA model was printed with water-soluble material using a fused deposition modeling (FDM) printer [16,23,24]. The melting temperature of the water-soluble material was 220 °C and the wire diameter was 1.75 mm. The maximum printing error of the negative model was less than 0.1 mm. Second, the negative model was coated with a water-soluble glue. The coating glue functioned as a slice separator of the negative model as the glue dried out. Both the water-soluble material and glue were mainly composed of polyvinyl alcohol (PVA), but they had different physical properties. The dried water-soluble glue will not stick to the external filler silicone, but the water-soluble material will. Third, the coated negative model was put into an optical glass box that was then carefully filled with transparent liquid silicone Sylgard 184 [25]. Before it was poured into the box, the silicone was needed to be vacuumed to remove bubbles in it. If not, the fabricated model wall would absorb tiny bubbles, which then seriously polluted measurement results. The liquid silicone was solidified by curing at room temperature over 48 h. Finally, the negative model and the coating glue were washed with cold water. The fabricated experimental model is shown in Figure 3. The compliance of the human airway tract and the mucus attached to the inner wall [26,27] were not considered in the present experimental model for the sake of accurate PIV measurements.

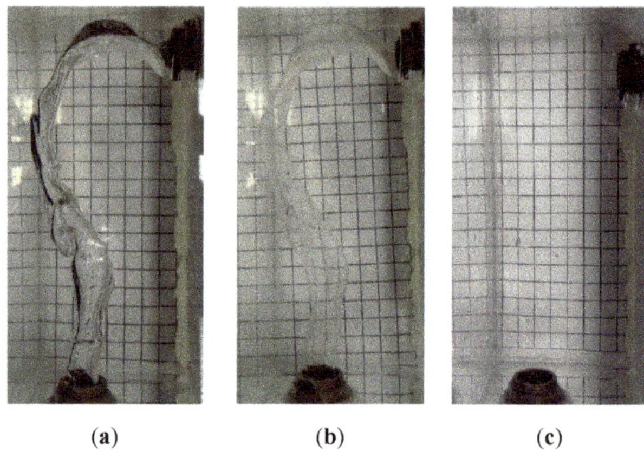

(a)   **(b)**   **(c)**

Figure 3. Optical index matching between ETA model and working fluids: (**a**) air; (**b**) water; (**c**) mixture of water and glycerol.

2.2. Optical Index Matching

In optical measurements of the flow inside the airway model, the optical index matching between the working fluid and model wall is crucial [20,28–31]. Good index matching can eliminate optical distortion to ensure measurement accuracy. After a series of priori tests with different mixtures of water and glycerol as working fluid, the mixture composed of 60.7% glycerol and 39.3% water [32] reached the best optical index matching. For comparison, Figure 3a–c presents optical index matching with air, water and water-glycerol mixture, respectively, as working fluids in the model. Clearly, the best optical index match was achieved with the water-glycerol mixture. As shown in Figure 3c, the lines in the quadrille paper sheet do not suffer from deformation as one looks through the model [20].

2.3. Oscillatory Respiratory Flow Rate

In the present unsteady measurement, the realistic normal male breathing flow rate transients at rest were used to determine the modeling flow rate transients. Statistical data of the breathing flow rate transients compiled by a Beijing medical institution from hundreds of supine male adults were used. Similar flow rate transient curves were also confirmed in Kim and Chung [21] and Huang and Zhang [33]. The tide volume (V_T) here was chosen to be 500 mL. The case where the inspiratory air volume is a little larger than the expiratory one [34] was not considered in the present test rig. The typical breathing cycle period at rest is within 3–5 s; 5 s were chosen due to the severe status of the OSAHS patient.

To mimic the realistic flow physics and respiratory physiology, the similarities of Womersley number ($\alpha = L/2 \cdot \sqrt{2\pi \cdot f_b / v}$) and Reynolds number ($Re = V \cdot L/v$) were kept between the breathing air and the modeling water-glycerol liquid [23,29,32,35]. The two similarity criteria ensured flow similarity such as the same relationship of pressure drop coefficient vs. flow coefficient. The Womersley number is a dimensionless number in biofluid mechanics representing the ratio of transient inertial force to viscous force and is an important similarity criterion for the oscillatory respiratory flows. Here, the Re number can be defined as $Re = 4f_b \cdot V_T / (L \cdot v)$, as mentioned by Janke et al. [23,24]. In these definitions, L, V_T, f_b and v refer to the inlet hydraulic diameter of the tract, tide volume, frequency of the oscillatory respiration, and kinematic viscosity of the working fluid, respectively. The kinematic viscosity (v) of the water-glycerol mixture and breathing air was 8.933×10^{-6} m^2/s and 15.03×10^{-6} m^2/s, respectively, at room temperature during the experiments. In this oscillatory respiratory flow, the maximum Reynolds numbers during inspiration and expiration phases were 1442 and 1149, respectively, and the Womersley number was 2.89. To maintain the similarity of Reynolds number and Womersley number, the breathing cycle period ($1/f_b$) of 5 s was tuned to be 8.5 s in the present in-vitro experimental model where the working fluid was taken as the water-glycerol liquid. The breathing flow rate transients as well as those employed in the experiment are shown in Figure 4. Under such oscillatory boundary condition, roles of inlet and outlet of this model would switch periodically. At the inspiration stage, the boundary at the mouth side can be considered as the inlet, and the boundary at the trachea side was the outlet. While at the expiration stage, the contrary would be the truth.

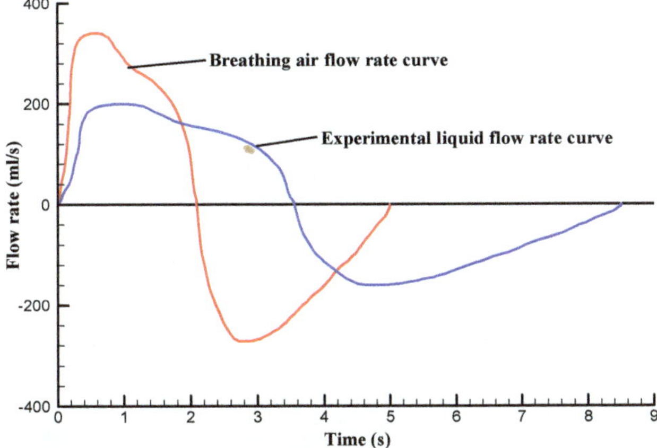

Figure 4. Human breathing and modelling flow rate transients.

To mimic the modeling flow rate transients as shown in Figure 4, several apparatuses and procedures were carefully designed in this research. The schematic and photo of the experimental setup are presented in Figure 5. A programmable logic controller (PLC), servo-motor, servo-motor driver, gear box, bent axle, piston, air exhausting device, reservoir and V_T monitor (Figure 5a) were employed to provide precise control over piston movements and thus experimental flow rate transients. Figure 5b,c presents a photo of the main apparatus and experimental rig, respectively. First, a PLC program was developed and run on a personal computer to transfer the PLC input digital signal into the PLC output analog voltage signal. Second, step by step, the driver of the servo-motor drove the movements of the servo-motor. The duration of the movement in each step was 20 ms. Thus, a total of 425 step movements were precisely controlled over the present single respiratory cycle (8.5 s). Third, the gear box was used to increase the servo-motor torque output, which eventually actuated the bent axle and the piston. Lastly, the air exhausting device was used to remove potential air bubbles absorbed into the test rig. The apparatus of the reservoir and V_T monitor was mainly made of a graduated cylinder, as shown in the Figure 5b. This cylinder can not only collect working fluids and add tracer particles, but also monitor the output V_T. The liquid level in the graduated cylinder would fluctuate up and down, following simultaneously with the oscillatory experimental boundary. By recording the locations of the highest and lowest liquid levels in the graduated cylinder, the output V_T was monitored.

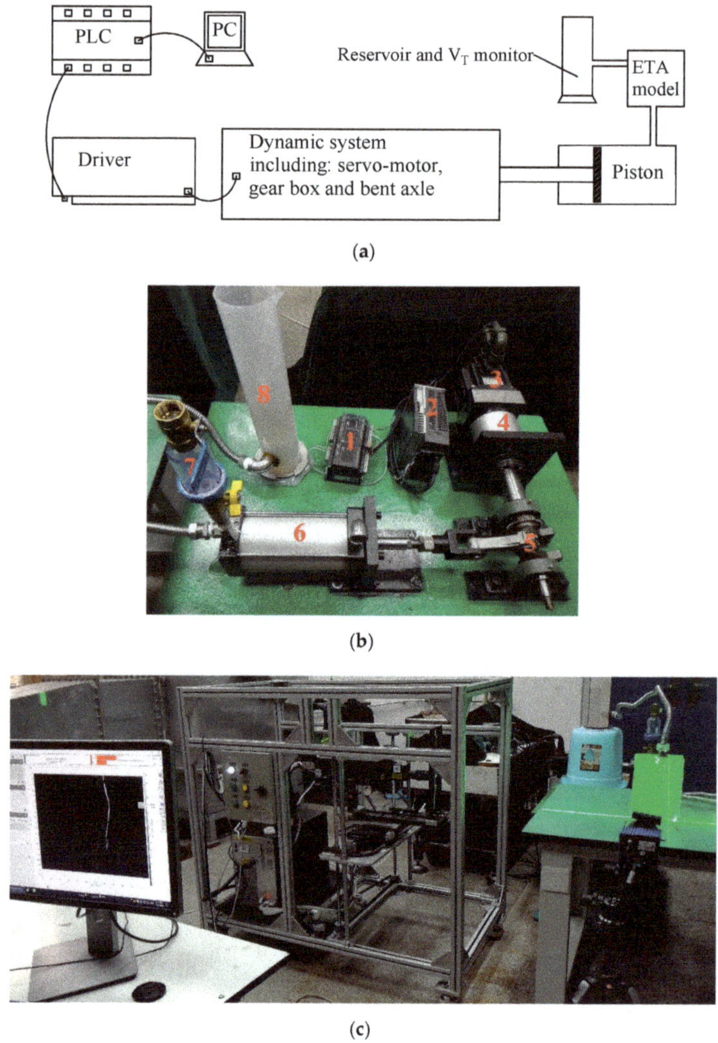

(a)

(b)

(c)

Figure 5. Experimental rig: (**a**) schematic; (**b**) main apparatus (except ETA tract and PIV system):1 PLC; 2 driver; 3 servo-motor; 4 gear box; 5 bent axle; 6 piston; 7 air exhausting device; 8 reservoir and V_T monitor; (**c**) photo of experimental rig.

3. PIV Techniques for the ETA Model with OSAHS

The other goal of this research was to apply the above experimental rig to measure the oscillatory respiratory flow velocity in an ETA model with OSAHS. The LaVision 2D-2C (two-dimensional, two-component) PIV system was used to measure the unsteady flow velocity at the two representative sections, i.e., section A-A and section B-B, as shown in Figure 2a,b.

A double pulse yttrium-aluminum-garnet (YAG) laser and a high-speed charge-coupled device (CCD) camera were employed to obtain digital images. A laser (Evergreen Corp.) with power up to 600 mJ per pulse was used to provide double laser sheets with frequency of 15 Hz, wavelength of 532 nm and thickness of approximately 1 mm to induce fluorescence of tracers. The CCD camera (Imager SX 4M) was used to store digital images. The camera features an interline transfer chip with progressive-scan low-noise readout, high spatial resolution of 2360 × 1776 pixels (4 Mpixel, $5.5 \times 5.5~\mu m^2$ per pixel size) in 12-bit mode, and fast frame rates up to 31 Hz. Here, 15 Hz frame rates were used to match the frequency of the laser pulse.

Determining the distinctive time (dt) delay of two subsequent particle images (double shutter) in the PIV measurements of oscillatory flow is an important task since the dt setting must be adapted to the maximum flow speed as well as the flow reversal condition [32]. Here, dt was determined according to the CCD spatial resolution, field of view of the camera, estimated average velocity, and target pixel shift of tracing particles in double shutter, which were 2360 × 1776 pixels, $15 \times 11.3~cm^2$, 0.7 m/s, and 7 pixels, respectively. The dt and the duration of exposure time of particle images were determined to be 500 μs and 200 μs, respectively. The length of camera image (11.3 cm) was smaller than that in ETA model (approximately 20 cm); consequently, one PIV measurement of A-A section had to be divided into two sub-sections. The two digital images obtained from the two sub-sections measurements at the same phase were then stitched to form a whole PIV measurement image. Within the present modeling cycle (8.5 s), 128 groups of double shutter images were continuously acquired with the PIV system and controlled with a synchronizer. All steps of control, acquisition and post processing were carried out in a DaVis 10.0.5 software environment [36].

The reflected light is very harmful to both the camera CCD itself and the CCD acquisition quality of near-wall flows. To avoid light reflection from walls, good optical index-matching, suitable tracer fluorescence and a long-pass glass filter were employed [16]. The working fluid with good optical index-matching not only eliminates optical distortions, but also reduces wall light reflection. Fluorescent tracers were used to scatter 590 nm wavelength light under 532 nm wavelength laser irradiation. Using the long-pass glass filter before the camera, the reflected light with wavelength below 550 nm will be filtered out, thus greatly alleviating wall reflection.

The long-pass glass filter and fluorescent tracers used in PIV measurement will not distort measured results. The long-pass glass filter had more than 99% transmittance of the scattered light with wavelength above 550 nm. The mean diameter d_p and density ρ_p of fluorescent tracers were 4 μm and 3.4 g/cm^3, respectively. The response time (τ_p) of tracing particles defined as $d_p^2 \rho_p/(18\mu)$ was calculated as 2.93×10^{-7} s [30]. The dynamic viscosity μ of the water-glycerol mixture was 1.03×10^{-2} Pa·s.

4. Results and Discussion

In a respiration cycle, four representative phases were highlighted to present PIV measurements of oscillatory flow velocity fields at two typical A-A and B-B sections, as shown in Figure 6. Locations of A-A and B-B sections are defined in Figure 2a,b. The four selected phases represent the reversal phase of expiration-inspiration (phase one), the peak phase of inspiration (phase two), the reversal phase of inspiration-expiration (phase three), and the peak phase of expiration (phase four), respectively. The measured flow velocity contours were colored with the magnitude of absolute velocity (ABS velocity). The measured results will be presented and discussed in the following two sub-sections for reversal phases (phases one and three) and peak phases (phases two and four). For a clear presentation, local views at B-B sections were deliberately enlarged. In order to quantitatively analyze measured results, the main-stream velocity transients at four cross-sections were also presented. The locations and the main-stream velocities profiles of the four cross-sections are shown in Figures 2d and 7, respectively.

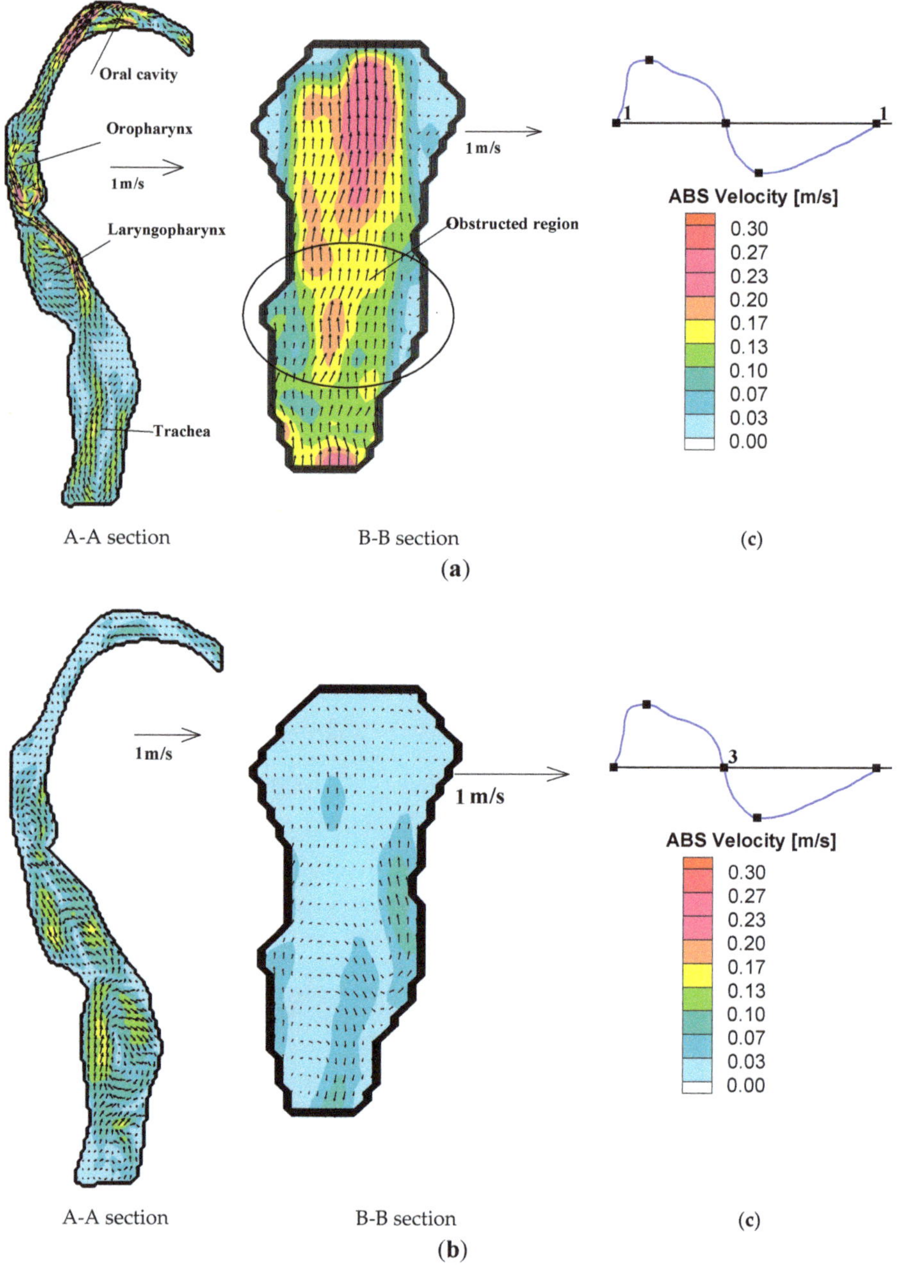

Figure 6. *Cont.*

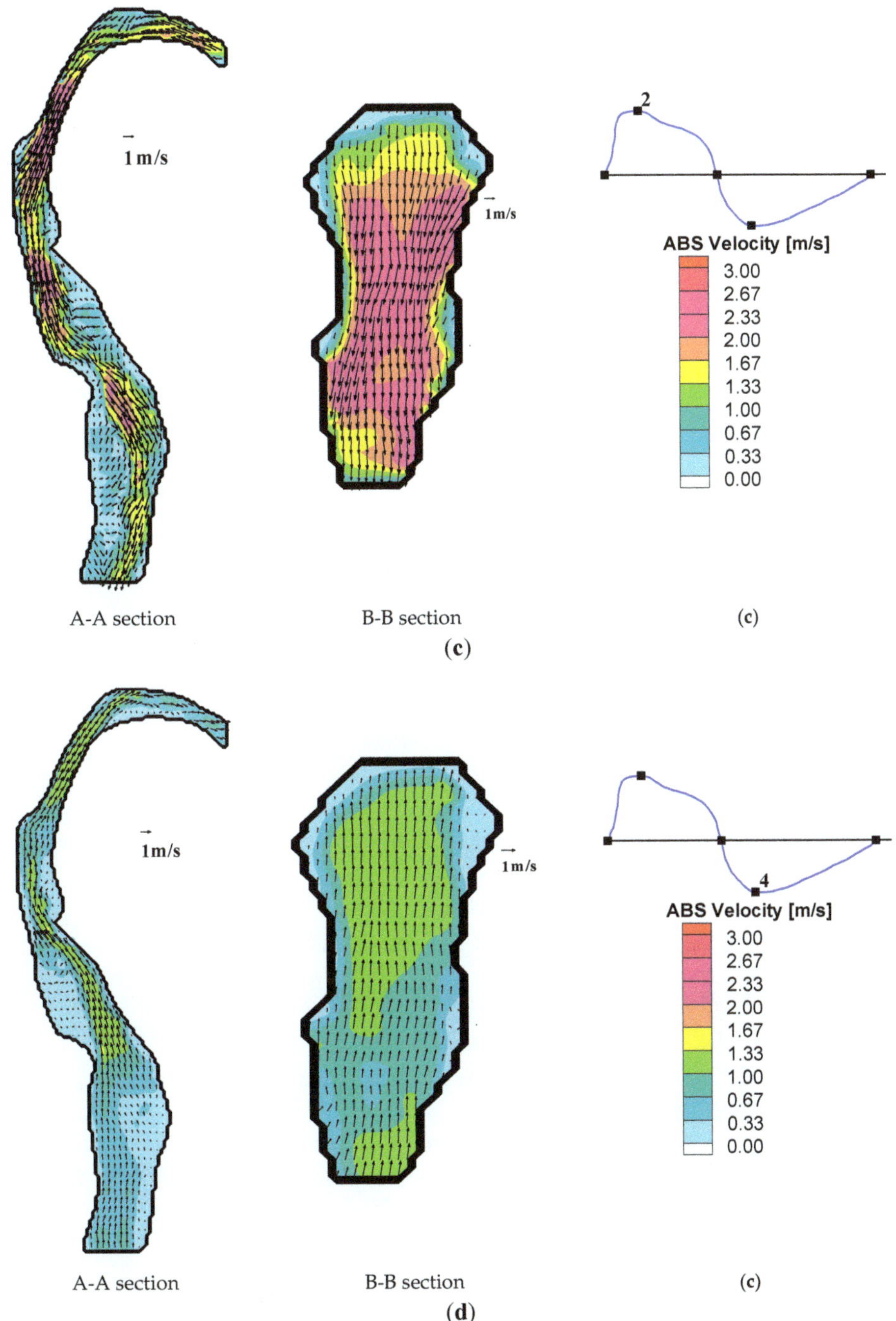

Figure 6. Measured results at four representative phases at A-A section and B-B section: (**a**) phase one: expiration-inspiration reversal transient; (**b**) phase three: inspiration-expiration reversal transient; (**c**) phase two: peak inspiration transient; (**d**) phase four: peak expiration transient.

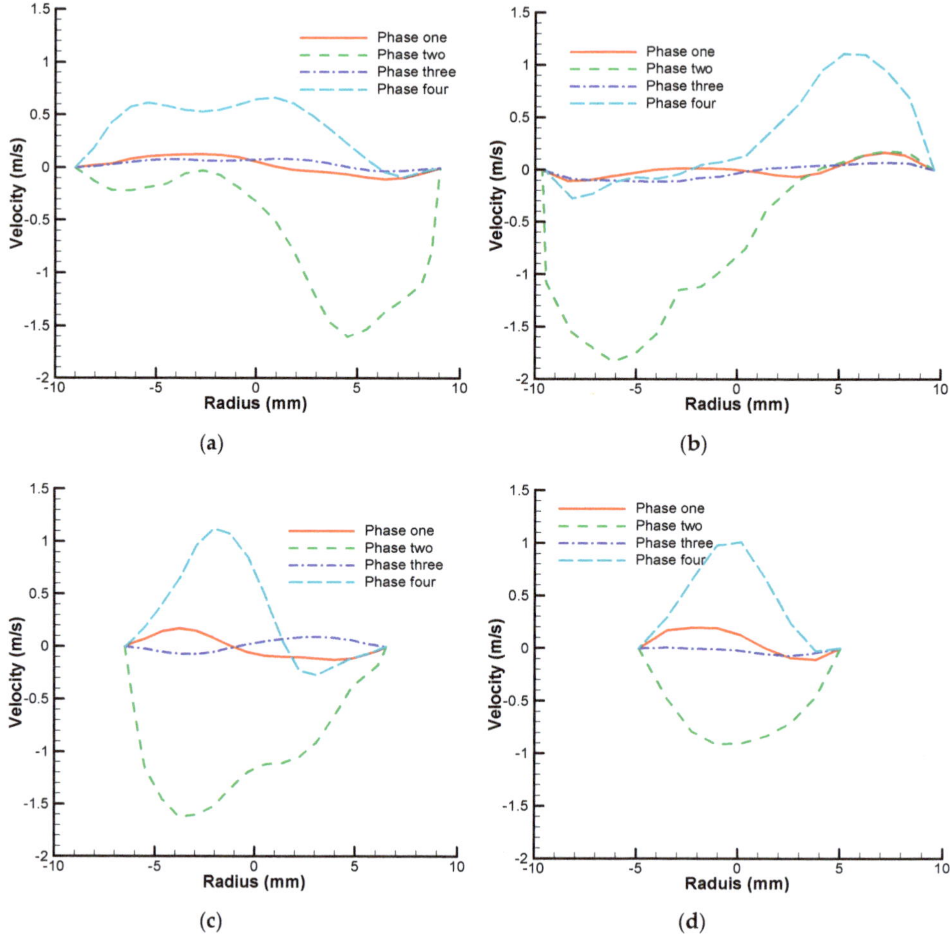

Figure 7. Measured main-stream velocities at four cross-sections at four representative phases: (**a**) cross-section 1; (**b**) cross-section 2; (**c**) cross-section 3; (**d**) cross-section 4. Locations of these four cross-sections are defined in Figure 2d.

4.1. Reversal Phases

At phase one, the respiratory flows are undergoing transition from expiration to inspiration (Figure 6a). The flow patterns presented are irregular both at the A-A section and the B-B section. The reversal flow and jet flow structures can be clearly observed at the A-A section. The interaction between the expiration and inspiration flows leads to strong flow reversal at the oral cavity and oropharynx region, which is hardly considered as either expiration or inspiration pattern. However, the local flow structure at the B-B section mainly corresponds to the expiration pattern with a large portion of expiration flow at the mainstream region, except only a small portion of reversal flow at the near wall region. The flow pattern at the expiration-inspiration phase is more complex than that observed in Adler and Brücker [22], who claimed that the flow at this phase could be regarded as expiration flow. This difference may be caused by more realistic oscillatory flow rate boundary conditions used in this research rather than the simplified sine-law flow rate boundary conditions used in Adler and Brücker [22]. Such jet-reversal flow structure is an indicator of complex expiration-inspiration transition at the quasi-static conditions where the net respiratory flow rate is zero while the unsteady inertial force and thus the respiratory flow acceleration and velocity are not zero.

We can also see that compared with the quasi-static condition at phase one, the jet-reversal flow structure occurs even more obviously at phase three (Figure 6b), due to the milder transition from inspiration to expiration (Figure 4). Neither an inspiration nor an expiration pattern is the dominating flow pattern at phase three. The jet-reversal flow structure extends into the oral cavity, oropharynx, laryngopharynx and trachea (Figure 6b, A-A section), and the widely spread reversal flow vortexes at the obstructed region (Figure 6b, B-B section). The inspiration-expiration pattern is very different from that in Adler and Brücker [22], who considered the flow at this phase to can be regarded as inspiration flow.

From the reversal transients, we could conclude that the realistic oscillatory respiratory boundary condition and the realistic model have a significant effect on respiration pattern, and associated flow structures. This observation can be clearly confirmed in Figure 7. Although the main-stream velocity profiles at reversal phases one and three are relatively small, their patterns are not similar to those at inspiration phase two and expiration phase four. Although quite few numerical and experimental studies focused on the reversal transients, the present PIV measurements confirm that the respiration patterns at the reversal phases can neither be regarded as the pattern of inspiration nor that of expiration due to the occurrence of jet-reversal flow and quasi-static condition.

4.2. Peak Phases

At peak inspiratory transient (phase two), the inspiration flows form an obvious jet-wake structure and a strong flow reversal (Figures 6c and 7b). Jet-wake flow structure, which often occurs in the human trachea and bronchi due to wall curvature [7], also appears at this phase in the ETA model with OSAHS. High-speed jet fluids accumulate at the dorsal side of the laryngopharynx region, while the anterior side of the trachea (Figure 6c, A-A section, Figure 7a,b). The narrowed tract does have a significant impact on inspiration flow at this phase with a clear injection phenomenon. As shown in the enlarged vector diagrams at the phase of peak inspiration, the jet-wake structure together with the flow reversal enhances fluid mixing and heat transfer, which benefits the velocity, temperature and humidity conditioning of the inhaled air as confirmed by Kim et al. [37].

At the peak expiration phase (phase four), however, the expiration flow shows very different flow patterns (Figure 6d). The high speed jet fluids accumulate at the anterior side of the laryngopharyngeal region and the dorsal side of the trachea (Figure 6d, A-A section, Figure 7a,b), which are at opposite sides of the tract at the peak inspiration phase (Figure 6c, A-A section). Such differences may be due to the combined effect of convection and viscous motions in the curved tract and complex wall attachments. The injection phenomenon caused by the narrowed region resembles the numerical observation in Liu et al. [12]. It should be noted that we used the oropharyngeal model in our research rather than the nasopharyngeal model used in Liu's study. The flow patterns at the trachea region are very different with the results in Wang et al. [38] using a simplified model and the results in Cui et al. [14] using an idealized model. In those studies, the expiration flow patterns at the trachea region were quite regular and presented straight tube-like flow. The differences may be caused by the more realistic model used in this research.

From the results above, we can clearly conclude that the inspiration and expiration flow patterns are quite different from each other. High speed jet fluids appeared at the opposite sides of the oropharynx, laryngopharynx and trachea regions, which was not highlighted in the previous similar numerical studies [6,14,19,26,37]. Expiration process is not a simple reversal process of inspiration.

Considering that the patient's nasal passageway was completely blocked, we believe that the collapse of the passageway most likely occurs at the peak inspiration phase at the obstructed region, thus resulting in apnea. There are two main reasons for this: firstly, the effect of obstructed tract on oscillatory respiratory flow fields at peak phases is much more apparent than that at the reversal phases. The occurrence of injection phenomenon at the narrowing obstructed area at the peak inspiration phase indicates very large pressure differences over this short, obstructed distance [13]. Secondly, at inspiration phase, the driving force induced by the movements of the thoracic cavity and abdominal cavity is a negative pressure. The combination of large pressure difference across the obstructed area and negative pressure at the peak inspiration phase would more easily induce collapse of passageway wall and eventual blockage of the tract, thus resulting in apnea. This condition worsens and becomes more dangerous for the deep-sleeping OSAHS patient, whose passageway tissue walls become very loose.

Overall, we presented a temporal and spatial evolution of the oscillatory respiratory flow patterns within a CT-scanned model with OSAHS using PIV measurements. It should be recognized that the calibrated CFD (computational fluid dynamics) method can provide even more flow details that are hardly, if not impossible, captured with PIV techniques [39], such as pressure drop and airway resistance in the CT-scanned complex geometry. Airway compliance is another important issue in human tract [40] that was not modeled in the present study. However, the present work contributes to an experimental modeling of the complex oscillatory respiratory flow by using PIV measurements and can be regarded as a step-stone for an even deeper understanding of OSAHS airway flows.

5. Conclusions

In the present research, particular efforts were made to experimentally mimic the respiratory flow conditions in an in-vitro experimental model to measure the oscillatory respiratory flow velocity inside a CT-scanned ETA model with OSAHS by using PIV technique. Four representative phases were highlighted to clarify the respiratory flow patterns and the associated flow structures in the ETA model.

With realistic non-sinusoidal respiratory flow rate and CT-scanned ETA model, the respiratory flow patterns at the reversal phases can neither be regarded as the pattern of inspiration nor that of expiration.

Flow patterns at the inspiratory and expiratory phases are very different. High-speed jet flows generally appear at the opposite sides of the oropharyngeal, laryngopharyngeal and tracheal regions at the inspiration and expiration phases. The expiration process is not a simple reversal process of inspiration.

The respiratory flow patterns and the associated flow structures have clinical implications for breathing tract blockage. The area of stenosis is apt to be blocked at the peak inspiration phase due to potential wall collapse.

To the best of the authors' knowledge, this has been the first research devoted to measuring oscillatory respiratory flows inside CT-scanned ETA model with OSAHS. The PIV measurements of unsteady flow velocity will be used to calibrate the computational fluid dynamics method, which can be further used to capture even more detailed flow fields.

Author Contributions: Funding acquisition, C.Z., Y.J.; project administration, Y.J.; formal analysis, Z.Z.; investigation, Z.Z.; data curation, Z.Z.; writing—original draft preparation, Z.Z.; writing—review and editing, Y.J. and C.Z. All authors have read and agreed to the published version of the manuscript.

Funding: This research was financially supported by the National Key Research and Development Project of China under Grant No. 2016YFB0200901, National Natural Science Foundation of China under Grant No. 51176146, and Shaanxi Key Research and Development Project under Grant No. 2018KWZ-01.

Conflicts of Interest: The authors declare that for this study they have no conflicts of interest.

References

1. Kesson, A.M. Respiratory virus infections. *Paediatr. Respir. Rev.* **2007**, *8*, 240–248. [CrossRef]
2. Lavie, L.; Lavie, P. Molecular mechanisms of cardiovascular disease in OSAHS: The oxidative stress link. *Eur. Respir. J.* **2009**, *33*, 1467–1484. [CrossRef] [PubMed]
3. Nithiarasu, P.; Sazonov, I.; Yeo, S.Y. Scan-Based Flow Modelling in Human Upper Airways. In *Patient-Specific Modeling in Tomorrow's Medicine*; Springer: Berlin/Heidelberg, Germany, 2011; pp. 241–280.

4. Huang, E.I.; Huang, S.Y.; Lin, Y.C.; Lin, C.M.; Lin, C.K.; Huang, Y.C.; Su, J.A. Increasing Hypopnea in Sleep Breathing Disturbance Improves Postoperative Oxygen Saturation in Patients with Very Severe Obstructive Sleep Apnea. *Appl. Sci.* **2020**, *10*, 6539. [CrossRef]

5. Wu, M.F.; Huang, W.C.; Chang, K.M.; Lin, P.C.; Kuo, C.H.; Hsu, C.W.; Shen, T.W. Detection Performance Regarding Sleep Apnea-Hypopnea Episodes with Fuzzy Logic Fusion on Single-Channel Airflow Indexes. *Appl. Sci.* **2020**, *10*, 1868. [CrossRef]

6. Taherian, S.; Rahai, H.; Lopez, S.; Shin, J.; Jafari, B. Evaluation of human obstructive sleep apnea using computational fluid dynamics. *Commun. Biol.* **2019**, *2*, 423. [CrossRef] [PubMed]

7. Zhu, Z.; Zhang, C.; Zhang, L. Experimental and numerical investigation on inspiration and expiration flows in a three-generation human lung airway model at two flow rates. *Resp. Physiol. Neurobiol.* **2019**, *262*, 40–48. [CrossRef]

8. Liu, Y.; So, R.; Zhang, C.H. Modeling the bifurcating flow in a human lung airway. *J. Biomech.* **2002**, *35*, 465–473. [CrossRef]

9. Zhang, C.H.; Liu, Y.; So, R.M.C.; Phan-Thien, N. The influence of inlet velocity profile on three-dimensional three-generation bifurcating flows. *Comput. Mech.* **2002**, *29*, 422–429. [CrossRef]

10. Liu, Y.; So, R.M.C.; Zhang, C.H. Modeling the bifurcating flow in an asymmetric human lung airway. *J. Biomech.* **2003**, *36*, 951–959. [CrossRef]

11. Lu, M.; Liu, Y.; Ye, J.; Luo, H.Y. Large Eddy simulation of flow in realistic human upper airways with obstructive sleep. *Procedia Comput. Sci.* **2014**, *29*, 557–564. [CrossRef]

12. Liu, X.; Yan, W.; Liu, Y.; Choy, Y.S.; Wei, Y. Numerical Investigation of Flow Characteristics in the Obstructed Realistic Human Upper Airway. *Comput. Math. Methods Med.* **2016**, *10*. [CrossRef] [PubMed]

13. Song, B.; Li, Y.; Sun, J.; Qi, Y.; Li, P.; Li, Y.; Gu, Z. Computational fluid dynamics simulation of changes in the morphology and airflow dynamics of the upper airways in OSAHS patients after treatment with oral appliances. *PLoS ONE* **2019**, *14*, e0219642. [CrossRef] [PubMed]

14. Cui, X.; Wu, W.; Ge, H. Investigation of airflow field in the upper airway under unsteady respiration pattern using large eddy simulation method. *Resp. Physiol. Neurobiol.* **2020**, *279*, 103468. [CrossRef] [PubMed]

15. Kim, S.K.; Chung, S.K. An investigation on airflow in disordered nasal cavity and its corrected models by tomographic PIV. *Meas. Sci. Technol.* **2004**, *15*, 1090–1096. [CrossRef]

16. Im, S.; Heo, G.E.; Jeon, Y.J.; Sung, H.J.; Kim, S.K. Tomographic PIV measurements of flow patterns in a nasal cavity with geometry acquisition. *Exp. Fluids* **2013**, *55*. [CrossRef]

17. Wu, H.; Wang, M.; Wang, J.; An, Y.; Wang, H.; Huang, Y. Direct visualizations of air flow in the human upper airway using in-vitro models. *Sci. China Life Sci.* **2019**, *62*, 235–243. [CrossRef] [PubMed]

18. Cui, X.; Gutheil, E. Three-dimensional unsteady large eddy simulation of the vortex structures and the mono-disperse particle dispersion in the idealized human upper respiratory system. *J. Aerosol Sci.* **2017**, *114*, 195–208. [CrossRef]

19. Cui, X.; Gutheil, E. Large eddy simulation of the flow pattern in an idealized mouth-throat under unsteady inspiration flow conditions. *Resp. Physiol. Neurobiol.* **2018**, *252–253*, 38–46. [CrossRef]

20. Evans, H.B.; Castillo, L. Index-matched measurements of the effect of cartilaginous rings on tracheobronchial flow. *J. Biomech.* **2016**, *49*, 1601–1606. [CrossRef]

21. Kim, S.K.; Chung, S.K. Investigation on the respiratory airflow in human airway by PIV. *J. Vis. Jpn.* **2009**, *12*, 259–266. [CrossRef]

22. Adler, K.; Brücker, C. Dynamic flow in a realistic model of the upper human lung airways. *Exp. Fluids* **2007**, *43*, 411–423. [CrossRef]

23. Janke, T.; Schwarze, R.; Bauer, K. Measuring three-dimensional flow structures in the conductive airways using 3D-PTV. *Exp. Fluids* **2017**, *58*. [CrossRef]

24. Janke, T.; Koullapis, P.; Kassinos, S.C.; Bauer, K. PIV measurements of the SimInhale benchmark case. *Eur. J. Pharm. Sci.* **2019**, *133*, 183–189. [CrossRef]

25. Wright, S.F.; Zadrazil, I.; Markides, C.N. A review of solid–fluid selection options for optical-based measurements in single-phase liquid, two-phase liquid–liquid and multiphase solid–liquid flows. *Exp. Fluids* **2017**, *58*. [CrossRef]

26. Bates, A.J.; Schuh, A.; Amine-Eddine, G.; McConnell, K.; Loew, W.; Fleck, R.J.; Woods, J.C.; Dumoulin, C.L.; Amin, R.S. Assessing the relationship between movement and airflow in the upper airway using computational fluid dynamics with motion determined from magnetic resonance imaging. *Clin. Biomech.* **2019**, *66*, 88–96. [CrossRef] [PubMed]

27. Le, T.B.; Moghaddam, M.G.; Woodson, B.T.; Garcia, G.J.M. Airflow limitation in a collapsible model of the human pharynx: Physical mechanisms studied with fluid-structure interaction simulations and experiments. *Physiol. Rep.* **2019**, *7*, e14099. [CrossRef] [PubMed]
28. Aycock, K.I.; Hariharan, P.; Craven, B.A. Particle image velocimetry measurements in an anatomical vascular model fabricated using inkjet 3D printing. *Exp. Fluids* **2017**, *58*. [CrossRef]
29. Lizal, F.; Jedelsky, J.; Morgan, K.; Bauer, K.; Llop, J.; Cossio, U.; Kassinos, S.; Verbanck, S.; Ruiz-Cabello, J.; Santos, A.; et al. Experimental methods for flow and aerosol measurements in human airways and their replicas. *Eur. J. Pharm. Sci.* **2018**, *113*, 95–131. [CrossRef]
30. Raffel, M.; Willert, C.E.; Scarano, F.; Kähler, C.J.; Wereley, S.T.; Kompenhans, J. *Particle Image Velocimetry: A Practical Guide*, 3rd ed.; Springer International Publishing: Cham, Switzerland, 2018.
31. Vanselow, C.; Stöbener, D.; Kiefer, J.; Fischer, A. Particle image velocimetry in refractive index fields of combustion flows. *Exp. Fluids* **2019**, *60*. [CrossRef]
32. Große, S.; Schröder, W.; Klaas, M.; Klöckner, A.; Roggenkamp, J. Time resolved analysis of steady and oscillating flow in the upper human airways. *Exp. Fluids* **2007**, *42*, 955–970. [CrossRef]
33. Huang, J.; Zhang, L. Numerical simulation of micro-particle deposition in a realistic human upper respiratory tract model during transient breathing cycle. *Particuology* **2011**, *9*, 424–431. [CrossRef]
34. West, J.B. *Respiratory Physiology: The Essentials*, 10th ed.; Lippincott Williams & Wilkins: Philadelphia, PA, USA, 2012.
35. Womersley, J.R. Method for the calculation of velocity, rate of flow and viscous drag in arteries when the pressure gradient is known. *J. Physiol.* **1955**, *127*, 553. [CrossRef] [PubMed]
36. Beresh, S.J. Comparison of PIV data using multiple configurations and processing techniques. *Exp. Fluids* **2009**, *47*, 883–896. [CrossRef]
37. Kim, M.; Collier, G.J.; Wild, J.M.; Chung, Y.M. Effect of upper airway on tracheobronchial fluid dynamics. *Int. J. Numer. Methods Biomed. Eng.* **2018**, *34*, e3112. [CrossRef] [PubMed]
38. Wang, Y.; Liu, Y.; Sun, X.; Yu, S.; Gao, F. Numerical analysis of respiratory flow patterns within human upper airway. *Acta Mech. Sinica PRC* **2009**, *25*, 737–746. [CrossRef]
39. Jalal, S.; Nemes, A.; Van de Moortele, T.; Schmitter, S.; Coletti, F. Three-dimensional inspiratory flow in a double bifurcation airway model. *Exp. Fluids* **2016**, *57*. [CrossRef]
40. Lucey, A.D.; King, A.J.; Tetlow, G.A.; Wang, J.; Armstrong, J.J.; Leigh, M.S.; Paduch, A.; Walsh, J.H.; Sampson, D.D.; Eastwood, P.R.; et al. Measurement, reconstruction, and flow-field computation of the human pharynx with application to sleep apnea. *IEEE Trans. Biomed. Eng.* **2010**, *57*, 2535–2548. [CrossRef]

Publisher's Note: MDPI stays neutral with regard to jurisdictional claims in published maps and institutional affiliations.

© 2020 by the authors. Licensee MDPI, Basel, Switzerland. This article is an open access article distributed under the terms and conditions of the Creative Commons Attribution (CC BY) license (http://creativecommons.org/licenses/by/4.0/).

Article

Experimental and Numerical Study on Performance Enhancement by Modifying the Flow Channel in the Mechanical Chamber Room of a Home Refrigerator

Dong Kyun Kim

Division of Mechanical Engineering, Tongmyong University, 428 Sinseon-ro, Nam-gu, Busan 48520, Korea; kimdk@tu.ac.kr; Tel.: +82-10-7402-0744

Received: 6 July 2020; Accepted: 8 September 2020; Published: 9 September 2020

Featured Application: Design of high performance refrigerator.

Abstract: Recently, many refrigerators of high have been designed so as to improve the performance of the refrigerator and to economize on energy. There are many methods of improving the total efficiency of the refrigerators using a methodological approach. In this experimental and numerical study, three methods—namely, modifying the flow channel and the operating conditions and changing the positions of the fan at the mechanical chamber room (MCR)—are used to analyze the performance improvement of the refrigerator. These methods of changing the flow channel in an MCR of a refrigerator are, in many ways, used as a regulation of the upper flow region of the condenser, especially in this research. Modifying the refrigerators' MCR is carried out by changing the shape of the cover back machine (CBM). We compute the fluid flow in a refrigerator MCR by computational fluid dynamics (CFD) techniques. The commercial code ANSYS CFX 16.2 is used for this computation. The regulation flow region shows a reduction in power consumption of about 0.75% per month. The result of the operating conditions of the fan at MCR is relative to the flow rate for reducing power consumption. The changing positions are also relative to the length of the reducing power consumption. Moreover, with the results obtained by CFD, we understand the flow structure in a refrigerators' MCR for various types of CBM. The results also show that the efficiency of the refrigerator is improved by 1.2%.

Keywords: mechanical chamber room; performance enhancement; flow channel; cover back machine; domestic refrigerator

---

1. Introduction

Some of the common household appliances are refrigerators, washing machines, clothes dryers, electric ovens, etc. [1]. Among the household appliances, the refrigerator is the most common one due to its indispensable nature and hence, is the most manufactured [1,2]. Currently, there are over a billion domestic refrigerators worldwide [2]. The energy consumption of household appliances increases in proportion to its usage. Among these appliances, the refrigerator is one of the most energy-consuming devices [2]. It is estimated that more than 45% of the world food production would go waste if not for the refrigerators' cold storage and distribution [2]. Thus, improving their energy efficiency, efficient energy consumption, and prolonged life becomes imperative [1,2]. There are different mechanisms to increase the efficiency of the refrigerator, namely, by improving the compressor performance, thermal isolation, heat exchanger design, implementation of optimum control in refrigerator operation [2], improvements in cabinets (such as insulation, improved gaskets), improvement in refrigeration systems (such as fan motors, high efficiency compressors), improved heat exchangers, advanced defrost mechanisms, etc. [1].

Appl. Sci. **2020**, *10*, 6284; doi:10.3390/app10186284　　　　　　　　　　www.mdpi.com/journal/applsci

Environmental problems, such as global warming, have precipitated a large number of studies—on various factors—that affect the power consumption of a refrigerator to improve their efficiency [3]. On the evaluation of the efficiency and performance of refrigerators, Dmitriyev and Pisarenko investigated the change in the performance of a refrigeration system using a Phase-change Material (PCM) cooling agent [4]. Laguerre et al. investigated the change in the temperature of freezing and refrigeration chambers in a food storage refrigeration system according to its storage temperature and usage frequency in an empirical manner [5]. Sergio Ledesma and J M Belman-Flores mathematically studied the thermal effect of the fresh food compartment with a change in glass shelve locations in a domestic refrigerator. The moving variance and the finite difference approximation were used to analyze the thermal distribution and the temperature change rate, respectively, in their study. The results showed that the use of thermal maps to analyze the mean and variance of the temperature was an important tool to assess a good combination of shelf locations [6]. M Rasti et al. experimentally studied the energy efficiency enhancement of a domestic refrigerator by replacing R134a with R436A without any mechanical alteration. The results showed that by replacing the R436A, the On time and energy consumption was reduced by 13% and 5.3%, respectively, for a day, and the optimum charge for R436A was reduced to 48% when compared to the charge of R134a [7]. Shikalgar and Sapali experimentally studied the energy and exergy analysis of a domestic refrigerator by comparing the hot-wall air-cooled (HWAC) and box type shell and tube water-cooled (BCTWC) condensers. The results showed that the coefficient of performance and the exergy efficiency of BCTWC was increased by 18–20% and 6.89–9.13%, respectively, when compared to the HWAC. In addition, the energy consumption of the BCTWC was reduced by 17% in comparison to HWAC [8]. Belman-Flores et al. experimentally and numerically studied the temperature stratification of the fresh food and freezer compartments of the domestic refrigerator with bottom mount configuration. The results showed that the proposed design had a good distribution of airflow and hence, uniform temperature at the fresh food compartment. Furthermore, in comparison to the original design, the new design showed less ON stages for the compressor [9]. Antonio and Afonso conducted a comparison study between the computational fluid dynamics (CFD) and Artificial Neural Network (ANN) modeling of the air temperature fields inside the refrigerator cabins. The results showed that the ANN model produced a low absolute error of 0.8 K in comparison to 1 K of the CFD model. Moreover, the temperature field obtained with the ANN model was more refined in comparison to the CFD model [10]. A review of advances in domestic refrigerators can be had from [1,2,11]. However, there are very few studies on the improvement in refrigeration performance according to the internal structure of a refrigeration chamber and its conditions that affect the performance of a refrigerator [12–15].

The purpose of this study is to improve the total efficiency of the refrigerator by maintaining the temperature inside the mechanical chamber room (MCR) at a constant level or by decreasing it—based on the improvement of the flow line and the operating condition of the cooling fans—experimentally. In addition, to optimize the flow line inside the MCR by changing the shield shape of the cover back machine (CBM) based on computational flow analysis.

2. Factors Affecting the Efficiency of the Refrigerator

The refrigerator consists of a front and rear (F and R) MCR. This chamber is called as a cover back machine (CBM). The MCR is installed at the rear of the refrigerator. Figure 1 shows the component arrangement of the MCR in the refrigerator. It consisted of a compressor, condenser, suction pipe, discharge pipe, cooling fan, guide fan, fan motor, capillary tube, a dryer. The suction pipes were installed at the entrance of the compressor, where a cooling agent moved. The discharge pipes were installed at the exit of the compressor.

In general, the F and R chambers were maintained at an inner temperature of −18 °C and +3 °C, respectively, where and they showed differences in the outer temperature as 48 °C and 27 °C, respectively, for an ambient temperature of +30 °C [16]. As a normal operation in a refrigerator, the compressor is operated as an On/Off control, and the cooling fan is operated at the same time.

A cover back machine (CBM) is installed to prevent the inflow of mechanical noise and other foreign objects into the internal space. The cooling fan is used to cool down the condenser. The cooling fan causes an internal flow inside the refrigerator.

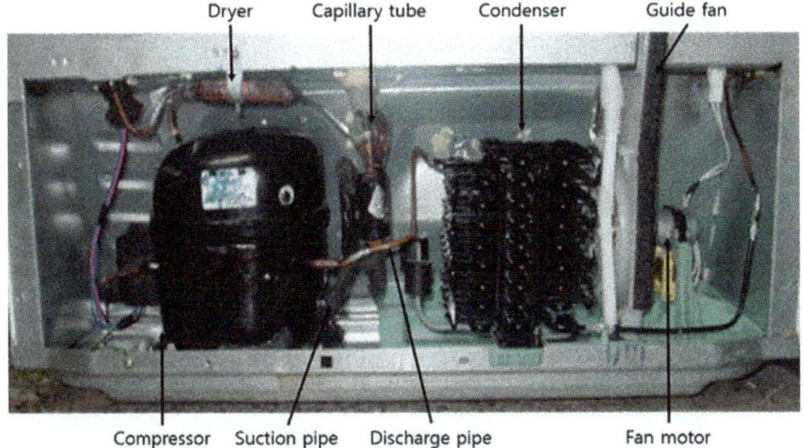

Figure 1. Component arrangement of the mechanical chamber room (MCR).

The operational losses, additional device (parasitic) losses, load losses in insulation used to isolate the inflow of external air, input losses in the On/Off control cycles, thermal load losses in the fan motors, input losses in the On/Off control of a compressor, heat exchange efficiency in a condenser, pressure decrease in the pipeline, loss in an evaporator for removing frosts are some of the factors that lead to the inefficiency of a refrigerator [16–19]. In addition, within the MCR, the heat source generated from the compressor, condenser, fans, and motors influence the decrease in the efficiency of the whole cycle of the refrigerator. Moreover, if the heat release is less than the heat released from within the MCR, the temperature of the system will be increased, leading to a decrease in the efficiency of the refrigerator. Thus, a cooling process is applied to the surface of the condenser using cooling fans, and the flow line in the MCR is rearranged to decrease the temperature. However, it is necessary to improve such a flow line to increase the efficiency as there are lots of resistant factors in the chamber due to the complicated flow line.

The internal flow inside the refrigerator caused by a cooling fan is discharged to each vent of the CBM through the forced convection generated by the fan as the compressor condenses refrigerant vapor. The performance of heat transfer inside the MCR caused by such forced convection affects its co-efficient of performance (COP). The COP is the refrigeration effect of a compression work, and it can be calculated as follows [4,14].

$$COP = \frac{\text{Cooling Efficiency}}{\text{Compressor work}} = \frac{Q_L}{AW} \tag{1}$$

As the vent temperature of a compressor and the condensation temperature of a condenser are decreased at a p-h diagram to improve the COP [18–20], the refrigerant condensed by the condenser may improve the refrigeration effect of COP due to the pressure reduction by an expansion valve. Furthermore, in a convectional heat transfer, by cooling the fan, the heat transfer coefficient of h depends on the internal temperature of the MCR and the air temperature around the vent of the CBM near the cooling fan. Because an effective shield in the vent of the CBM at the front of each major component significantly affects the COP, the convectional heat transfer coefficient, and power consumption, it becomes an important factor in the flow inside the MCR and its flow line [3,13].

3. Background Theory

The Carnot cycle applied to a refrigerator consists of an isentropic process that corresponds to the work of a compressor, static pressure heat release process in a heat exchanger, expansion process in a capillary tube, and static pressure heat suction process in an evaporator [21]. In general, the static pressure heat release and heat suction process are reversible processes that show a decrease in its efficiency as the temperature difference between high and low temperature regions in a heat exchange process. In addition, it plays a role in decreasing the refrigeration capacity due to the increase in the compression rate of the compressor and the decrease in volume efficiency.

Regarding the solution for solving these problems, although an expansion for the section of heat exchanges is to be recommended, it represents significant difficulties in the structure of the MCR and the space for installing this chamber for the present refrigerator [22]. Therefore, it is the most ideal way to improve the efficiency of a refrigerator by decreasing the temperature in an heat exchanger through changing the flow line using cooling fans, increasing the surface heat transfer coefficient, and decreasing the compression rate of a compressor.

Figure 2 shows the schematic of the reverse Carnot cycle. In the figure, h is enthalpy; h_1 and h_2 are the initial and final enthalpy state of the compressor, respectively. The numbers 1, 2, 3, and 4 denote the initial enthalpy state of the compressor. The numbers 1′, 2′, 3′, and 4′ denote the improved enthalpy state of the compressor. $h_{2'}$ is the improved enthalpy state of the compressor. P is the pressure. AW $(= h_2 - h_1)$ is the difference between the two enthalpy states. Q_H and Q_L are the high and low heat flux, respectively. Thirty degrees denotes the low temperature of the system, and 40 °C; 80 °C denotes the high temperature of the system. In the operating condition, as shown in Figure 2, the amount of heat transfer and COP can be obtained using the following equations:

$$\text{COP} = \frac{\text{Cooling Efficiency of Evaporator}}{\text{Power Consumption of Compressor}} = \frac{Q_L}{W}, \tag{2}$$

$$Q_C = KA\Delta T = m_{ref}\Delta h, \tag{3}$$

$$Q_L = h_2 - h_1 \Rightarrow h_{2'} - h_{1'}, \tag{4}$$

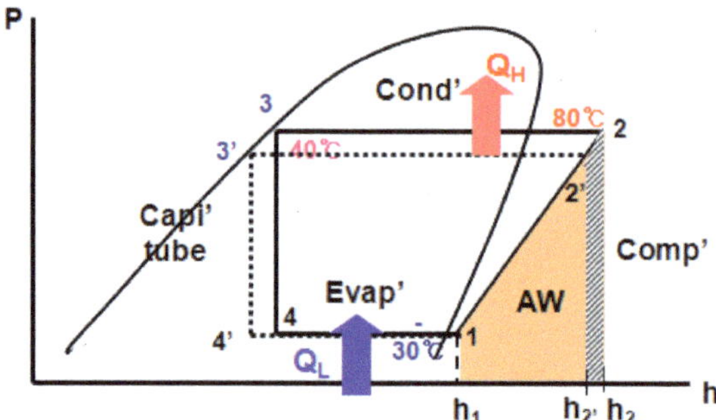

Figure 2. Schematic diagram for the reverse Carnot cycle.

The equations above are used to investigate changes in the efficiency of the whole system through improving the heat transfer performance by reducing the temperature difference between high and low temperature regions [23].

4. Experimental Conditions

Figure 3 shows the arrangement of the experimental apparatus used in this study. The experiment was carried out in a constant temperature and humidity chamber to maintain uniform conditions. Two specimens, #A and #B, were used under the same room conditions to ensure the accuracy of the experiment. In addition, the system consisted of a logger that configures 60 channels for measuring the temperature inside the MCR and the surface temperature of the compressor and condenser, a power meter that measures the power of these specimens, and a static voltmeter that was able to provide a uniform voltage [21].

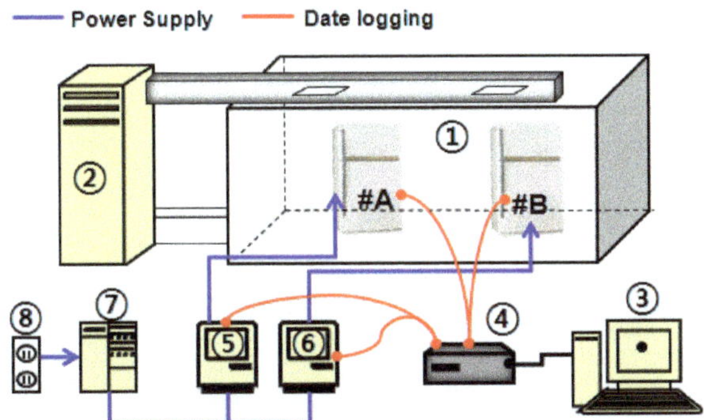

Figure 3. Arrangement of experimental apparatus.

The specification of the constant temperature and humidity chamber and instruments are presented in Table 1. In the experimental conditions, the temperature inside the constant temperature, and humidity chamber and relative humidity were configured as a specific range based on the standard for the test of a refrigerator. Furthermore, a T-Type thermocouple was connected to the logger, the surfaces of the compressor and condenser installed inside the refrigeration chamber and MCR, and inside the MCR to measure the temperature [22,23].

Table 1. Specifications of the apparatus.

No.	Name	Specifications
1	Chamber room	Temperature 30 ± 1 °C, Humidity 75 ± 5%
2	Chamber controller	Temperature 10–35 °C, Humidity 60–95%
3	Computer	Workstation, 16 Core, 128 Gb
4	Data logger	DS-600 (YOGAWA 60 Channel), Thermocouple (T-type)
5	Power meter for #A	WT-200 (YOKOGAWA)
6	Power meter for #B	WT-200 (YOKOGAWA)
7	Automatic Voltage Regulator (AVR)	3KVA, Output V: 220 V ± 2%
8	Power supply	Voltage: 220 V, Frequency: 60 Hz

In addition, the power that was applied to each specimen was measured using a voltmeter in which the output port of the voltmeter was directly connected to the logger. Figure 4 shows the compressor On/Off cycle of the defrost state of the refrigerator. From the figure, it can be seen that the defrost peak occurred twice. The first peak occurred initially at 1 h, and the second peak occurred during the 15th h. The figure also shows the temperature of the refrigerator ranging from 0 °C to 20 °C. As shown in Figure 4, the internal temperature of the MCR and the surface temperature and power of the compressor and condenser were measured by configuring a stable region between the processes for removing frost. The measurement was repeated five times for each condition and determined as an ensemble average.

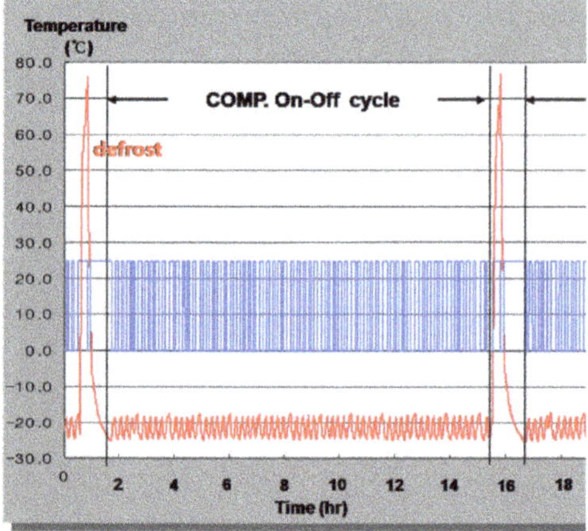

Figure 4. Compressor On/Off cycle for defrost state.

5. Numerical Analysis

5.1. Governing Equations and Numerical Solutions

This study was performed under the assumption that the fluid flow type inside the MCR represents a three-dimensional steady state, incompressible, and turbulent flow. The physical properties of the working fluid were kept at a constant level for temperature and pressure for numerical calculation. Due to the disturbance of the fluid by the cooling fan, the flow in the machine room will behave in a complex manner. Therefore, the standard $k - \varepsilon$ turbulence model that represents excellent estimation performance in an internal flow analysis that has been widely adopted was used to perform flow analysis in the refrigerator machine room [9,10,24]. The governing equations were composed of the continuous equation, the momentum equation, the turbulent kinetic energy equation, and the turbulent kinetic energy dissipation rate equation, and are as shown in the following Equations (5)–(10) [25]:

$$\frac{\partial u_i}{\partial x_i} = 0, \tag{5}$$

$$\frac{\partial u_i}{\partial t} + u_j \frac{\partial u_i}{\partial x_j} = -\frac{1}{\rho}\frac{\partial p}{\partial x_j} + \nu \frac{\partial^2 u_i}{\partial x^2_j}, \tag{6}$$

$$\nu_t = \frac{C_\mu k^2}{\varepsilon}, \tag{7}$$

where ν_t is the turbulent viscosity, C_μ is the constant. k and ε can be obtained from the following equations for turbulent kinetic energy and turbulent dissipation rate, respectively

$$\frac{\partial u_j k}{\partial x_j} = \frac{\partial}{\partial x_j}\left\{\left(\nu + \frac{\nu_t}{\sigma_k}\right)\frac{\partial k}{\partial x_j}\right\} + P_k - \varepsilon, \tag{8}$$

$$\frac{\partial u_j k}{\partial x_j} = \frac{\partial}{\partial x_j}\left\{\left(\nu + \frac{\nu_t}{\sigma_\varepsilon}\right)\frac{\partial \varepsilon}{\partial x_j}\right\} + \frac{\varepsilon}{k\rho}(C_{\varepsilon 1}P_k - C_{\varepsilon 2}\varepsilon), \tag{9}$$

Here, $C_{\varepsilon 1}$, $C_{\varepsilon 2}$, σ_ε, and σ_k are model constants. P_k is the turbulence generation term according to viscosity and buoyancy and is defined as follows:

$$P_k = \nu_t \frac{\partial u_i}{\partial x_j}\left(\frac{\partial u_i}{\partial x_j} + \frac{\partial u_j}{\partial x_i}\right), \tag{10}$$

where the coefficients were defined as follows.

$$C_\mu = 0.09, \sigma_k = 1.0, C_{\varepsilon 1} = 1.44, C_{\varepsilon 2} = 1.92, \tag{11}$$

5.2. Computational Domain and Grid System

In this study, the total number of grid elements was 1,350,000. To investigate the change in the result value according to the number of grids, the grid elements were gradually increased from a small number. In the case of 551,000 grid systems, changes in flow distribution trends and internal velocity values in the fan and condenser were reduced (grid number 1,350,000), and changes in numerical calculation values, such as flow and velocity, became constant in more than 2,239,250 grid elements. Therefore, 1,350,000 grid systems were adopted in this study, as shown in Figure 5. Figure 6a,b shows the front and rear view of a computational grid system of the MCR. From Figure 6a,b, the arrangements of the compressor, condenser, and the CBM can be visualized. The internal velocity was measured between the compressor and condenser, as shown in Figure 6a.

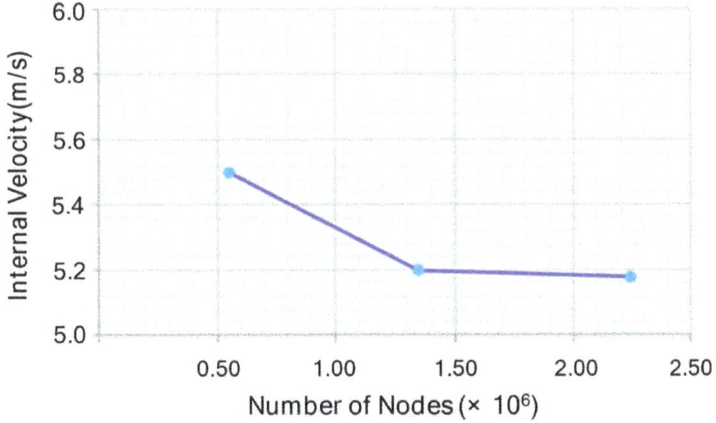

Figure 5. Grid dependency of flow at MCR.

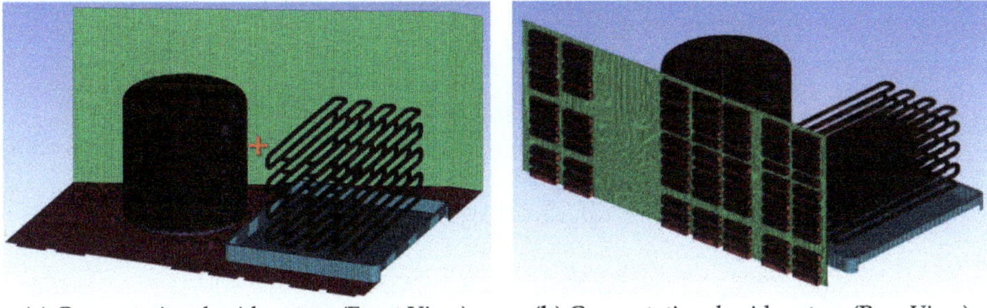

(**a**) Computational grid system (Front View)   (**b**) Computational grid system (Rear View)

Figure 6. Computational grid system.

The governing equations were a type of continuously composed governing equations and were analyzed using the commercial CFD software ANSYS-CFX 16.2 [26,27]. The elements used in the calculation were a type of hexahedral elements, about 1,350,000, as shown in Figure 6. Moreover, a type of non-staggered grid system that determines pressure and velocity at different points was used to perform a linked calculation in pressure and velocity [5]. The CFX integrated coupled solver (pressure and velocity) was used as the solution method in this study.

5.3. Boundary Condition

The discharge from the MCR was configured as an atmospheric condition, and the inflow by air through a fan was applied to the calculation by transforming the components measured using laser Doppler velocimetry (LDV), such as the velocity components of the fan along the axial, rotational, and radial directions of a cooling fan into X, Y, and Z components, respectively, in the Cartesian coordinate system [12,28]. The ideal air gas model was used in this study. The root mean square (RMS) convergence criterion was set to 10^{-4}. The following equations are approximate equations for each test point using the velocity components calculated by using the LDV. The equation can be used to calculate the velocity for each point of the fan.

$$V(a) = -145232X^3 + 880.56X^2 + 11.54X - 0.35, \tag{12}$$

$$V(\theta) = 45170X^3 + 4474X^2 - 38.64X + 0.49, \tag{13}$$

$$V(r) = 64420X^3 - 1775.73X^2 - 29.55X + 0.69, \tag{14}$$

where $V(a)$ is the velocity component along the axial direction, $V(\theta)$ is the velocity component along the circumferential direction, and $V(r)$ is the velocity component along the radial direction, and they correspond to X, Y, and Z components, respectively, in the Cartesian coordinate system.

6. Experimental Results and Discussion

6.1. Power Consumption and Inner Temperature of Chamber Room

Figure 7 shows the power consumption for various temperature conditions of standard KS C 9305. From the Figure, it can be seen that the cause of an increase in power consumption was due to the increase in the chamber room temperature. As the temperature was raised by 1 °C, the power consumption was increased by 2 kW/month. The value of the power consumption can be calculated using the linear fitting function Equation (15).

$$y = 2.06x - 12.67, \tag{15}$$

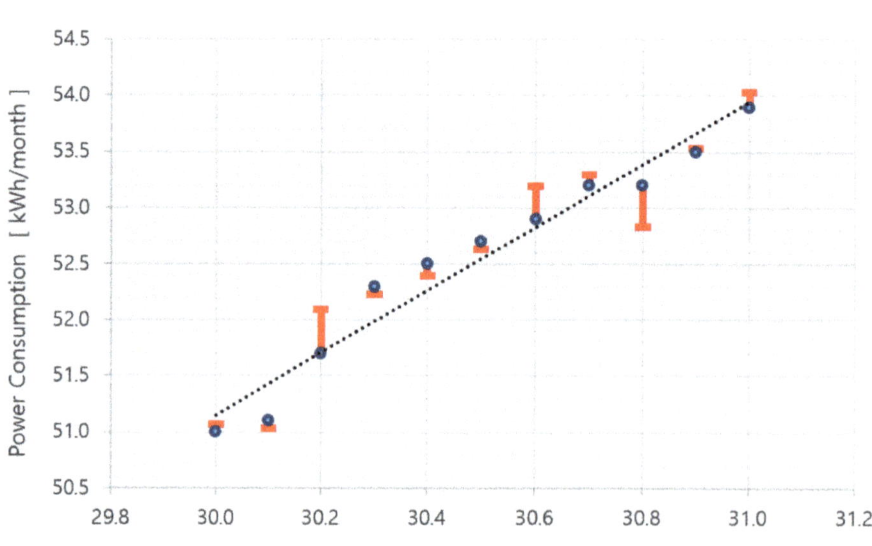

Figure 7. Monthly power consumption for various temperatures in the chamber room.

6.2. Improvement of MCR Flow Line

The characteristics of the MCR will cause heat transfer due to the forced convection if the flow line in the chamber is improved. By improving the flow line, it was possible to ensure more wind was received by the condenser by adjusting the flow line of the upper space of the condenser. In the improvement in the flow line, as shown in Figure 8, a box was attached to the upper wall of the MCR in which a 15 mm gap was left on the upper side of the condenser to avoid contact with it not to generate an insulation effect on its surface. Furthermore, the space between the compressor and the condenser where the flow finally passes through was configured as a curved surface to present a smooth flow.

(a) Original structure

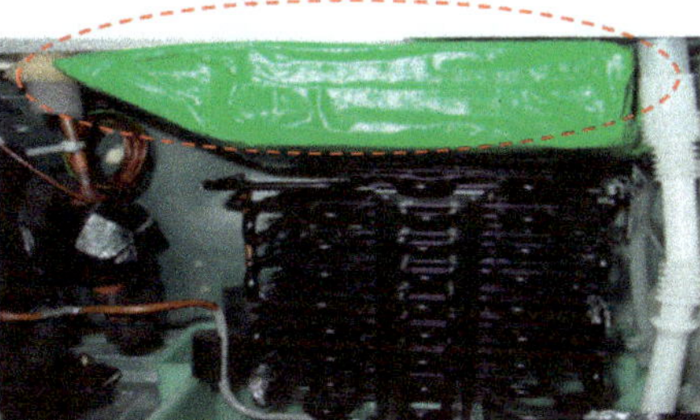

(b) Improved structure

Figure 8. Original and modified model of the machine room.

The temperature of the compressor and condenser installed inside the MCR for the original model was 53.9 °C and 43.2 °C, respectively. The temperature of the compressor and condenser for the improved model was 53.7 °C and 43.0 °C, respectively. Regarding the distribution of the average temperature of the specimens #1 and #2 inside the MCR in the improved model compared to that of the conventional model, the condenser showed a decrease in the temperature of 0.2 °C, and the compressor represented a decrease in the temperature of 0.2 °C. Thus, it represented an effect that decreased the monthly power consumption by 0.2 kWh, in which the reduction rate was 0.4%. By improving the internal flow line, it showed a decrease in the temperature of the whole surface of the condenser. Moreover, it is expected that the exit of the compressor that corresponds to discharge pipes also represented a decrease in the temperature. It is due to the fact that the flow, which had a high-speed element, progressed to the compressor via the condenser, except for the flow that intermittently passed through it in the resistance determined by the shape of the condenser by optimizing its upper space for the flow that progresses to the compressor. In addition, it was verified that the temperature in the condenser and compressor decreased due to the improvement of the flow line inside the MCR. Table 2 shows the reduction in power for the modified model when compared to that of the original model.

Table 2. Reduction in power of original and modified models.

Item / Case	Power Original (kWh/Month)	Power Modified (kWh/Month)	Reduction (%)
Case #1	50.1	49.7	0.8
Case #2	51.4	51.0	0.7

6.3. Change in Position of the Cooling Fan

The hot air in the MCR is discharged using a cooling fan. Thus, the airflow and internal flow rate can be varied according to the position of such a cooling fan, and this leads to present a difference in the heat transfer performance. As illustrated in Figure 9, the experiment was performed by adjusting the position of the cooling fan from a maximum of 40 mm to a minimum of 20 mm, with an interval of 5 mm referenced on the condenser.

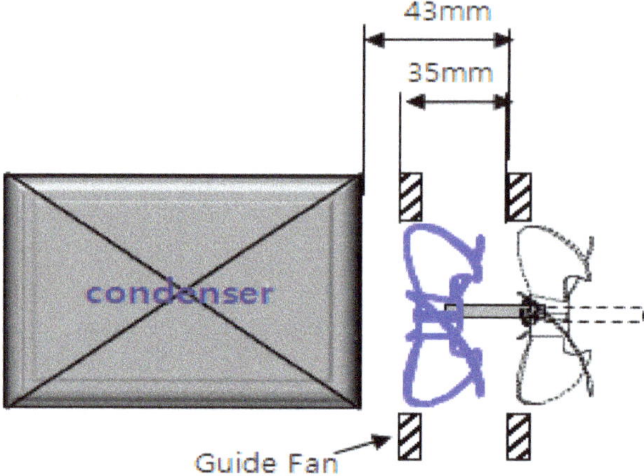

Figure 9. Displacement of the cooling fan position.

In the experimental results of the monthly power consumption, as shown in Figure 10, the power consumption decreased when the position of the fan was closer to the condenser. Thus, it can be seen that the interval determined by 20 mm represented the highest efficiency. However, it is necessary to configure the position of the fan by considering the structure of the MCR as it is not possible to reduce the distance due to its structure excessively.

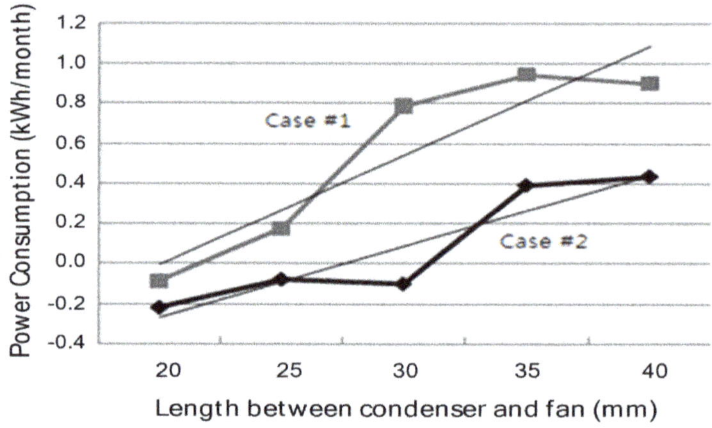

Figure 10. Monthly power consumption for various lengths between the condenser and the fan.

6.4. Optimization of Fan Flow Rate

The airflow discharged from the fan inside the MCR becomes an essential factor for determining the performance of the whole heat transfer. In addition, the determination of an optimum condition for the fan operation is very important to improve the whole efficiency of the refrigerator because the power consumption operated by the fan according to the airflow represents large differences.

The airflow of the fan was obtained under the same conditions for a single product/system. The measurement method of the airflow for a single product/system was measured after adjusting motor revolutions per minute (RPM) of the motor. The RPM could be adjusted by controlling the frequency and voltage of the motor. This study applied a voltage-controlled adjusting method. The airflow of the fan was measured according to the position of each fan as an average value in which 20 points were determined for measuring the airflow. Furthermore, the airflow converted as the measured value was presented, as shown in Figure 11.

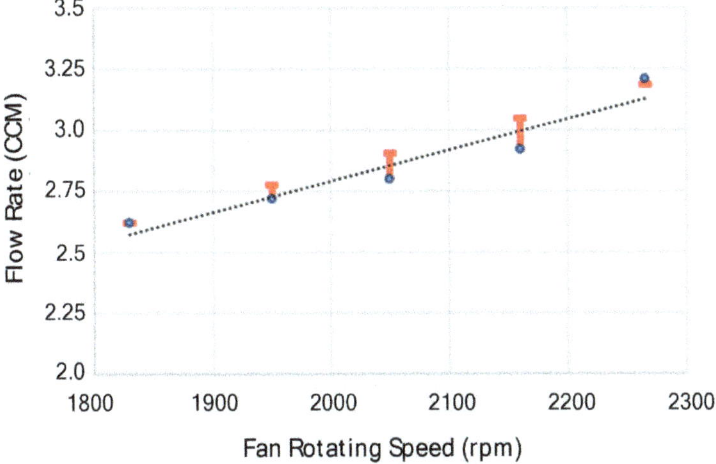

Figure 11. Fan flow rate for various fan rotating speeds.

Figure 12 illustrates the results of the measurement of power consumption according to the airflow of the fan. The power consumption decreased as the airflow of the fan increased. It brought a decrease in the temperature of the compressor and condenser inside the MCR due to the increase in the airflow of the fan. Moreover, the decrease in such temperature increased the efficiency of the whole system. It can be seen that the results agree with the decrease in the temperature of the compressor and condenser as presented in the above two subsections that increased the efficiency of the refrigerator.

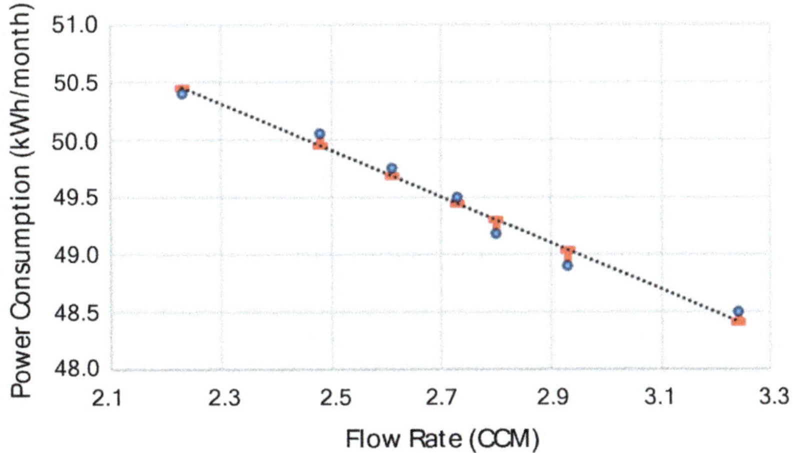

Figure 12. Power consumption for various flow rates of the cooling fan.

7. Numerical Results and Discussion

7.1. Flow Patterns of Existing Refrigerator

Figure 13a–c show the original CBM and the flow pattern of a model that applied conventional CBM. In the flow pattern, the internal air that cannot be discharged through the left vent of the compressor moved up to the upper section of the compressor and circulated to the upper section of the condenser. Then, the internal air was discharged to the upper left vent of the compressor according to the rise in the circulated air that moved up to the uppermost section.

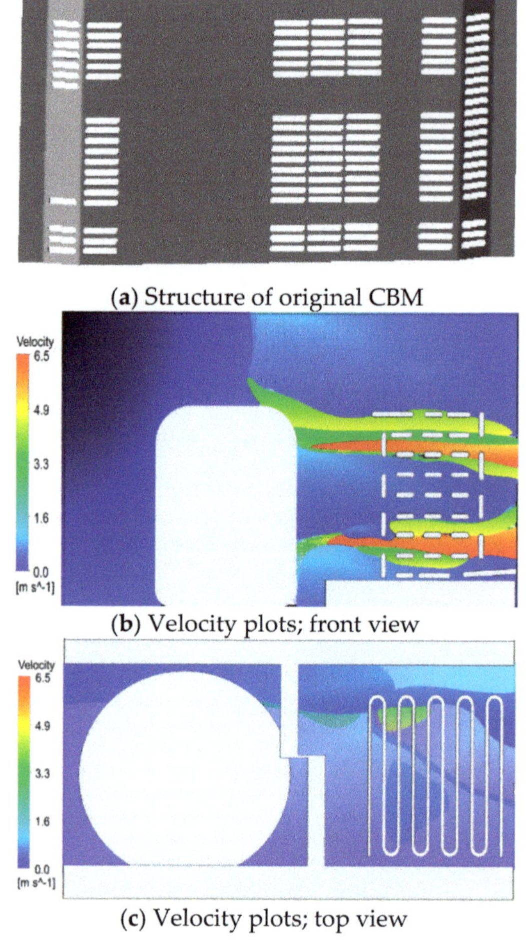

(**a**) Structure of original CBM

(**b**) Velocity plots; front view

(**c**) Velocity plots; top view

Figure 13. Original cover back machine (CBM) and flow pattern in the machine room of the refrigerator.

In the flow pattern at the space between the compressor and the condenser, the air passed from the condenser collided to the right side of the compressor, and that generated the circulated flow to the upper section of the condenser, and a whirlpool-like circulation was formed. As a result, it caused an increase in temperature at the entrance line of the condenser.

7.2. Selection of Modification Models

In the results of the flow analysis for the existing models, it was necessary to ensure that the left vent for the flow that circulates at the left section of the compressor, the upper vent for the upper flow circulation, and the vent for the whirlpool and flow circulated due to the wall at the space between the compressor and the condenser. In addition, the vent was to be cut-off due to the internal flow from around the condenser to the compressor. The modified models were selected, as shown in Figure 14.

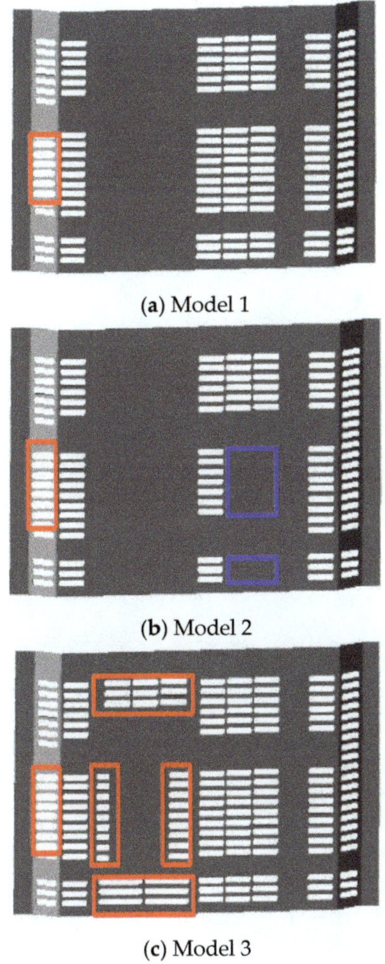

(**a**) Model 1

(**b**) Model 2

(**c**) Model 3

Figure 14. Modified structure of the CBM.

7.3. Numerical Simulation of the Modification Models

Figure 15 illustrates the velocity field at a vertical section of +40 mm to the CBM referenced around the cooling fan. From the Figure, it can be seen that there was a smooth flow around the compressor in the modified model (c) due to the presence of a vent at the rear of the compressor when compared to models (a) and (b). In addition, it can be seen that the flow passed through the upper section of the compressor due to the fact that the flow stopped up at the vent of the condenser. In addition, model (d) shows lots of flows at the upper and lower sections of the compressor according to the installation of vents at the rear and around the compressor when compared with the model (c). In addition, there were smooth flows due to the discharge of flow that was generated at the space between the compressor and the condenser and was congested at the upper section of the compressor.

Figure 16 illustrates the velocity field at a vertical section of −20 mm to the CBM around the cooling fan. As shown in the figure, regarding the models (b–d) that ensure the vents at the left section of the compressor, it can be seen that the flow progressed around the compressor, whereas model (d) represents the most active movement. This shows that all modified models represented smooth flow around the compressor when compared to the original model.

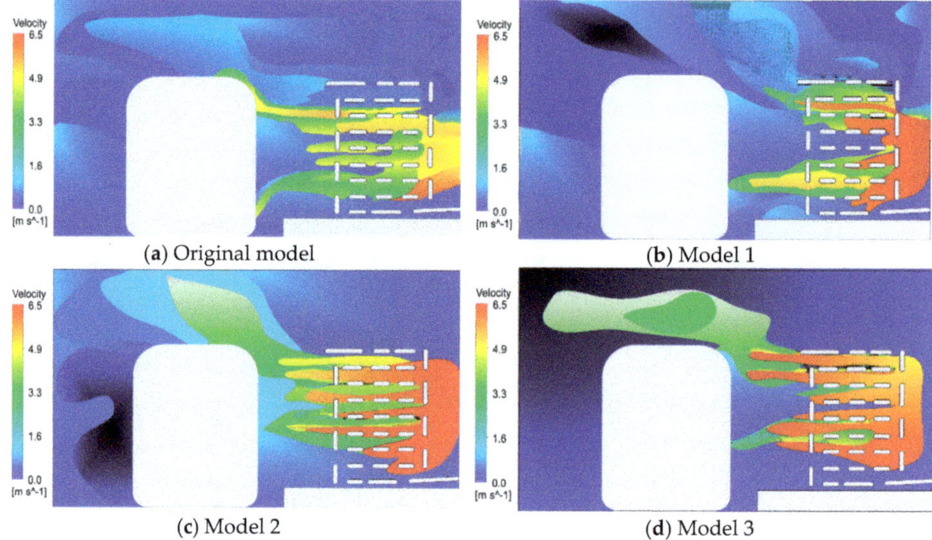

Figure 15. Velocity plots of vertical section at +40 mm position from the center—front view.

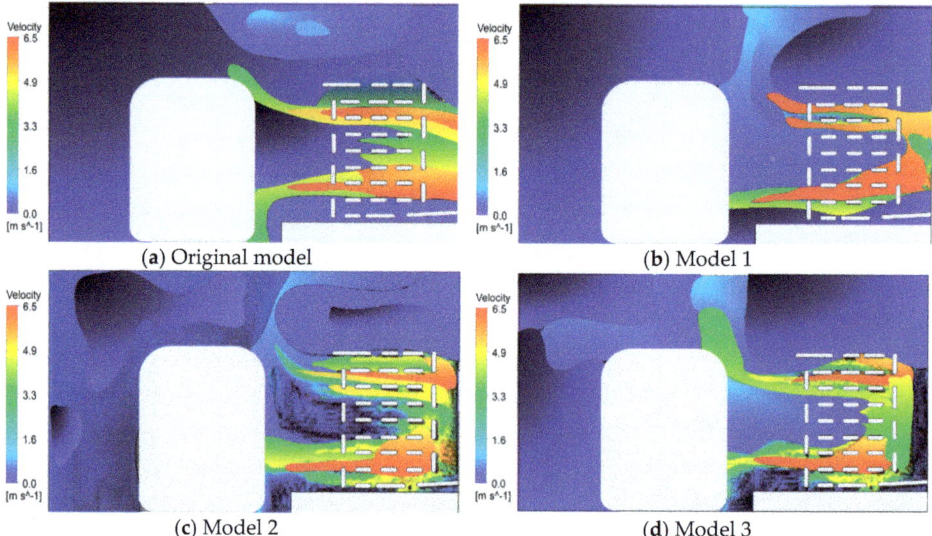

Figure 16. Velocity plots of vertical section at −20 mm position from the center—front view.

Figure 17 illustrates the velocity field at a horizontal section of +20 mm to the CBM around the cooling fan. From the Figure, it can be seen that the models (b–d) showed the discharge of flows to the lower section of the compressor, whereas model (c) showed the flow that progressed to the compressor due to the cut-off of the vent of the condenser. Moreover, it can be seen that in model (d), flow was generated at the upper section due to the influence of the right upper vent of the compressor.

Figure 18 illustrates the velocity field at a horizontal section of −40 mm to the CBM around the cooling fan. From the Figure, it can be seen that the flow was discharged to the left section of the compressor. In the case of model (d), the flow was discharged as a vent was determined at the lower section of the compressor. In addition, because the horizontal sections in Figures 17 and 18 were not a normal symmetry but an axial symmetry, the sign of + to the horizontal center of the cooling fan and the—sign inside the MCR can be considered as the flow to the CBM.

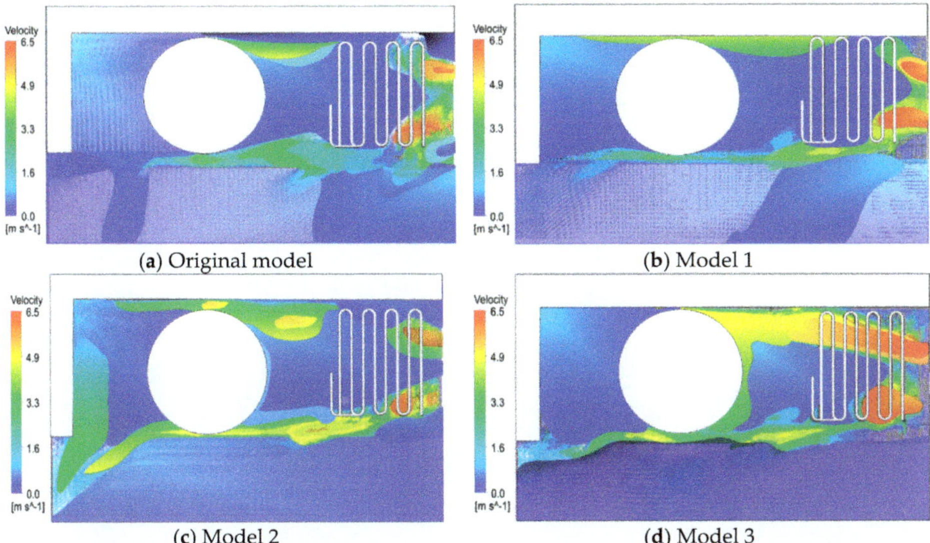

Figure 17. Velocity plots of horizontal section at +20 mm position from the center—top view.

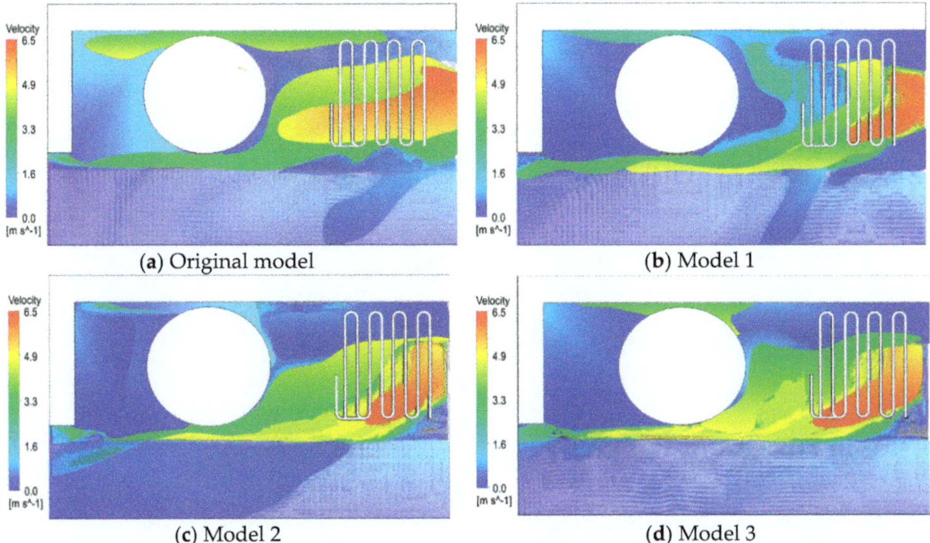

Figure 18. Velocity plots of horizontal section at −40 mm position from the center—top view.

Figure 19 and Table 3 represents a mass flow rate for the original and modified CBMs. Models 1 and 2 CBMs showed a high mass flow rate as the vent was generated at the left side of the compressor, and Model 3 CBM showed a high mass flow rate at the section of the compressor. Thus, it can be considered that the flow around the compressor is actively generated when compared to that of the existing models.

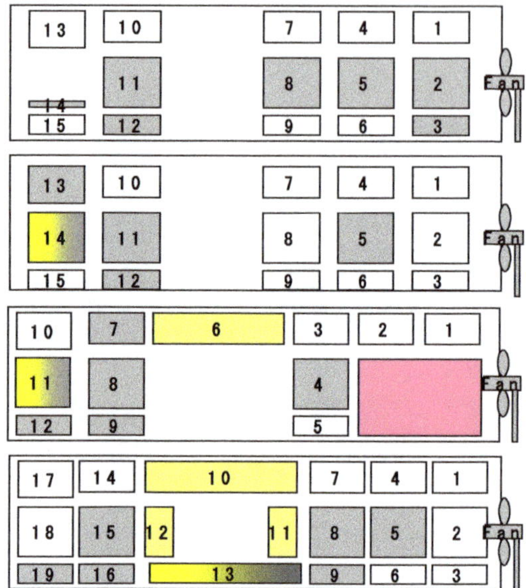

Figure 19. Mass flow rates of the original and modified CBMs.

Table 3. Mass flow rate (kg/s) of the original and modified cover back machine (CBM).

Outlet.	Existing Model	Model 1	Model 2	Model 3
1	0.60	0.73	0.22	0.28
2	1.70	0.33	0.22	0.47
3	1.25	0.37	0.22	0.28
4	0.51	0.72	1.53	0.28
5	1.15	1.47	0.93	1.23
6	0.58	0.65	0.97	0.55
7	0.36	0.65	1.45	0.28
8	1.39	0.90	1.47	2.31
9	0.36	0.94	1.59	1.18
10	0.88	0.82	0.91	0.33
11	1.36	0.55	1.06	0.54
12	1.75	2.08	1.29	0.54
13	0.65	1.09		2.03
14	1.83	1.03		0.97
15	1.03	0.27		2.05
16				2.37
17				0.89
18				0.89
19				1.82

Since the modified models represented more smooth flows around the compressor when compared with the existing models, the decrease in the temperature at the compressor could be estimated. The temperature of the compressor and condenser installed inside the MCR for the original model was 54.5 °C and 43.5 °C, respectively. The temperature of the compressor and condenser for the improved model 3 were 52.0 °C and 42.7 °C, respectively. The temperature of the modified model 3 showed a decrease of about 2.5 °C and 0.8 °C at the compressor and condenser, respectively. As noted in Table 4, according to the decrease in the temperature, the power consumption of the refrigerator decreased by about 1.2% when compared to that of the existing models. Thus, the improvement in the flow line by modifying it in the MCR increases the efficiency of power consumption.

Table 4. Reduction in power for recent and modified models.

Item Case	Power Recent Model (kWh/Month)	Power Modified Model (kWh/Month)	Reduction (%)
Case #1	49.9	49.3	1.3
Case #2	52.8	52.3	1.1

8. Conclusions

An experimental and numerical study was carried out to analyze the performance improvement of a domestic refrigerator. In the results of the measurement of power consumption under the conditions of flow line control in the MCR and fan operation, the following conclusions can be summarized as follows:

The improvement of the flow line decreased the power consumption by 0.4% due to the decrease in the temperature of the compressor and condenser inside the MCR. The power consumption decreased in proportion to the distance between the fan inside the MCR and the condenser. Because the decreased power consumption was larger than the increase in the power consumed by the operation of the fan according to the airflow, the power consumption proportionally decreased according to the increase in the airflow.

For the modified Model 3 that improved the flow line around the compressor, it showed a smooth flow due to the lowest congested section. In the calculation of the mass flow rate for the existing and modified CBMs, the mass flow rate at the vent showed a high level in the model that vents around the compressor. In the smooth airflow inside the MCR in the modified CBM, it decreased power consumption by about 1.2% due to a decrease in the temperature of the section of the compressor (shell).

Funding: This research was funded by the Tongmyong University Research Grants 2017.

Acknowledgments: This research was supported by the Tongmyong University Research Grants 2017(2017A017).

Conflicts of Interest: The authors declare no conflict of interest.

References

1. Bansal, P.; Vineyard, E.; Abdelaziz, O. Advances in household appliances—A review. *Appl. Therm. Eng.* **2011**, *31*, 3748–3760. [CrossRef]
2. Belman-Flores, J.; Barroso-Maldonado, J.M.; Rodríguez-Muñoz, A.P.; Camacho-Vázquez, G. Enhancements in domestic refrigeration, approaching a sustainable refrigerator—A review. *Renew. Sustain. Energy Rev.* **2015**, *51*, 955–968. [CrossRef]
3. James, S.J.; Evans, J. The temperature performance of domestic refrigerators. *Int. J. Refrig.* **1992**, *15*, 313–319. [CrossRef]
4. Dmitriyev, V.; Pisarenko, V. Determination of optimum refrigerant charge for domestic refrigerator units. *Int. J. Refrig.* **1984**, *7*, 178–180. [CrossRef]
5. Laguerre, O.; Ben Amara, S.; Moureh, J.; Flick, D. Numerical simulation of air flow and heat transfer in domestic refrigerators. *J. Food Eng.* **2007**, *81*, 144–156. [CrossRef]
6. Ledesma, S.; Belman-Flores, J. Mathematical application to analyze the thermal behavior of a domestic refrigerator: Influence of the location of the shelves. *Int. J. Refrig.* **2017**, *74*, 362–370. [CrossRef]
7. Rasti, M.; Hatamipour, M.S.; Aghamiri, S.F.; Tavakoli, M. Enhancement of domestic refrigerator's energy efficiency index using a hydrocarbon mixture refrigerant. *Measurement* **2012**, *45*, 1807–1813. [CrossRef]
8. Shikalgar, N.; Sapali, S. Energy and exergy analysis of a domestic refrigerator: Approaching a sustainable refrigerator. *J. Therm. Eng.* **2019**, *5*, 469–481. [CrossRef]
9. Belman-Flores, J.; Gallegos-Muñoz, A.; Puente-Delgado, A. Analysis of the temperature stratification of a no-frost domestic refrigerator with bottom mount configuration. *Appl. Therm. Eng.* **2014**, *65*, 299–307. [CrossRef]
10. António, C.C.; Afonso, C. Air temperature fields inside refrigeration cabins: A comparison of results from CFD and ANN modelling. *Appl. Therm. Eng.* **2011**, *31*, 1244–1251. [CrossRef]

11. James, S.; Evans, J.; James, C. A review of the performance of domestic refrigerators. *J. Food Eng.* **2008**, *87*, 2–10. [CrossRef]

12. Rouaud, O.; Havet, M. Computation of the airflow in a pilot scale clean room using K-ε turbulence models. *Int. J. Refrig.* **2002**, *25*, 351–361. [CrossRef]

13. Kattakayam, T.A.; Srinivasan, K. Thermal performance characterization of a photovoltaic driven domestic refrigerator. *Int. J. Refrig.* **2000**, *23*, 190–196. [CrossRef]

14. Jung, D.; Kim, C.-B.; Song, K.; Park, B. Testing of propane/isobutane mixture in domestic refrigerators. *Int. J. Refrig.* **2000**, *23*, 517–527. [CrossRef]

15. Vineyard, E.A.; Sand, J.; Miller, W. Refrigerator-freezer energy testing with alternative refrigerants. *NASA STI Recon Tech. Rep. N* **1989**, *89*, 27037.

16. Farley, K.; Alissi, M.S.; Schoenhals, R.J.; Ramadhyani, S. Effects of ambient temperature and control settings on thermal performance and energy consumption of a household refrigerator-freezer. *ASHRAE Trans.* **1987**, *93*, CONF-870620.

17. *2009 ASHRAE Handbook of Fundamentals*; American Society of Heating, Refrigerating and Air-Conditioning Engineers Inc.: Atlanta, GA, USA, 2009.

18. Bejan, A. *Convection Heat Transfer*; John Wiley & Sons: Hoboken, NJ, USA, 2013.

19. Cengel, Y.A.; Boles, M.A. *Thermodynamics: An Engineering Approach 6th Editon (SI Units)*; The McGraw-Hill Companies Inc.: New York, NY, USA, 2007.

20. Ashrae, P.C. *ASHRAE Handbook of Fundamentals*; American Society of Heating, Refrigeration and Air Conditioning Engineers Inc.: Atlanta, GA, USA, 2005.

21. Amara, S.B.; Laguerre, O.; Charrier-Mojtabi, M.-C.; Lartigue, B.; Flick, D. PIV measurement of the flow field in a domestic refrigerator model: Comparison with 3D simulations. *Int. J. Refrig.* **2008**, *31*, 1328–1340. [CrossRef]

22. Björk, E.; Palm, B. Performance of a domestic refrigerator under influence of varied expansion device capacity, refrigerant charge and ambient temperature. *Int. J. Refrig.* **2006**, *29*, 789–798. [CrossRef]

23. Zhou, Q.; Pannock, J.; Radermacher, R. Development and testing of a high-efficiency refrigerator. *ASHRAE Trans.* **1994**, *100*, 1351–1358.

24. Bayer, O.; Ruknettin, O.; Paksoy, A.; Aradag, S. CFD simulations and reduced order modeling of a refrigerator compartment including radiation effects. *Energy Convers. Manag.* **2013**, *69*, 68–76. [CrossRef]

25. R. ANSYS. *ANSYS CFX-Solver Theory Guide*; Release 16.2; ANSYS Inc.: Canonsburg, PA, USA, 2015.

26. Afonso, C.; Matos, J. The effect of radiation shields around the air condenser and compressor of a refrigerator on the temperature distribution inside it. *Int. J. Refrig.* **2006**, *29*, 1144–1151. [CrossRef]

27. Fukuyo, K.; Tanaami, T.; Ashida, H. Thermal uniformity and rapid cooling inside refrigerators. *Int. J. Refrig.* **2003**, *26*, 249–255. [CrossRef]

28. Hoang, M.L.; Verboven, P.; De Baerdemaeker, J.; Nicolai, B.M. Analysis of the air flow in a cold store by means of computational fluid dynamics. *Int. J. Refrig.* **2000**, *23*, 127–140. [CrossRef]

© 2020 by the author. Licensee MDPI, Basel, Switzerland. This article is an open access article distributed under the terms and conditions of the Creative Commons Attribution (CC BY) license (http://creativecommons.org/licenses/by/4.0/).

Article

Investigation of Fluid-Structure Interaction Induced Bending for Elastic Flaps in a Cross Flow

Tayyaba Bano [1], Franziska Hegner [2], Martin Heinrich [1] and Ruediger Schwarze [1,*

[1]   Institute of Mechanics and Fluid Dynamics (imfd), TU Bergakademie Freiberg, Lampadiusstr. 4,
     09599 Freiberg, Germany; Tayyaba.bano@extern.tu-freiberg.de (T.B.);
     Martin.Heinrich@imfd.tu-freiberg.de (M.H.)
[2]   Human-Computer Interaction Group (HCIG), Technische Hochschule Ingolstadt (THI), Esplanade 10,
     85049 Ingolstadt, Germany; franziska.hegner@thi.de
*   Correspondence: Ruediger.Schwarze@imfd.tu-freiberg.de

Received: 30 July 2020; Accepted: 31 August 2020; Published: 5 September 2020

Abstract: With the recent increase in the design of light and flexible structures, numerical investigations of fluid and structure together play a significant role in most engineering applications. Therefore, the current study presents an examination of fluid-structure interaction involving flexible structures. The problem is numerically solved by a commercial software ANSYS-Workbench. Two-way coupled three-dimensional transient simulations are carried out for the flexible flaps of different thicknesses in glycerin for a laminar flow and Reynolds number ranging from 3 < Re < 12. The bending line of the flaps is compared with experimental data for different alignments of the flaps relative to the fluid flow. The study reports the computation of the maximum tip-deflection and deformation of flaps fixed at the bottom and mounted normal to the flow. Additionally, drag coefficients for flexible flaps are computed and flow regimes in the wake of the flaps are presented. As well, the study gives an understanding on how the fluid response changes as the structure deforms and the model is appropriate to predict the behavior of thick and comparatively thinner flaps. The results are sufficiently encouraging to consider the present model for analyzing turbulent flow processes against flexible objects.

Keywords: fluid-structure interaction; flexible flaps; bending line; deformation

---

1. Introduction

Fluid-structure interaction (FSI) states the mutual interaction of a structure with a flowing fluid and develops the fluid and solid motions and the dependency of these motions on each other. Such kind of problems involve, the velocity and pressure of fluid effects and are affected by the structure deformation. In recent years, it has become a prominent characteristic of various engineering projects because design involves lighter and more flexible structures. The conduct of these structures with a fluid generates multi physical situations which are usually complex to handle. Numerical calculations play an important role in most engineering areas and have become more challenging due to the prediction of different flow regimes which cannot be seen in experiments (EXP) but it is still a challenge to have accurate numerical simulation of a FSI problem because of strong fluid-structure coupling. FSI problems are ever present in nature for example flapping wings, flags [1–9], aquatic animals [10], flexible vegetation [11–13], in hemodynamics [1,14] and human hair and hair sensors [15]. Thin structures are of great interest for microfluidic procedures in engineering and biomedical applications [16–18] and structural dynamics [18–20].

For investigating fluid motion together with the structure, extensive theoretical, experimental and numerical studies across flexible plates are carried out in the context of plate flapping, vibration, flutter instability, the effect of wake on plate stability and the estimation of natural frequency for

Appl. Sci. **2020**, *10*, 6177; doi:10.3390/app10186177   www.mdpi.com/journal/applsci

plates exposed to axial flow directions [21–25]. Results indicate that the flapping amplitude of the plate is mainly controlled by the elastic modulus and density ratio whereas frequency is a function of length-thickness ratio and density ratio. Also with an increase in Reynolds number, the plate indicates the transfer of symmetric to asymmetric flapping [21]. The forcing amplitude is the major influence on the plate resonance while flow velocity and plate bending rigidity indicate only the minor effect providing the forcing frequency is normalized with the natural frequency of the plate [10].

Aside from flat plates in axial flow direction, research work is offered to examine a vertical bottom-fixed flat plate in a laminar boundary layer. A study representing the Reynolds number, structure-to-fluid mass ratio and bending of the plate, respectively, determine the dynamic behavior of the flexible plate. Based on this, a two-dimensional (2D) numerical study [26] presents the dynamic response of a flexible plate engrossed in laminar condition for Reynolds numbers of 100, 200, 400 and 800 by an immersed boundary method (IBM). The dynamics of the plate indicate that the plate is stable for a low Reynolds number; however, when the Reynolds number reaches 800, the plate starts vibrating. The effect of bending rigidity, ranging from 0.1–1.6, demonstrates the periodic vibrations of the plate at low values and chaotic vibration when bending rigidity approaches between 0.2 and 0.4, then switching to periodic vibration again for a bending rigidity of 0.8 and 1.6. In comparison with the flapping flag [27], the plate shows a considerably different response in terms of frequency and vibrations.

The interaction of severely deformed structures with their environment is significant in many life science applications and technical systems. Nevertheless, precise and stable numerical methods for illustrating such problems are still exceptional. Tian et al. [28] developed a coupling between an existing immersed boundary flow solver together with a nonlinear finite element (FE) solver. Furthermore, they evaluate six practical applications of FSI specifically for three dimensions (3D), which primarily have large deformations. Considering a flexible plate as one of the test examples, the free-end deflection and drag coefficients measurements for Reynolds numbers ranging from 100–1600 with and without the effect of gravity and buoyancy forces are presented. Initially, the results of the baseline case with experimental data [29] are compared and then new simulations are offered for future benchmark cases. Moreover, the plate test case is used as a model to investigate the dynamic interaction of aquatic plants. The method overall leads to a numerical stability for all test cases.

The dynamic changes in the structure could possibly affect many aspects of fluid forces and present interesting fluid performance. Therefore, it is an important aspect for multiphase flow industrial applications. To compute such problems, tip-displacement of a rubber beam in a two-phase flow of air and oil was examined using a two-way coupled FSI [30]. A benchmark case [31] was simulated and previously published data of simulations and experiments were compared in terms of beam tip-deflection. The difference between 2D and 3D simulations has been clearly explained as well.

A scheme that has treated the deflection line of single and tandem beam configurations is presented by Axtmann et al. [32]. It is attained by interacting both experiments and direct numerical simulations (DNS) for Reynolds numbers ranging from 1–60 in a viscous fluid. Bending is calculated via beam bending theory. Bending lines of simulations and experiments demonstrate a good agreement for both single and tandem arrangements. A prediction model for cross flow velocity is offered. Moreover, local drag coefficient along pillar is computed and an empirical model based on drag coefficient is tested and proposed. Furthermore, a two-way coupled FSI study for different rows of flexible flaps is conducted to investigate the wavering behavior at the tip of flaps and development of vortices in the gap among the flaps. Experiments have been performed for water and glycerin against the silicon rubber flaps, reaching the maximum Reynolds number of 120, and the resulting tip-deflection of flaps is compared experimentally and numerically [5].

Investigating transient directional deformations of flexible obstacles by means of different numerical algorithms has been a subject of various studies involving the relevance of the extended arbitrary Langrangian Eulerian (ALE) method [33] and monolithic solvers [34,35]. The test problems presented [33] can be used as a benchmark for the formulation of numerical problems involving 2D

FSI with immersed structures that have large displacements. While 2D benchmark cases, free-surface flows and test cases of an elastic obstacle and plate have been presented for monolithic approaches and validated with other numerical methods [34,35]. Similarly, for the analysis of hemodynamic problems, a new and more sophisticated computational formulation named immersed structural potential method (ISPM), based on the original IBM, is presented. The novel method accounts for the computation of FSI force field at the fluid mesh and description of cardiovascular tissue scheme by viscoelastic fiber-reinforced constitutive models. A new time-integration method is also introduced for the computation of a deformation gradient tensor. Three advancement cases of deformable cylinders and flapping membranes are presented. The study is applicable for highly complex FSI problems incorporating large deformations under pulsating flows and sinking and floating rigid bodies [36]. To describe the deflection of flaps and moving boundaries, the fluid- solid interface-tracking/interface-capturing technique (FSITICT) is applied to the FSI problem with ALE interface tracking and Eulerian interface tracking. The method is demonstrated by test FSI cases. First, test computations are performed for two flapping membranes and the second test case is composed of one flapping membrane and a movable, elastic arterial-structure. The flapping membranes and fluid are modelled in fully Eulerian coordinates in both test cases, while the elastic structure in the second case is computed in Lagrangian coordinates. The overall objective is to reduce the computational cost and test the applicability of the method for large deformations in order to analyze different cardiac cycles [37].

In the light of the above survey, it is concluded that the previous research efforts are made to encounter the dynamics of flapping structures in uniform flows. Prediction of the bending line and tip-displacement of a flexible flap fixed at the bottom in 3D is rarely seen in the literature. With the simple geometry of a flap in a rectangular fluid domain, the problem is fundamental in nature and can be used in applications that coincide with the flow, forcing functions and the solid boundary conditions. Therefore, this study emphasizes on the deformation characteristics of the flaps over time, flap bending positions and different flap orientations. The focus of this study is to explore the motion of an elastic flap immersed in a rectangular fluid domain and to analyze the bending of flexible structures in 3D under laminar flow conditions.

This paper is organized as follows-first, an insight into experimental procedure is presented then a computational model is demonstrated in terms of different flap thicknesses and positions. Later, the response of flexible flaps is analyzed in terms of bending and deformation in addition, the results of flap bending are offered in comparison with the experimental work. The effect of the different clamping angles is discussed as well. Similarly, the computation of drag coefficient at various clamping positions is a part of the current research. Flow regimes are revealed at various time steps and velocity contours in the wake are shown. Finally, some discussions are presented.

2. Material and Method

A numerical two-way FSI model is presented for validating the towing-tank experimental data for flexible flaps. For the validation purposes, transient structure (flap) and unsteady flow are modeled using a commercial software ANSYS-Workbench. Each field is demonstrated separately by defining its geometry, properties, mesh, boundary conditions and respective solvers. Coupling is introduced between individual domains to establish a two-way FSI.

The experimental and numerical setups, settings and conditions are presented in the following subsets.

2.1. Experiments

A summary of the experimental data is given in the present section, whereas detail can be found in the reference [32]. Flaps of different thicknesses are fixed in the center of a plate and immersed into the working fluid.

Flaps are designed with length of 100 mm, width of 20 mm and thicknesses of 5 and 10 mm. Several measurements for the deformable flaps are taken by varying the flow conditions (Reynolds number, Re = 3 to 12) and flap arrangements that is, at clamping angles from 0° to 90° and inclined angle of 45°. The bending of the flaps was recorded for each case. As the effective Reynolds number is small and viscous forces are dominating, only bending response is observed in the experiments instead of vortex shedding, which would cause periodic vibrations.

2.2. Numerical Model

FSI modeling mainly deals with the information about how the fluid domain acts as a function of structure and how structural domain behaves as a function of fluid domain, at the interface, information is shared between the structural and fluid solver thus founding FSI [38]. In the present case, an elastic structure was surrounded by incompressible Newtonian fluid flow. $\Omega_f(t)$ is defined as a time dependent fluid domain and $\Omega_s(t)$ is defined as a time dependent structural domain. It is assumed that for all the time t, $\Omega_f(t)$ is entirely occupied by the fluid and $\Omega_s(t)$ is entirely occupied by the structure. Besides, $\Gamma_o(t) = \Omega_f(t) \cap \Omega_s(t)$ is an interface where the elastic structure interacts with the fluid. A schematic of FSI in terms of structural domain, fluid domain and interface for a present case of an elastic flap immersed in a rectangular fluid domain is shown in Figure 1.

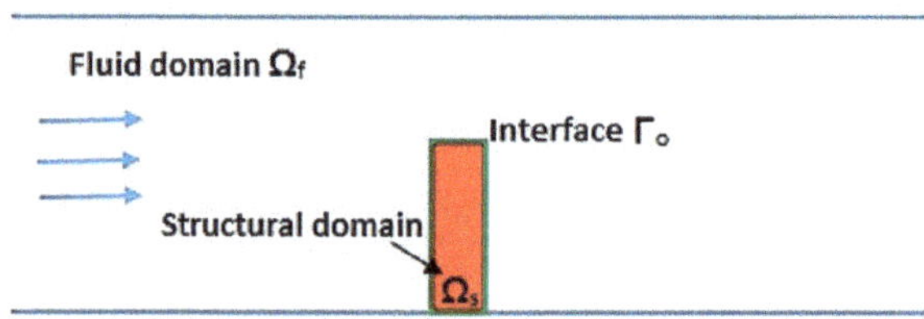

Figure 1. A schematic of FSI in the present work.

For the established FSI solution, the kinematic, dynamic and geometric conditions at the common interface of fluid and structure should be satisfied, numerically at the interface $\Gamma_o(t)$, the following conditions are assumed for all the time t:

$$v_i^s = v_i^f \tag{1}$$

$$\sigma_{ij}^s n_i = \sigma_{ij}^f n_i. \tag{2}$$

Here, v^s and v^f represent the structure and fluid velocity constituting the kinematic stability. Dynamic stability is attained by the continuous normal stresses of solid σ^s and fluid σ^f at the interface. n being the unit vector normal at the interface and the subscript i indicates its direction. According to the geometric condition stability fluid-solid domain always match and do not overlap [38].

Equations of motion for the structure and fluid are expressed as follows:

Structure equations

The flexible flap is modeled as an elastic continuum therefore the balance of linear momentum is given by:

$$\rho^s \dot{v}_i^s = \sigma_{ij,j}^s \text{ in } \Omega_s(t). \tag{3}$$

For linear elastic material, the linear Hook's law is followed by the structural stress as:

$$\sigma_{ij}^s = 2H\varepsilon_{ij} + \lambda\varepsilon_{kk}\delta_{ij}, \tag{4}$$

where ρ^s is the structural density and $\sigma_{ij,j}^s$ is the structural stress. The subscript ij of the stresses in Equations (3), (4), (6) and (7) represents the components of stress tensor whereas the subscript j, indicates the differential term involved in the linear elastic model.

The structural stress is a function of strain ε and the Lamé parameters H and λ, which are defined as:

$$\varepsilon_{ij} = \tfrac{1}{2}\left(u_{i.j} + u_{j,i}\right)$$
$$H = \frac{E}{2(1+v)}$$
$$\lambda = \frac{Ev}{(1+v)(1-2v)},$$

where u represents the displacement, E and v are Young's modulus and Poisson ratio, respectively [38–40].

Fluid equations

The governing equations for time-dependent fluid regime are modeled by incompressible continuity and Navier-Stokes equations in terms of velocity v^f, static pressure P and fluid density ρ^f as follows [38–40]:

$$\text{div } \dot{v}_i^f = 0 \text{ in } \Omega_f(t) \tag{5}$$

$$\rho^f \dot{v}_i^f = \sigma_{ij,j}^f \quad \text{in } \Omega_f(t). \tag{6}$$

The fluid stress for an incompressible fluid is simplified to:

$$\sigma_{ij}^f = -P\delta_{ij} + \tau_{ij}$$
$$\tau_{ij} = \tfrac{2}{3}\mu\left(e_{ij} - \delta_{ij}e_{kk}\right) \tag{7}$$
$$e_{ij} = \left(v_{i.j}^f + v_{j,i}^f\right).$$

The dynamic mesh motion for an FSI problem is defined in terms of arbitrary movement of the grid points with respect to their frame of reference by considering the convection of the material points. The material derivative in terms of the movement of fluid mesh can be expressed as [40]:

$$\frac{dv^f}{dt} = \frac{\partial v^f}{\partial t} + \left(v^f - U\right) \cdot \nabla v^f, \tag{8}$$

where U is the mesh movement of fluid, also $U = K\,U_{sur}$ representing K as a stiffness matrix of an elastic media and U_{sur} being the fluid dynamics surface mesh movement.

Modeling of structure and fluid in ANSYS-Workbench are explained in the following sections:

2.2.1. Structural Model

The solid model is designed in 'Transient Structural' (ANSYS-Mechanical) by a (FE) analysis, afterwards the geometry is defined in the ANSYS Design Modeler. To estimate the flow as close as possible to the experiments, all dimensions and parameters are needed to be precisely defined according to the experimental setup and can be seen in Table 1.

Analogous to the experiments, the structural model is considered as a flexible flap that is, the bottom wall is fixed, while all other walls are free. The fixed wall is assigned as 'Fixed support' whereas the remaining part of the flap is given the 'Fluid Solid Interface.' The flap is meshed by hexahedral elements and are depicted in Figure 2a. Specifying 'w = 0.02 m' being the width of the flap and height as 5 w, the flap is positioned at 7.5 w (center) of the plate and 6.25 w from the side walls, whereas the distance of the flap from outlet is specified as 75 w. Accordingly, the rectangular flow domain is indicated as (150w, 20w, 12.5w). The structural model and FSI domain are shown in Figure 2a–c respectively. Part (a) of the figure represents the flap, its dimensions, mesh and transient conditions, (b) indicates the bending definitions and (c) shows the collected computational domain of fluid and structure.

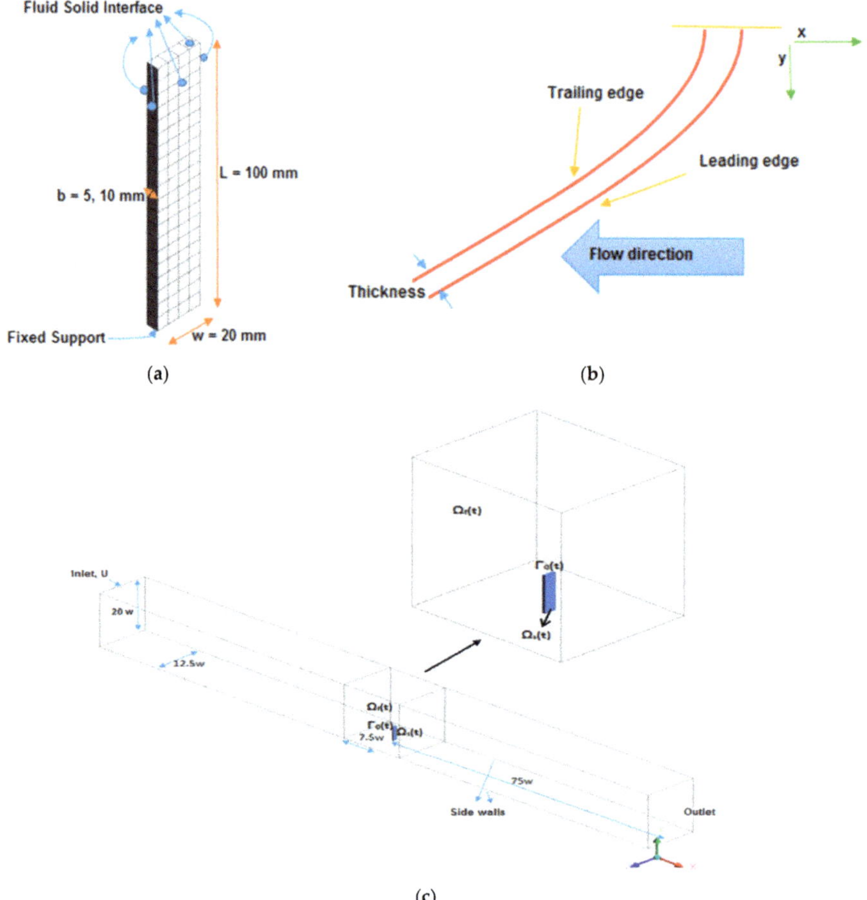

Figure 2. (**a**) Flap configuration; (**b**) Bending definitions; (**c**) Computational domain.

2.2.2. Computational Model

The fluid domain is modeled by ANSYS-FLUENT for flaps immersed in a glycerin environment as depicted in Figure 2c. Because of the low Reynolds number range, a laminar fluid model is chosen. For an effective simulation in comparison to the experiments, a fixed coordinate system is used. The constant flow velocity as 'velocity inlet' is given for the input conditions, plate and flap are assigned as walls with a 'no slip' boundary condition while the top walls, remaining bottom parts and sides are assigned as 'slip walls' and 'pressure outlet' is defined for 0 Pa at the outlet conditions. A non-dimensional Reynolds number Re is defined as $Re = vw/v$ based on the flow velocity v, flap width w and fluid kinematic viscosity v. The inlet flow velocities, v, ranging from 0.125–0.5 m/sec are given as input values based on the identified Reynolds number. The engineering parameters used for the computations are listed in Table 1. Due to subsonic incompressible flow field, pressure based solver with coupled scheme is used and second order implicit transient formulation is employed for modeling the unsteady flow.

Table 1. Engineering parameters and dimensions.

Fluid-Glycerin	Density, ρ^f	1220 kg/m^3
	Dynamic viscosity, μ	1 kg/m s
Structure-Flap	Length, L	100 mm
	Width, w	20 mm
	Thickness, b	5 mm, 10 mm
	Density, ρ^s	1030 kg/m^3
	Young's modulus, E	1.23 MPa
	Poisons ratio, v	0.3
	Shear modulus, G	0.473 MPa

Five mm thick flap:

Initially, simulations were performed for a flap of 5 mm thickness. To consider mesh independence, three different meshes were tested to explore the effect of mesh resolution on a flap bending. This mesh resolution was needed to compensate for the effect of computational time such as the computational time increasing after coupling the computational fluid dynamics (CFD) and FEA models [41]. Bending line convergence for different meshes was attained with coarser to finer: mesh 1: 52,486 elements, mesh 2: 101,641 elements and mesh 3: 137,693 elements. So, mesh 3 was chosen for the final solution, although this number of elements may vary since flap deforms and dynamic meshing takes place. Bending convergence is shown in Figure 3 whereas different meshes before and after flap deformations are depicted in Figure 4. Moreover, the difference between results obtained with mesh 2 and 3 is small, showing that grid convergence is attained. The difference between mesh 2 and 3 in terms of bending x and height y is 0.53% and 1.1% respectively.

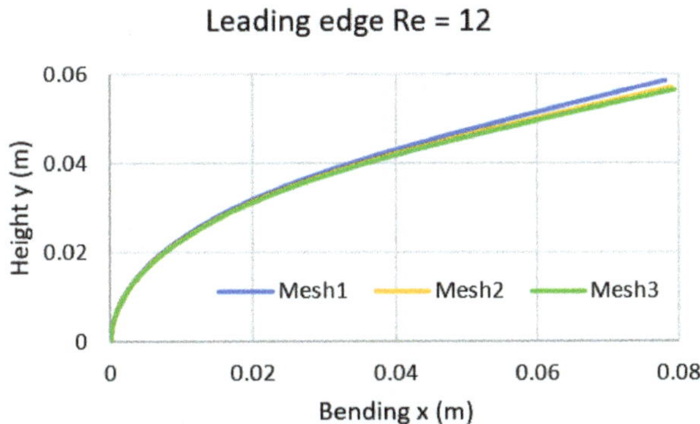

Figure 3. Bending line comparison of different mesh resolutions at Re = 12.

As the large fluid domain is present, it is only desirable to have high mesh resolution around the flap and the dynamic mesh region in order to have influence of pressure and deformation gradient. For a 3D dynamic mesh domain in Fluent, a structured mesh cannot be applied [41]. For that reason, a tetrahedron unstructured mesh was chosen around the flap, though the rest of the fluid domain was designed as a structured mesh. To simulate the flap movement, both smoothing and remeshing methods were used in the dynamic mesh region such that the displacement of the flap was large compared to the grid size. Dynamic meshing is needed for both surface and volume mesh in 3D cases and it is necessary to retain element quality for both surface and volume meshes otherwise simulations may result in errors and computational differences [41]. Considering remeshing parameters, mesh scale information was followed and a minimum cell size of 6×10^{-4} m and a maximum cell size of 0.04 m were used as input values. In addition, the interface is allocated as 'system coupling' in dynamic mesh zones under dynamic mesh settings with a cell thickness of 0.6 mm along the wall. To provide a balance among stable mesh movement and total computational time, a time step Δt of 0.001 s is used for the coupled FSI problem for 2 s of physical flow time where the flap attained the steady-state condition.

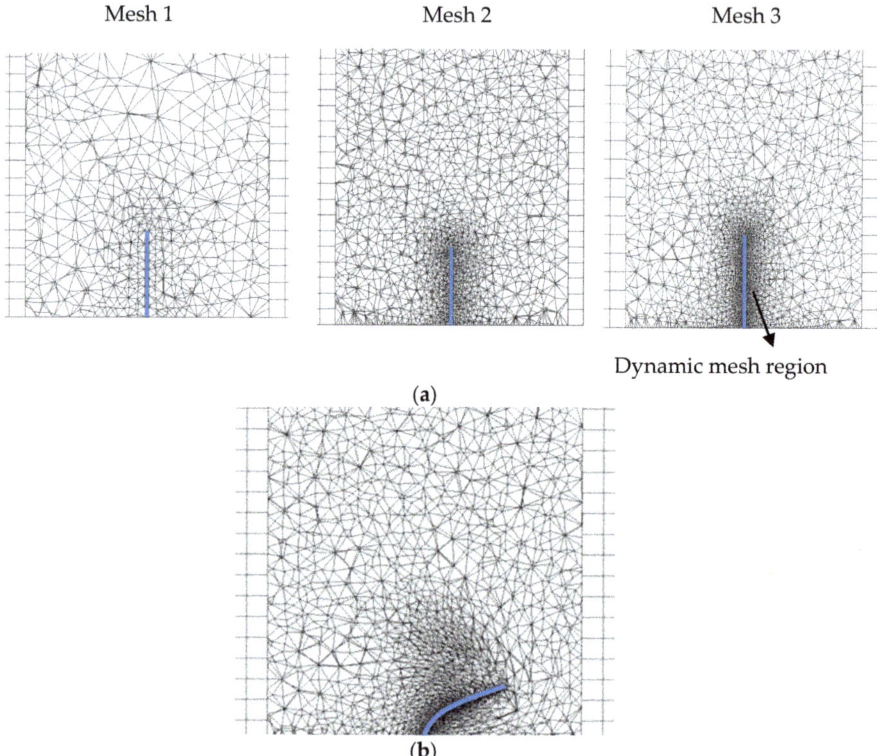

Figure 4. (a) Different meshes used in the simulations, with increasing number of elements from left to right at initial time step; (b) deformed mesh 3 after 2 s.

Ten mm thick flap:

In order to meet the experimental conditions, FSI analysis for a 10 mm thick flap is performed for different flap orientations, that is, varying the clamped angle from 0° to 90°-for the angle of 0° the flap width is facing the flow, correspondingly angles are changed from 0°–90°. At 90°, the flap is rotated completely and is parallel to the flow direction. Analysis is also performed for an inclined angle of 45°. The 10 mm flap is modeled by the same structural mesh conditions as a 5 mm thick flap. Also for the fluid domain, the same mesh parameters are employed. Though the total number of elements vary in the range of 137,000–140,000 for different angles. The same boundary conditions and models are specified for 10 mm as described for the 5 mm thick flap. However, minimum and maximum length scales of remeshing differ, depending upon the flap positioning. Flap orientations designed in Design Modeler are shown in Figure 5a,b where (a) shows the different clamping positions for angle 0°–90° and (b) indicates inclined position of the flap with the angle of 45°.

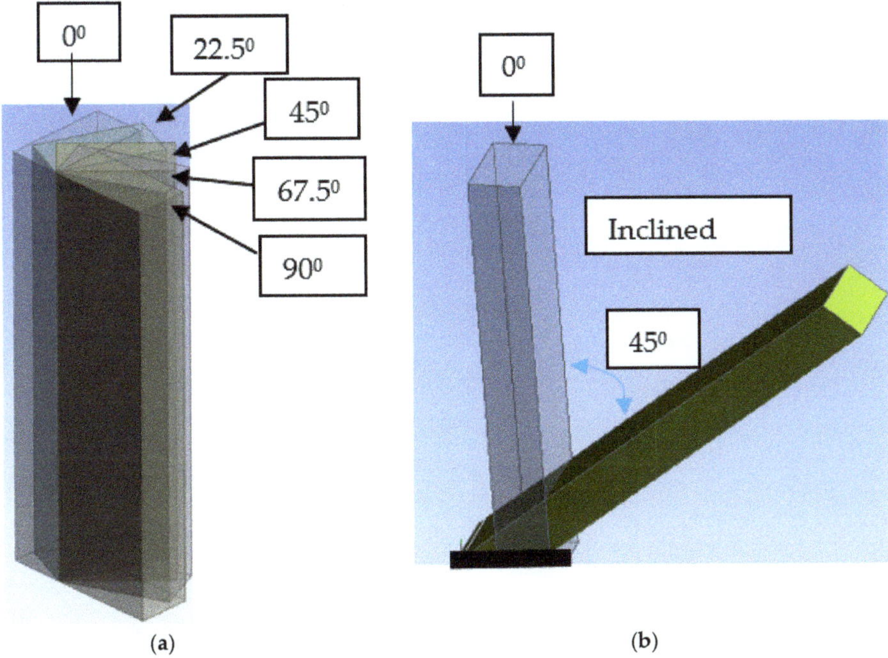

(a)

(b)

Figure 5. (**a**) Flap positioning at different clamping angles angle 0°–90°; (**b**) Inclined angle.

3. Results and Discussions

3.1. 5 mm Thick Flap

A comparison between FSI and the experimental bending of a flap at different Reynolds numbers is expressed in Figure 6a–c. For maximum flow velocity of v = 0.5 m/sec (Re = 12) the flap bending is near to 1/2 of the flap length that is, y/L ≈ 0.5, for both FSI and experimental results. The simulated bending line represents the deflection of the flap in terms of x and y coordinates after reaching a steady-state. The flap bending increases with an increase in Reynolds number. However, a non-linear connection between flap deflection and Reynolds number is identified. Computational results can be taken for the whole flap length, though for the simplicity of measurements, five displacement points are measured along the length of the flap in experiments. In simulations, the x-coordinate bends to a maximum of x = 0.079 m and x = 0.076 m for leading edge and trailing edge respectively while in experiments, it is measured to be x = 0.070 m for leading edge and x = 0.067 m for the trailing edge. The results from experiments and simulations show the same bending behavior and are in good agreement. For the highest velocity, bending curves for both leading and trailing edge are shown in the Figure 6a,b, whereas the remaining results are focused only on leading edge of the flap.

Time-dependent total deformation curve of an elastic flap is processed by ANSYS-Transient Structural. Total deformation consists of directional deformations, d, in all three directions that is, X, Y and Z for a given flap. Resultant deformation varies with time from smallest to highest Reynolds number till it reaches its maximum value and then becomes constant. The deformation curves for all four Reynolds numbers at a physical flow time of 2 s after reaching steady-state can be seen in Figure 7 while Figure 8 indicates the flap visualization versus time till reaching steady-state for a Reynolds number of 9. The deformation is influenced by the magnitude of inflow velocity that is, maximum structural deformation occurs at the tip of the flap and has a least value for Reynolds 3 and increases by 0.0449 m reaching the peak value of 0.0982 m for a Reynolds number of 12. Besides, less computational time is required by the flap for the higher velocity flow in converging to the steady-state value. Thus the FSI model consists of the observations for the maximum deflections offered by the flap.

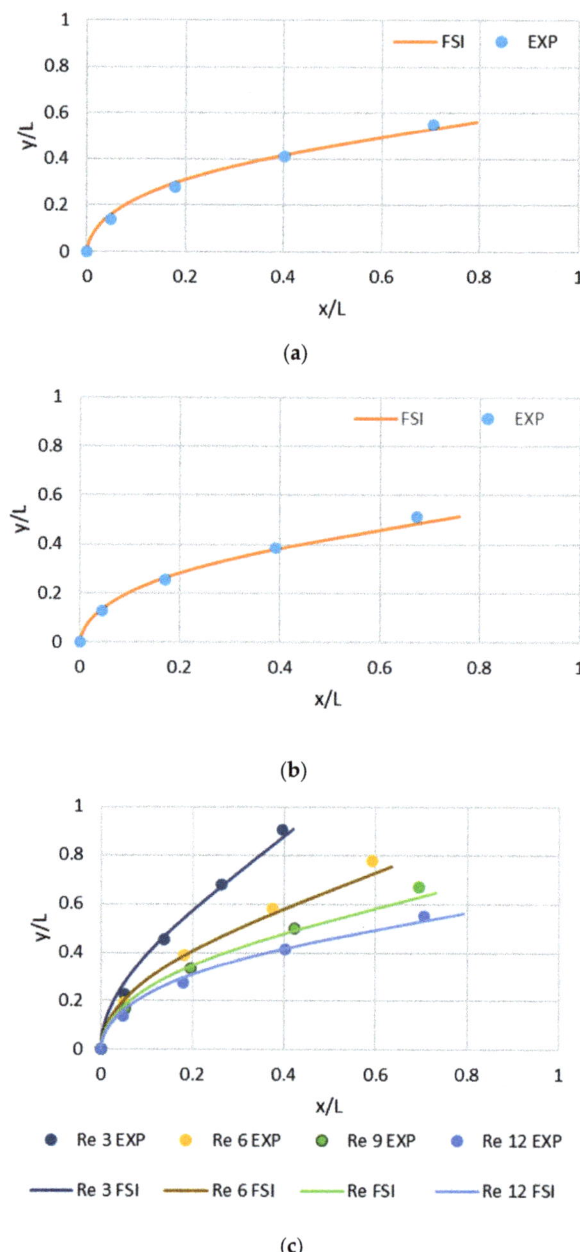

Figure 6. Flap bending; (**a**) Re = 12 leading edge; (**b**) Re = 12 trailing edge; (**c**) Leading edge as a function of Reynolds number, thickness 5 mm.

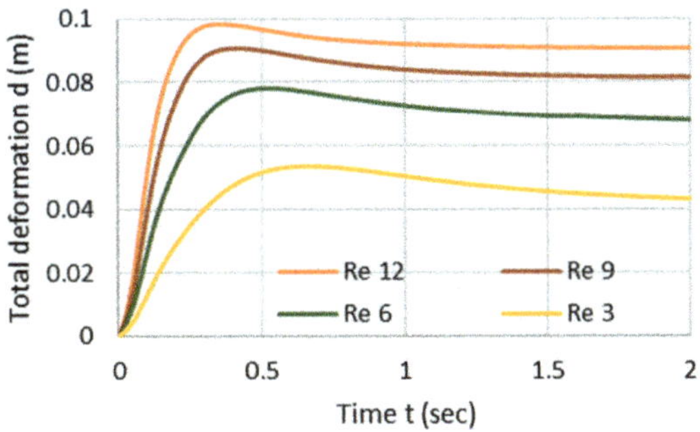

Figure 7. Deformation verses time as a function of Reynolds number.

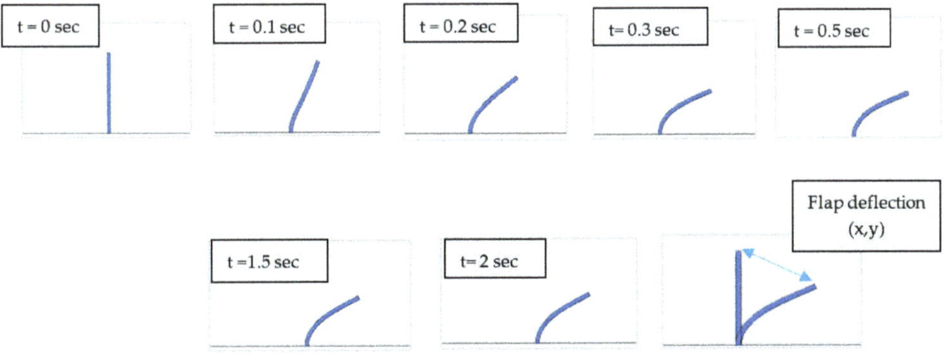

Figure 8. Flap visualization at time t, Re = 9.

3.2. 10 mm Thick Flap

As described in Section 2.2.2, a flap of 10 mm thickness is used to simulate experimental data with different clamping orientations. For the highest Reynolds number of 12, the maximum deformation for a 10 mm thick flap is 0.53 times the 5 mm thickness for angle 0°. However, it continues to decrease with increase in the clamping angle and shows the least at 90° which is 0.37 times the deformation at clamping angle of 0°. Similarly, as the Reynolds number and the clamping angle increases, the x-coordinate of flap bending decreases indicating the minimum flap deflection with highest clamping angle (see Figure 9b) that is, the maximum x-coordinate for angle 0° is $\Delta x = 0.039$m whereas for 90° it bends to x = 0.011 m. Similarly, a small flap deflection is also observed for inclined angle of 45°. Deformation at each position can be seen in Table 2 as well. Likewise, the comparison of experimental and computational results at various flap positions are shown in Figures 9 and 10 respectively.

Table 2. Deformation across flap orientations.

Angle (°)		Maximum Transient Deformation d (m)
Clamping angle	0	0.0526
-	22.5	0.0519
-	45	0.0463
-	67.5	0.0316
-	90	0.0196
Inclined	45	0.0247

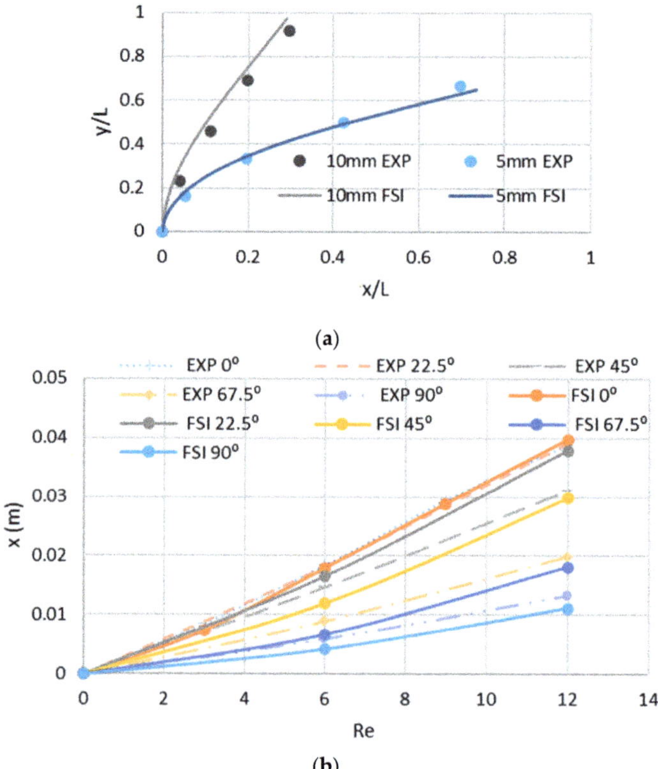

(a)

(b)

Figure 9. (**a**) Bending comparison at different thicknesses, Re = 9 clamping angle 0°; (**b**) Bending verses Reynolds number for different clamping angles, thickness 10 mm.

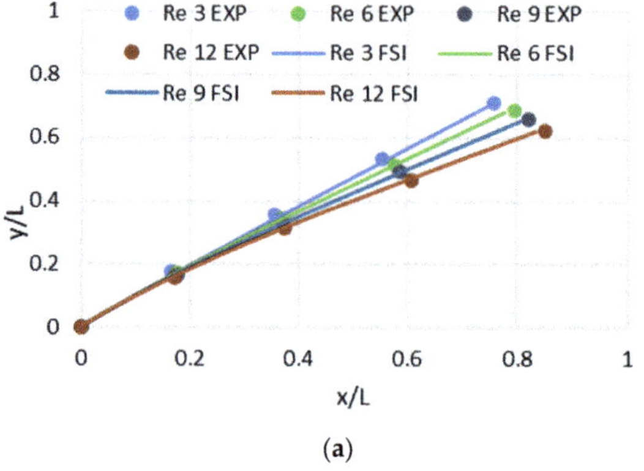

(a)

Figure 10. *Cont.*

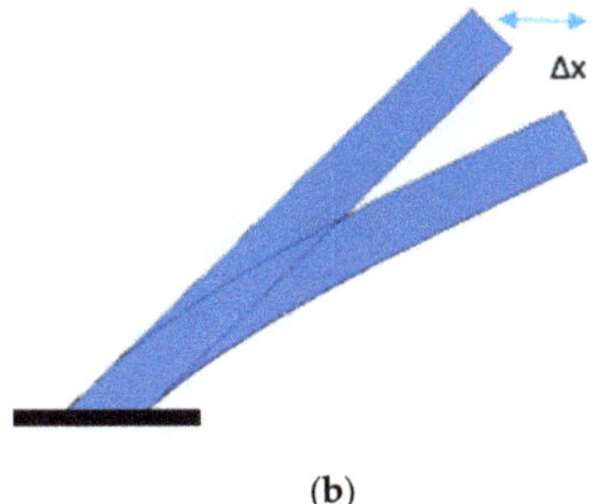

(b)

Figure 10. Inclined angle 45°: (**a**) Flap bending; (**b**) Representation of flap deflection with respect to its initial state, thickness 10 mm

3.3. Drag Coefficient

For flow around solid bodies, the drag coefficient is defined as [42]:

$$C_D = \frac{F_D}{\frac{1}{2}\rho v^2 S},$$

where F_D is the drag force exerted by the fluid on the body and ρ is the fluid density, S is the body area perpendicular to the flow direction and v is the velocity of the flow. For the present simulations, the drag coefficient, C_D, for three representative flexible flap cases is computed, that is, for a flap with a thickness of 10 mm and the clamping angles of 0°, 90° and the inclined angle of 45° at a Reynolds number of 12. The frontal area of the flap facing the flow is defined for drag calculations meant for a current 3D case. Figure 11 represents the variation of drag coefficients versus time. The mean C_D during t = 1.5–2 s for clamping angle of 90° is 3.61 and 1.95 for clamping position of 0°. Whereas for an inclined angle of 45° it is smaller than the other two cases and gives a mean of 1.45. It appears that the more bended flap is a smaller obstacle to the flow and therefore reduces the resistance from the mainstream. Moreover, the drag coefficient values are comparable with a benchmark case of a flexible flat plate in a cross flow [28].

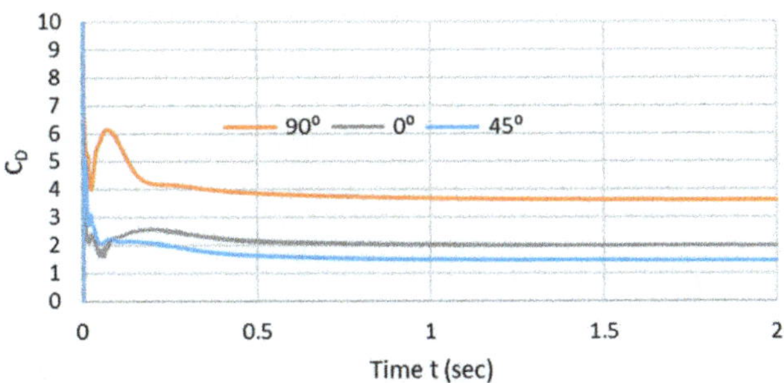

Figure 11. C_D-comparison for different clamping positions.

3.4. Flow Regimes

The velocity evaluation along with flap position for thickness 5 and 10 mm intended for clamped angle 0° and Reynolds 12 is displayed in the present section at various time steps. As discussed above (Section 3.2) and shown in Figure 12, the first two time steps at t = 0.05 s and 0.1 s corresponds to the initial deflection of the flap for both thicknesses. Nevertheless, the maximum deformation reached by the 10 mm thick flap is less for the 5 mm thick flap and attained in less time, whereas, for 5 mm thickness, the flap continues to bend and attain steady-state at approximately t = 1 s.

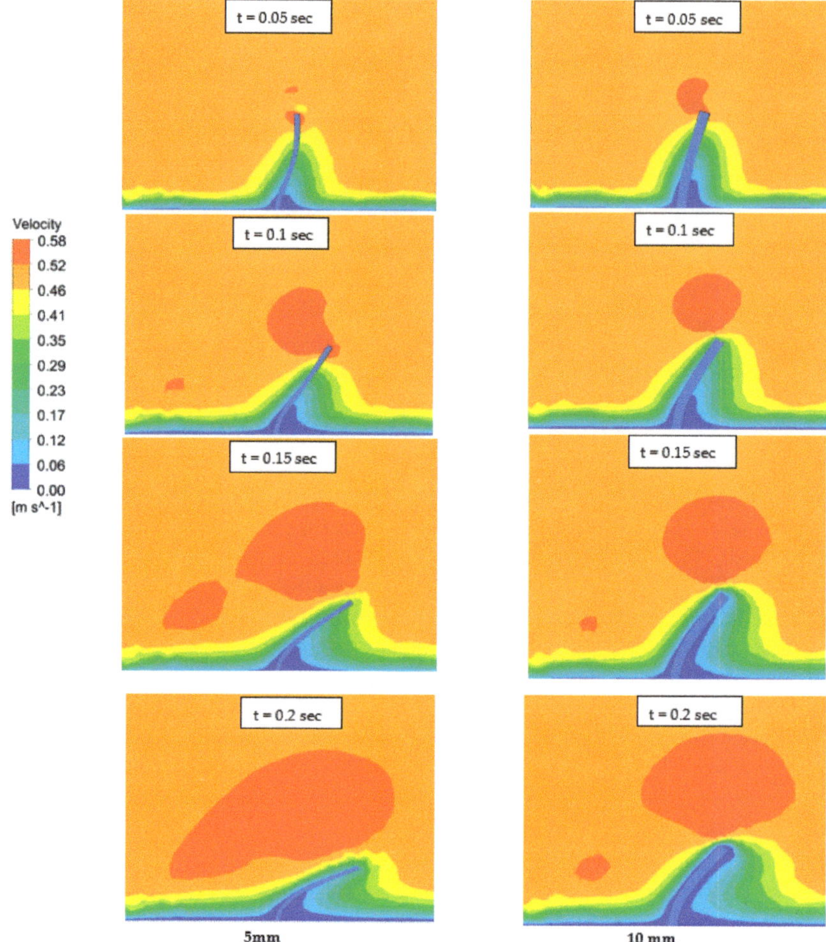

Figure 12. Flap movement of different thicknesses at various time steps.

The steady-state 3D configurations of clamping angle 0°, 90° and inclined angle 45° at maximum inlet velocity is shown in Figure 13a–c presenting the region behind the flap where the transient bending converges to a steady-state and no flow fluctuation occurs downstream of the flap. The position of the flexible flap for the specified angles has changed from being straight to deformed. Flap wake indicates the absence of vortex shedding as observed in experiments because of a low Reynolds number range. Moreover, the flow past the flap is shown and compared for Reynolds numbers of 6 and 12 for clamping angle of 90° in Figure 13d,e and also for 5 mm thick flap representing the flow for minimum and maximum Reynolds number in Figure 13f,g, respectively. The higher Reynolds number leads to the higher velocity of the flow at the flap-tip and further downstream in the wake of the flap. Besides, the maximum bending of the flap in Figure 13g bends the streamlines near the flap downstream.

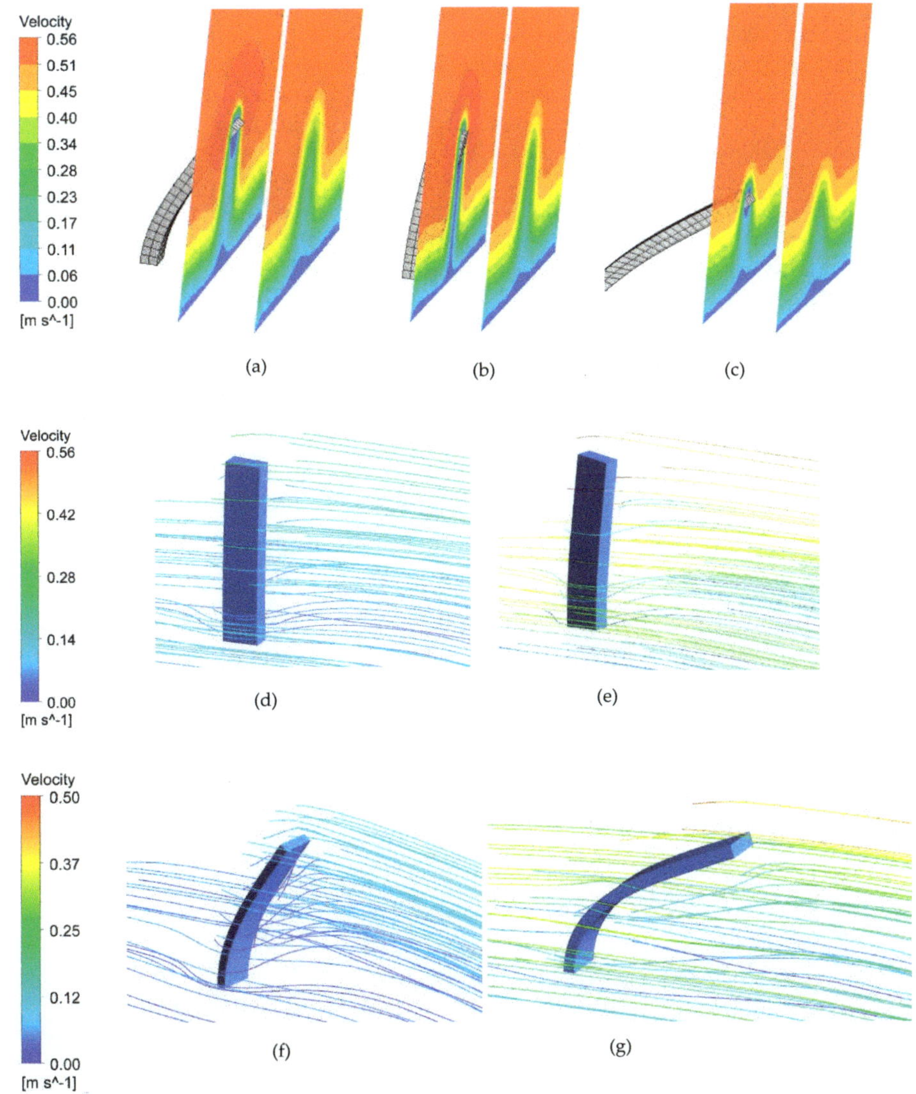

Figure 13. Comparison of flap wake; 10 mm Re = 12 (**a**) angle 0°; (**b**) 90°; (**c**) Inclined 45°; (**d**) angle 90° Re = 6; (**e**) angle 90° Re = 12; (**f**) 5mm Re = 3; (**g**) Re = 12.

4. Discussion

The present investigation focuses on the numerical investigation of flexible flaps in a glycerin environment for different thicknesses and flap orientations. The deformation of the flaps was measured to observe the various flap clamping positions under laminar flow and also the maximum tip-deflection of the flap was observed. For both experimental and simulation conditions, the flaps settle down to a steady-state and the bending patterns attained by current simulation model are very close to that observed in the experiments. The focus of the present FSI simulations is to design a 3D interaction between structure and fluid in order to narrow the literature gap for flexible structures. The results indicate that flap bending increases with the increase in Reynolds number in the range of

3 < Re < 12. However, for a different clamping position of 0° to 90° indicates minimum flap deflection with the highest clamping angle. Correspondingly, for an inclined position of 45° less deflection is perceived as compared to its undeformed shape. The objective is to judge the flap positions and flow regimes at varying computational time and to observe transient deformation response under different flow velocities. Likewise, the drag coefficients are also predicted for various flap alignments. Henceforward, the numerical model is suitable to accurately predict the response of thick and relatively thinner structures.

The flap bending in the current work is observed for low Reynolds number; however, the bending and periodic nature of flap motion is schedule by another 3D FSI model for higher Reynolds number in case of water as a working fluid. For this purpose, experiments will be performed for thinner flaps and the numerical model will be helpful in understanding the flow physics around the flaps. This will be further helpful in designing the flexible structures in hemodynamic, biomedical and structural dynamics.

Author Contributions: Conceptualization, T.B.; Methodology, T.B. and F.H.; Software, T.B.; Validation T.B.; Formal Analysis, T.B. and F.H.; Investigation, T.B.; Writing—original draft preparation, T.B.; writing—review and editing, T.B., F.H., M.H. und R.S.; supervision, M.H. und R.S. All authors have read and agreed to the published version of the manuscript.

Funding: This research was funded to T.B. by a scholarship from the funds of the Freistaat Sachsen Germany, grant number Landesstipendium and to F.H. by Deutsche Forschungsgemeinschaft (DFG) Germany, grant number DFG 1494/25-1.

Conflicts of Interest: The authors declare no conflict of interest.

Nomenclature

The following symbols are used in the manuscript:

L	Flap length	(m)
w	Flap width	(m)
b	Flap thickness	(m)
Re	Reynolds number	
Ω	Domain	
Γ	Fluid-structure interface	
t	Time	(sec)
v	Velocity	(m s^{-1})
σ	Stress	(N m^{-2})
n	Unit vector	
ε	Strain defined in Equation (4)	(m m^{-1})
H, λ	Lamé parameters	
δ	Identity tensor defined in Equations (4) and (7)	
ρ	Density	(kg m^{-3})
u	Displacement	(m)
E	Young's Modulus	(N m^{-2})
ν	Poisson ratio	
P	Static pressure	
τ	Viscous stress tensor defined in Equation (7)	
e	Velocity gradient defined in Equation (7)	
U	Fluid mesh velocity	(m s^{-1})
K	Stiffness matrix	
μ	Dynamic viscosity	(kg m^{-1} s^{-1})
υ	Kinematic viscosity	(m^2 s^{-1})
G	Shear Modulus	(N m^{-2})
x, y	Bending coordinates	(m)

d	Deformation	(m)
F	Force	(N)
C	Coefficient	
S	Area	(m^2)

Superscripts

f	Fluid
s	Structure
0	Degree

Subscripts

o	Fluid-structure interface
sur	surface
i,j,kk	Indices notation
D	Drag

References

1. Tian, F.B.; Luo, H.; Zhu, L.; Liao, J.C.; Lu, X.Y. An efficient immersed boundary-lattice Boltzmann method for the hydrodynamic interaction of elastic filaments. *J. Comput. Phys.* **2011**, *230*, 7266–7283. [CrossRef] [PubMed]
2. Shelley, M.J.; Zhang, J. Flapping and Bending Bodies Interacting with Fluid Flows. *Annu. Rev. Fluid Mech.* **2011**, *43*, 449–465. [CrossRef]
3. Saeedi, M.; Wüchner, R.; Bletzinger, K.U. Fluid-Structure interaction analysis and performance evaluation of a membrane blade. *J. Phys. Conf. Ser.* **2016**, *753*, 102009. [CrossRef]
4. Nakata, T.; Liu, H. A fluid–structure interaction model of insect flight with flexible wings. *J. Comput. Phys.* **2012**, *231*, 1822–1847. [CrossRef]
5. Favier, J.; Li, C.; Kamps, L.; Revell, A.; O'Connor, J.; Brücker, C. The PELskin project—Part I: Fluid–structure interaction for a row of flexible flaps: A reference study in oscillating channel flow. *Meccanica* **2016**, *52*, 1767–1780. [CrossRef]
6. Sawada, T.; Hisada, T. Fluid-Structure Interaction Analysis of a Two-Dimensional Flag-in-Wind Problem by the ALE Finite Element Method. *JSME Int. J. Ser. A* **2006**, *49*, 170–179. [CrossRef]
7. Combes, S.A.; Daniel, T.L. Flexural stiffness in insect wings. II. Spatial distribution and dynamic wing bending. *J. Exp. Boil.* **2003**, *206*, 2989–2997. [CrossRef]
8. De Langre, E. Effects of Wind on Plants. *Annu. Rev. Fluid Mech.* **2008**, *40*, 141–168. [CrossRef]
9. Shelley, M.J.; Vandenberghe, N.; Zhang, J. Heavy Flags Undergo Spontaneous Oscillations in Flowing Water. *Phys. Rev. Lett.* **2005**, *94*, 094302. [CrossRef]
10. Paraz, F.; Eloy, C.; Schouveiler, L. Experimental study of the response of a flexible plate to a harmonic forcing in a flow. *Comptes Rendus Mécanique* **2014**, *342*, 532–538. [CrossRef]
11. Dijkstra, J.T.; Uittenbogaard, R.E. Modeling the interaction between flow and highly flexible aquatic vegetation. *Water Resour. Res.* **2010**, *46*. [CrossRef]
12. Sand-Jensen, K. Drag forces on common plant species in temperate streams: Consequences of morphology, velocity and biomass. *Hydrobiologia* **2008**, *610*, 307–319. [CrossRef]
13. Mohamed, A.A. Modeling coupling between eelgrass Zostera marina and water flow. *Mar. Ecol. Prog. Ser.* **2007**, *338*, 81–96. [CrossRef]
14. Selim, K.; Logg, A. Simulating Heart Valve Dynamics in FEniCS. In Proceedings of the 5th National Conference on Computational Mechanics (MekIT 2009), Trondheim, Norway, 26–27 May 2009.
15. Maschmann, M.R.; Ehlert, G.J.; Dickinson, B.T.; Phillips, D.M.; Ray, C.W.; Reich, G.W.; Baur, J.W. Bioinspired Carbon Nanotube Fuzzy Fiber Hair Sensor for Air-Flow Detection. *Adv. Mater.* **2014**, *26*, 3230–3234. [CrossRef] [PubMed]
16. Lambert, R.A.; Rangel, R. The role of elastic flap deformation on fluid mixing in a microchannel. *Phys. Fluids* **2010**, *22*, 52003. [CrossRef]
17. Rao, K.S.; Sravani, K.G.; Yugandhar, G.; Rao, G.V.; Mani, V. Design and Analysis of Fluid Structure Interaction in a Horizontal Micro Channel. *Procedia Mater. Sci.* **2015**, *10*, 768–788. [CrossRef]

18. Bogaers, A.; Kok, S.; Reddy, B.; Franz, T. An evaluation of quasi-Newton methods for application to FSI problems involving free surface flow and solid body contact. *Comput. Struct.* **2016**, *173*, 71–83. [CrossRef]

19. Yamada, T.; Yoshimura, S. Line search partitioned approach for fluid-structure interaction analysis of flapping wing. *Comput. Model. Eng. Sci.* **2008**, *24*, 51–60.

20. Fairuz, Z.M.; Abdullah, M.; Yusoff, H.; Abdullah, M. Fluid Structure Interaction of Unsteady Aerodynamics of Flapping Wing at Low Reynolds Number. *Eng. Appl. Comput. Fluid Mech.* **2013**, *7*, 144–158. [CrossRef]

21. Yu, Z.; Wang, Y.; Shao, X. Numerical simulations of the flapping of a three-dimensional flexible plate in uniform flow. *J. Sound Vib.* **2012**, *331*, 4448–4463. [CrossRef]

22. Tang, L.; Païdoussis, M.P. On the instability and the post-critical behaviour of two-dimensional cantilevered flexible plates in axial flow. *J. Sound Vib.* **2007**, *305*, 97–115. [CrossRef]

23. Tang, L.; Païdoussis, M.P. The influence of the wake on the stability of cantilevered flexible plates in axial flow. *J. Sound Vib.* **2008**, *310*, 512–526. [CrossRef]

24. Kerboua, Y.; Lakis, A.; Thomas, M.; Marcouiller, L. Vibration analysis of rectangular plates coupled with fluid. *Appl. Math. Model.* **2008**, *32*, 2570–2586. [CrossRef]

25. Eloy, C.; Lagrange, R.; Souilliez, C.; Schouveiler, L. Aeroelastic instability of cantilevered flexible plates in uniform flow. *J. Fluid Mech.* **2008**, *611*, 97–106. [CrossRef]

26. Wang, L.; Lei, C.; Tian, F. Fluid–structure interaction of a flexible plate vertically fixed in a laminar boundary layer over a rigid wall. In Proceedings of the 21st Australasian Fluid Mechanics Conference, Adelaide, Australia, 10–13 December 2018.

27. Connell, B.S.; Yue, D.K.P. Flapping dynamics of a flag in a uniform stream. *J. Fluid Mech.* **2007**, *581*, 33–67. [CrossRef]

28. Tian, F.B.; Dai, H.; Luo, H.; Doyle, J.F.; Rousseau, B. Fluid–structure interaction involving large deformations: 3D simulations and applications to biological systems. *J. Comput. Phys.* **2014**, *258*, 451–469. [CrossRef] [PubMed]

29. Luhar, M.; Nepf, H.M. Flow-induced reconfiguration of buoyant and flexible aquatic vegetation. *Limnol. Oceanogr.* **2011**, *56*, 2003–2017. [CrossRef]

30. Svenning, E.; Mark, A.; Edelvik, F. Simulation of a highly elastic structure interacting with a two-phase flow. *J. Math. Ind.* **2014**, *4*, 7. [CrossRef]

31. Botia-Vera, E.; Souto-Iglesias, A.; Bulian, G.; Lobovsky, L. Three SPH novel benchmark test cases for free surface flows. In Proceedings of the 5th ERCOFTAC SPHERIC Workshop on SPH Applications, Manchester, UK, 22–25 June 2010.

32. Axtmann, G.; Hegner, F.; Brücker, C.; Rist, U. Investigation and prediction of the bending of single and tandem pillars in a laminar cross flow. *J. Fluids Struct.* **2016**, *66*, 110–126. [CrossRef]

33. Basting, S.; Quaini, A.; Čanić, S.; Glowinski, R. Extended ALE Method for fluid–structure interaction problems with large structural displacements. *J. Comput. Phys.* **2017**, *331*, 312–336. [CrossRef]

34. Jodlbauer, D.; Langer, U.; Wick, T. Parallel block-preconditioned monolithic solvers for fluid-structure interaction problems. *Int. J. Numer. Methods Eng.* **2018**, *117*, 623–643. [CrossRef]

35. Ryzhakov, P.B.; Rossi, R.; Idelsohn, S.; Oñate, E. A monolithic Lagrangian approach for fluid–structure interaction problems. *Comput. Mech.* **2010**, *46*, 883–899. [CrossRef]

36. Gil, A.J.; Carreño, A.A.; Bonet, J.; Hassan, O. The Immersed Structural Potential Method for haemodynamic applications. *J. Comput. Phys.* **2010**, *229*, 8613–8641. [CrossRef]

37. Wick, T. Flapping and contact FSI computations with the fluid–solid interface-tracking/interface-capturing technique and mesh adaptivity. *Comput. Mech.* **2013**, *53*, 29–43. [CrossRef]

38. Richter, T. *Fluid-Structure Interactions: Models, Analysis and Finite Elements*; Springer Science and Business Media LLC: Cham, Switzerland, 2017; Volume 118.

39. Galdi, G.P.; Rannacher, R. *Fundamental Trends in Fluid-Structure Interaction*; World Scientific Pub Co Pte Lt: Singapore, 2010.

40. Hou, G.; Wang, J.; Layton, A.T. Numerical Methods for Fluid-Structure Interaction—A Review. *Commun. Comput. Phys.* **2012**, *12*, 337–377. [CrossRef]

41. Sederstrom, D. Methods and Implementation of Fluid-Structure Interaction Modeling into an Industry-Accepted Design Tool. Ph.D. Thesis, University of Denver, Denver, CO, USA, 2016. ProQuest Dissertations and Theses.
42. Ferziger, J.H.; Perić, M.; Street, R.L. *Computational Methods for Fluid Dynamics*, 4th ed.; Springer Nature Switzerland AG: Cham, Switzerland, 2020.

© 2020 by the authors. Licensee MDPI, Basel, Switzerland. This article is an open access article distributed under the terms and conditions of the Creative Commons Attribution (CC BY) license (http://creativecommons.org/licenses/by/4.0/).

Article

Numerical Simulation of the Oil Droplet Size Distribution Considering Coalescence and Breakup in Aero-Engine Bearing Chamber

Fei Wang *, Lin Wang *, Guoding Chen and Donglei Zhu

School of Mechanical Engineering, Northwestern Polytechnical University, Xi'an 710072, China;
gdchen@nwpu.edu.cn (G.C.); zhudonglei@mail.nwpu.edu.cn (D.Z.)
* Correspondence: wfei@mail.nwpu.edu.cn (F.W.); wanglin@nwpu.edu.cn (L.W.); Tel.: +86-187-2900-1815 (F.W.)

Received: 12 June 2020; Accepted: 13 August 2020; Published: 14 August 2020

Abstract: In order to improve the inadequacy of the current research on oil droplet size distribution in aero-engine bearing chamber, the influence of oil droplet size distribution with the oil droplets coalescence and breakup is analyzed by using the computational fluid dynamics-population balance model (CFD-PBM). The Euler–Euler equation and population balance equation are solved in Fluent software. The distribution of the gas phase velocity field and the volume fraction of different oil droplet diameter at different time are obtained in the bearing chamber. Then, the influence of different initial oil droplet diameter, air, and oil mass flow on oil droplet size distribution is discussed. The result of numerical analysis is compared with the experiment in the literature to verify the feasibility and validity. The main results provide the following conclusions. At the initial stage, the coalescence of oil droplets plays a dominant role. Then, the breakup of larger diameter oil droplet appears. Finally, the oil droplet size distribution tends to be stable. The coalescence and breakup of oil droplet increases with the initial diameter of oil droplet and the air mass flow increasing, and the oil droplet size distribution changes significantly. With the oil mass flow increasing, the coalescence and breakup of oil droplet has little change and the variation of oil droplet size distribution is not obvious.

Keywords: bearing chamber; breakup; coalescence; population balance model; two phase flow; oil droplet size distribution; computational fluid dynamics

---

1. Introduction

The bearing chamber is an important part in the aero-engine lubrication system. The lubricating oil is shed into the bearing chamber in the form of oil droplet via under-race lubrication method [1]. After the high-speed moving oil droplets collide with the wall, some of them are deposited to form the oil film, and others are splashed to form a large number of secondary oil droplets. Then, the complex air–oil two phase flow and heat transfer state of air, oil droplet, and oil film coexist in the bearing chamber. Because the oil droplet size is very small, many secondary oil droplets are suspended in the bearing chamber under the action of the air phase flow field. Furthermore, the interaction between oil droplets induces many dynamic events, such as collision, coalescence, deposition, evaporation, breakup, etc. The coalescence and breakup of oil droplets will affect the size distribution of secondary oil droplets and further restrict the research of flow and heat transfer in the bearing chamber. Moreover, the oil droplet size distribution directly affects the mixture state of oil and gas. Previously the Rosin–Rammler (R–R) distribution is widely used for the calculation of oil droplet size distribution in bearing chamber. However, experimental research indicates that R–R distribution is not consistent with the actual condition. Therefore, the study of droplet diameter distribution of the secondary oil droplets, which considers the coalescence and breakup of oil droplets, is of great significance in the aero-engine lubrication design.

At present, the research on coalescence and breakup of oil droplets mostly adopts the population balance model, which is earliest used in chemical process by Hulbert et al. [2]. Since then it has become a general method to analyze the particle diameter distribution of dispersed phase in multiphase system. The population balance model describes the time and space distribution of discrete phase by tracing the change of number density function of discrete phase with convection, diffusion and the source term of coalescence and breakup in the flow field. This method has been widely used because it remains part of the size information of discrete phase, and uses the averaging method to describe the motion of continuous phase, which reduces computation greatly. The core of population balance model lies in the profound understanding of coalescence and breakup mechanism of oil droplets. The coalescence and breakup of oil droplets is affected by the flow field around the droplets and the physical parameters (such as surface tension) of the oil droplet, which involves a very complicated multi-level and multiscale physical mechanism. The mechanism description of oil droplet breakup can be divided into three categories: (1) the droplet breakup caused by turbulence, (2) the droplet breakup caused by viscous shear force, and (3) the droplet breakup caused by the instability of the liquid–liquid interface. The coalescence mechanism of oil droplet is more complex. Moreover, the probability model is used more often, as the coalescence rate of oil droplet is equal to the product of collision frequency and coalescence efficiency between oil droplet. The collision mechanism of oil droplet can be divided into four categories: (1) collision caused by random turbulence fluctuation, (2) collision caused by buoyancy, (3) collision caused by shear force, and (4) collision induced by wave vortex. The coalescence efficiency model can be divided into three categories: energy model, velocity model, and liquid membrane drainage model.

During the past decades, many scholars have done many researches on the coalescence and breakup of bubbles/droplet and the size distribution of bubbles/droplet. Liao and Lucas et al. [3,4] reviewed the coalescence and breakup model of liquid, and introduced the research process of oil droplet coalescence and breakup mechanism. Ramkrishna et al. [5] investigated the effect of coalescence and breakup of oil droplet on the size distribution by using the population balance model, and further studied the flow behavior of liquid–liquid system from mechanism. Coulaloglou and Tavlarides et al. [6] assumed that the oscillating droplet will break up when the turbulent kinetic energy transmitted by the turbulent vortex to the droplet exceed the surface energy of the droplet, but the two droplets will coalesce when the contact time exceed the film discharge time of the two droplets. Tsouris and Tavlarides et al. [7] studied the droplet coalescence behavior in the turbulent using high-speed imaging technology and population balance model, proposed a novel collision frequency model, and compared the predicted droplet diameter with the experimental data. The results show that the model is applicable under different discrete phase fraction. Luo and Svendsen et al. [8] established a bubble breakup model, in which the bubble breakup rate is equal the product of the collision frequency with the turbulent vortex and the bubble breakup efficiency.

Regarding the solution of the population balance equation, Li et al. [9] used the quadrature method of moments (QMOM) to study the droplet breakup and coalescence behavior of the liquid–liquid dispersion system in the stirred reactor. The results show that the droplet coalescence behavior occurs with some difficultly in the low discrete phase fraction, due to the small probability of droplet collision. Li et al. [10] reviewed the CFD-PBM simulation in extraction column, and the principles and advantages and disadvantages of different solving methods of population balance model were compared. Xing et al. [11] used the experimental method and CFD-PBM coupled model to analyze the influence of viscosity in the bubble bed on the hydrodynamic behavior in the bubble area, such as the total gas holdup, the big and small bubble holdup, and the bubble size distribution. The results show that the CFD-PBM coupled model has a good prediction ability for the hydrodynamic behavior of the bubble bed in a wide range of gas velocities and viscosities. Wang et al. [12,13] used the population balance model to describe the bubble size distribution in the gas–liquid system, considered the different mechanism of bubble coalescence and breakup, established a relatively mature bubble coalescence and breakup model, and calculated the bubble size distribution in the gas–liquid system by numerical solution of

PBM. Zhao et al. [14] used a multi-Monte Carlo algorithm to simulate the collision and coalescence process of particles in nanoparticle flow, and the simulation results were in good agreement with the direct numerical simulation results. Chen et al. [15] analyzed the parameter relationship between the structural condition of bearing chamber and the oil droplet size distribution, clarified the oil droplet mass distribution based on the continuous oil drop size distribution, and introduced the concept of continuous oil droplet diameter into the analysis of oil droplet movement. Glahn et al. [16] measure the size distribution of oil droplet using the phase Doppler particle analyzer (PDPA) technology under various operating condition in the rotating disk chamber. In conclusion, the population balance theory has been widely applied in the analysis of coalescence and breakup between bubbles and droplets. However, no relevant research work has been done to consider the influence of coalescence and breakup on the oil droplet size distribution in the bearing chamber.

Therefore, based on the population balance model, the coalescence and breakup behaviors of oil droplets are described through the coalescence and breakup model of the oil droplet. Then, the air–oil two-phase flow model and the population balance model are solved by Fluent software. The numerical simulation of the coalescence and breakup processes of the oil droplet in a bearing chamber is performed. The air phase velocity distribution and the volume fraction of the oil droplets at different times are discussed. Moreover, the variation rules of oil droplet size distribution are obtained in different initial oil droplet diameter, air inflow, and oil inflow. The coalescence and breakup of oil droplets can change the size distribution of secondary oil droplets. Moreover, the oil droplet size distribution directly affects the mixture state of oil and gas. The research work in this paper can improve the accuracy of air–oil two-phase flow and heat transfer analysis, and provide reference for further research on oil mist concentration test in the aero-engine bearing chamber.

2. Numerical Model

2.1. Geometric Model and Physical Parameters

The geometric structure of the bearing chamber is shown in Figure 1. The structural parameters and values of the bearing chamber involved in the analysis mainly include the rotor radius of the bearing, $r_s = 60$ mm; the height of the bearing chamber, $h_b = 30$ mm; the width of the bearing chamber, $w_b = 30$ mm; and the diameter of oil scavenge d_s (scavenge is shown in Figure 2) and the air vent d_v are both 16 mm. Seal air flows into the bearing chamber from air inlet and discharges through vent. The gap of oil inlet is 2 mm. Moreover, the lubricating oil is shed into the bearing chamber from the clearance between the bearing outer ring and the cage (oil inlet in Figure 1) and discharges through scavenge. The equivalent size of oil inlet is 4 mm (oil inlet in Figure 2).

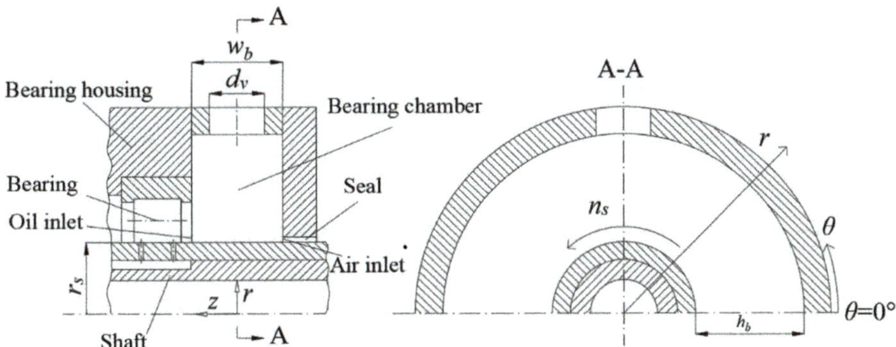

Figure 1. The structural diagram of aero-engine bearing chamber.

In the analysis, the oil brand [17] is aviation lubricating oil 4109. The oil density ρ_l is 926 kg/m^3, the oil dynamic viscosity μ_l is 0.007 Pa·s, and the oil surface tension coefficient σ_l is 0.035 N/m. The gas density ρ_g is 1.225 kg/m^3 and the air dynamic viscosity μ_g is 1.789 × 10^{-5} Pa·s.

2.2. Mesh Model and Grid Independence Verification

According to the structural parameters of the bearing chamber, Gambit software is used to build up the geometric model of the flow region, hexahedral grids are used to discrete the flow region, and the grid of air inlet and oil inlet is refined. The grid model is shown in Figure 2.

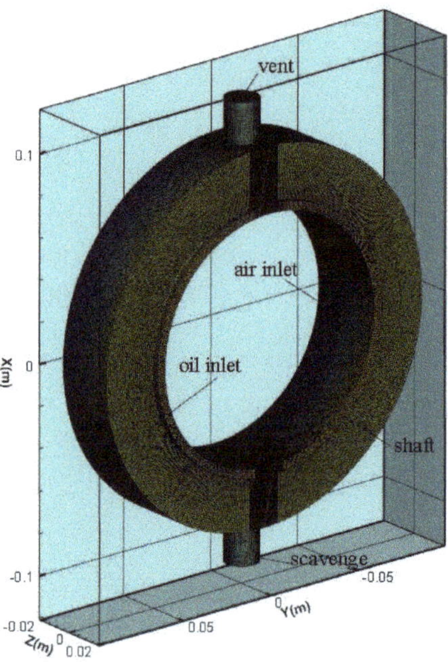

Figure 2. Grid model of the bearing chamber.

In order to eliminate the influence of mesh on the calculation, three groups mesh with different size are generated on the gambit platform, and the number of meshes is 487,585, 609,719, and 1,530,258, respectively. Under the condition of air inflow 12 g/s, the air phase velocities at four position are extracted on the section z = 0, and the four coordinate points are as follows; (−0.11,0,0), (−0.09,0,0), (0.09,0,0), (0.11,0,0). The calculation results are shown in Figure 3. From the figure, the velocities at four position obtained by three groups mesh are approximate. In order to save computing cost, the grid model of 609,719 is chosen in this paper.

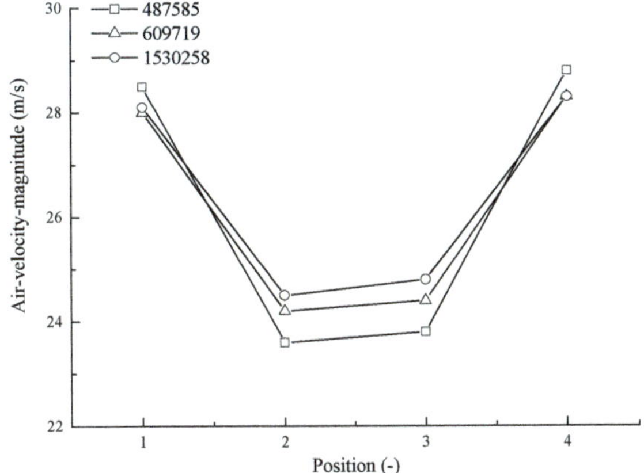

Figure 3. Verification of grid independence, the number of meshes is 487,585, 609,719, and 1,530,258, respectively, when air inflow is 12 g/s, the air phase velocities at four positions (−0.11,0,0), (−0.09,0,0), (0.09,0,0), and (0.11,0,0) in z = 0 plane.

2.3. Mathematical Model

In the air–oil two-phase flow, the dynamic events of oil droplet include collision, coalescence, breakup, deposition, etc. Under the combined effect of these dynamic events, the oil droplet diameter distribution changes with time. Under a series of physical actions, not only do the conservation of mass, momentum, and energy between oil droplet and air need to be considered, but also the population balance equation needs to describe the oil droplet diameter distribution. Especially, the characteristics of oil droplet collision, coalescence, and breakup are significant in the bearing chamber. It is necessary to focus on the change of oil diameter before and after coalescence and breakup. Therefore, the parameters, such as the volume fraction and velocity of each phase, are obtained by solving the Eulerian–Eulerian two-fluid model, and the dynamic events of oil droplet are described by using the population balance model. The oil droplet diameter distribution with the change of time and space is obtained by combining the two models. The Eulerian–Eulerian two-fluid model and the population balance model in this paper are as follows.

2.3.1. Two-Fluid Model

In two-fluid model [18], the air and oil phase are both treated as interconnected continuous media. As the volume occupied by one phase cannot be occupied by the other, the phase volume fraction is introduced. The volume fraction is a continuous function of space and time, and the sum of the volume fractions of each phase is 1. The volume fraction equation is as follows,

$$\alpha_q = \frac{V_q}{V}, \ 0 \le \alpha_q \le 1 \tag{1}$$

$$\sum_{q=1}^{n} \alpha_q = 1 \tag{2}$$

where V is the total volume of air oil two-phase, V_q is the volume of the q phase, α_q is the volume fraction of the q phase, and n has two phases: gas is labeled as phase 1 and oil droplet is labeled as phase 2. It should be noted that a two-fluid model can be considered a valid approximation only up to Stokes number ~0.1–0.2. A similar set of continuity equations and momentum equations can be derived for each phase.

The continuity equation is

$$\frac{\partial(\alpha_q \rho_q)}{\partial t} + \nabla \cdot (\alpha_q \rho_q \mathbf{U}_q) = 0 \tag{3}$$

where ρ_q is the density of the q phase; $\mathbf{U}_q$ is the velocity vector of the q phase.

The momentum equation is

$$\begin{aligned}
\frac{\partial(\alpha_q \rho_q \mathbf{U}_q)}{\partial t} &+ \nabla \cdot (\alpha_q \rho_q (\mathbf{U}_q \otimes \mathbf{U}_q)) + \nabla \cdot (\alpha_q \boldsymbol{\tau}_q) + \nabla \cdot (\alpha_q \mathbf{R}_q) \\
&= -\alpha_q \nabla p + \alpha_q \rho_q \mathbf{g} \pm \mathbf{M}_q
\end{aligned} \tag{4}$$

where $\boldsymbol{\tau}_q$ is the viscous-stress tensors of the q phase, p is the pressure share by gas phase and the liquid phase, $\mathbf{R}_q$ is the Reynolds stress tensors, $\mathbf{g}$ is the gravitational acceleration vector, and $\mathbf{M}_q$ is the interfacial force term between the gas phase and liquid phase. When q is oil droplet, $\mathbf{M}_q$ is positive; when q is air, $\mathbf{M}_q$ is negative. The term $\mathbf{M}_q$ is usually composed by three forces: drag, virtual mass, and lift. In present work, only the drag force is taken into account, whereas the others are neglected. The drag force is calculated as follows,

$$\mathbf{M}_q = \alpha_1 \alpha_2 \left(\frac{3}{4} C_D \frac{\rho_l}{d_{32}} |\mathbf{U}_r| \right) \mathbf{U}_r \tag{5}$$

where $\mathbf{U}_r = \mathbf{U}_g - \mathbf{U}_l$ is the slip velocity and $\mathbf{U}_g$, $\mathbf{U}_l$ are the average vector of the air and oil, respectively. C_D is the drag coefficient; d_{32} is the mean Sauter diameter, calculated as the ratio between the moments of order three and two of the droplet size distribution.

In the two-fluid model, the turbulent Reynolds stress should be closed. The Reynolds stress arises from the operation of Reynolds averaging on the momentum equation. In this paper, the k-ε turbulence model [19,20] is selected to close the turbulence term in two fluid model. The turbulence model k-ε can be represented by the following equations,

$$\frac{\partial(\alpha_q k)}{\partial t} + \nabla \cdot (\alpha_q k \overline{u_q}) = \nabla \cdot (\alpha_q \frac{\nu_t}{\sigma_k} \nabla k) + \alpha_q G_k - \alpha_q \varepsilon \tag{6}$$

$$\frac{\partial(\alpha_q \varepsilon)}{\partial t} + \nabla \cdot (\alpha_q \varepsilon \overline{u_q}) = \nabla \cdot (\alpha_q \frac{\nu_t}{\sigma_\varepsilon} \nabla \varepsilon) + \alpha_q \frac{\varepsilon}{k} (C_{k1} G_k - C_{k2} \varepsilon) \tag{7}$$

where k is turbulent kinetic energy; ε is turbulent energy dissipation term; ν_t is turbulent viscosity, which can be expressed as $C_\mu k^2 / \varepsilon$, C_{k1}, C_{k2}, C_μ, σ_k, and σ_ε are model constants; and G_k is the generating terms of turbulent kinetic energy.

2.3.2. Population Balance Model

The application of the population balance model [21] is used to calculate the oil droplet diameter distribution in the bearing chamber. The transport equation of the oil droplet number density function can be expressed as

$$\begin{aligned}
\frac{\partial}{\partial t}[(n(V,t))] + \nabla \cdot \left[\vec{u}(n(V,t)) \right] &= \underbrace{\frac{1}{2} \int_0^V a(V-V', V') n(V-V', t) n(V', t) dV'}_{\text{I}} \\
&- \underbrace{\int_0^\infty a(V, V') n(V, t) n(V', t) dV'}_{\text{II}} + \underbrace{\int_0^\infty b(V') \beta(V|V') n(V', t) dV'}_{\text{III}} - \underbrace{b(V) n(V, t)}_{\text{IV}}
\end{aligned} \tag{8}$$

In the right-hand side of the equation in turn are the source term (I) generated by coalescence, the sink term (II) generated by coalescence, the source term (III) generated by breakup, and the sink term (IV) generated by breakup. Where $n(V, t)$ is the number distribution function (1/m^6), which means the number of oil droplets in the range of V and $V + dV$ in unit volume is $n(V, t)dV$; $a(V, V')$ is the coalescence rate function (m^3/s) of oil droplets in volume V and V' due to collision. $b(V')$ is the breakup rate function (1/s) of oil droplets in volume V'. $\beta(V|V')$ is the probability distribution density function (1/m^3) of oil droplets in volume V, which breaks up from oil droplets in volume V'.

The population balance equation (PBE) is a hyperbolic integral-partial differential equation; only a few simple cases have an analytical solution, so it is necessary to use a numerical method to solve the PBE. At present, the usually used methods include the discrete method, moment method, and Monte Carlo method. Based on the calculation cost and accuracy, the discrete method is used to solve the population balance equation in this paper.

The basic idea of the discrete method is dividing the continuous diameter distribution of oil droplets into N discrete subintervals, and the size of all the droplets in the subinterval is equal to the node value g_i. Meanwhile, the distribution function of the oil droplet number can be approximately as follows,

$$n(V, t) = \sum_{i=1}^{N-1} N_i(t)\delta(V - g_k) \tag{9}$$

where N_i is the number of droplets in unit volume.

$$N_i(t) = \int_{V_i}^{V_{i+1}} n(V, t)dV \tag{10}$$

In most cases, the oil droplet size calculated by the coalescence and breakup model is not consistent with the node value of the subinterval, so it is necessary to allocate such droplets to the node value of the interval in a certain proportion, and the same time guarantee the conservation of the total oil droplet mass and the total number of oil droplet. Based on the above principles, Ramkishna proposes that the ratio of the newly generated droplets in the interval (g_i, g_{i+1}) to the node g_i and g_{i+1} is $\psi(V, g_i)$ and $\zeta(V, g_i)$, and satisfies that

$$\psi(V, g_i)g_i + \zeta(V, g_{i+1})g_{i+1} = V \tag{11}$$

$$\psi(V, g_i) + \zeta(V, g_{i+1}) = 1 \tag{12}$$

By substituting Equations (9), (11), and (12) into Equation (8) and through a series of algebraic transformations, the discrete droplet transport equation is given by Equation (13).

$$\frac{dN_i(t)}{dt} + \nabla \cdot [\vec{u}N_i(t)] = \underbrace{\sum_{\substack{j,k \\ g_{i-1} \leq (g_i + g_k) \leq g_{i+1}}}^{j \geq k} \left(1 - \frac{1}{2}\delta_{j,k}\right)x_{i,j,k}a(g_i, g_k)N_j(t)N_k(t)}_{\text{I}} \tag{13}$$

$$\underbrace{-N_i(t)\sum_{k=1}^{M} a(g_i, g_j)N_k(t)}_{\text{II}} + \underbrace{\sum_{k=1}^{M} \gamma_{i,k}b(g_k)N_k(t)}_{\text{III}}$$

$$\underbrace{-b(g_i)N_i(t)}_{\text{IV}}$$

where

$$x_{i,j,k} = \begin{cases} \dfrac{g_{i+1}-V}{g_{i+1}-g_i} & V_i \leq V \leq V_{i=1} \\[2mm] \dfrac{V-g_{i-1}}{g_i-g_{i-1}} & V_i \leq V \leq V_{i=1} \\[2mm] 0 & otherwise \end{cases} \tag{14}$$

After the droplet is broken up, the distribution proportion of each sub interval is expressed as

$$\gamma_{i,k} = \int_{g_i}^{g_{i+1}} \frac{g_{i+1}-V}{g_{i+1}-g_i}\beta(V,g_k)dV + \int_{g_i}^{g_{i+1}} \frac{V-g_{i-1}}{g_i-g_{i-1}}\beta(V,g_k)dV \tag{15}$$

Luo coalescence model [8] is adopted in this paper. The coalescence rate in model is defined as the frequency of new droplets generated after the collision of two droplets, and the volume of the two oil droplets is V_i and V_j. The coalescence rate is given by

$$a(V_i, V_j) = \omega_{ag}(V_i, V_j)P_{ag}(V_i, V_j) \tag{16}$$

where $\omega_{ag}(V_i, V_i)$ is the collision frequency; $P_{ag}(V_i, V_j)$ is the probability of coalescence. The collision frequency is obtained by

$$\omega_{ag}(V_i, V_j) = \frac{\pi}{4}(d_i + d_j)^2 n_i n_j \bar{u}_{ij} \tag{17}$$

where $\bar{u}_{ij}$ is the characteristic velocity of collision between two oil droplets with diameter d_i and d_j, and the number density is n_i and n_j, which is defined as

$$\bar{u}_{ij} = (\bar{u}_i^2 + \bar{u}_j^2)^{0.5} \tag{18}$$

$$\bar{u}_i = 1.43(\varepsilon d_j^2)^{1/3} \tag{19}$$

$$P_{ag}(V_i, V_i) = \exp\left\{-c_i \frac{\left[0.75(1+x_{ij}^2)(1+x_{ij}^3)\right]^{0.5}}{(\rho_2/\rho+0.5)^{0.5}(1+x_{ij})^3}We_{ij}^{1/2}\right\} \tag{20}$$

$$x_{ij} = \frac{d_i}{d_j} \tag{21}$$

$$We_{ij} = \frac{\rho_i d_i (\bar{u}_{ij})^2}{\sigma} \tag{22}$$

where c_i is constant, ρ_i is the density of each phase.

Laakkonen breakup model [22] is adopted in this paper, which considers that the breakup of oil droplet is affected by both surface tension and viscous force; the degree of influence depends on the magnitude of two forces; and the collision model can be expressed as

$$b(V_j) = C_2\varepsilon^{1/3}\mathrm{erfc}\left(\sqrt{C_3\frac{\sigma}{\rho_l\varepsilon^{2/3}d_j^{5/3}} + C_4\frac{\mu_l}{\sqrt{\rho_l\rho_g}\varepsilon^{2/3}d_j^{5/3}}}\right) \tag{23}$$

where $C_2 = 2.52$, $C_3 = 0.04$, $C_4 = 0.01$.

The probability distribution function of sub droplets is given by

$$\beta(V_i|V_j) = \frac{30}{V_j}\left(\frac{V_i}{V_j}\right)^2\left(1-\frac{V_i}{V_j}\right)^2 \tag{24}$$

The final breakup frequency is

$$\Omega_{br} = b(V_j)\beta\big(V_i\big|V_j\big) \tag{25}$$

2.4. Initial and Boundary Conditions

In the transient analysis, the boundary conditions include the inlet and outlet conditions of gas phase and lubricating oil, as well as the relevant wall conditions. The boundary conditions are treated as follows.

(1). Inlet boundary conditions: mass flow is adopted at the air inlet and oil inlet. At the air inlet, the initial air mass flow is 12 g/s. At the oil inlet, the initial oil mass flow is 20 g/s, and initial oil droplet diameter is 28 μm.

(2). Outlet boundary conditions: outlet boundary conditions adopt pressure-outlet and set the liquid reflux rate as 0 to ensure the stability of calculation.

(3). Wall boundary conditions: all walls adopt a no-slip velocity boundary condition. The shaft wall adopts a moving wall and the rotor speed is 16,000 r/min.

The calculation in this paper does not consider the effect of temperature. The oil droplets collision, coalescence, breakup, evaporation and deposition, nucleation and other dynamic events, as well as the geometry of the bearing chamber will affect the oil droplet size distribution in the bearing chamber. However, taking all the factors into account at the same time can make the analysis more complicated. Therefore, the calculation in this paper assumes that the temperature is constant and does not take into account the evaporation of oil droplets and the change of fluid properties.

2.5. Calculation Method

The coalescence and breakup model of the oil droplet in the bearing chamber is solved by using ANSYS Fluent 17.0. The 3D unsteady model is adopted, the first-order upwind scheme is used, and the phase Coupled SIMPLE algorithm modified by pressure and velocity is employed for solving the transient equation. The coalescence and breakup model are as follows. The coalescence model uses the Luo model; the breakup model uses Laakkonen breakup model, in which both surface tension and viscous force are considered; and the oil droplet breakup is combine constrained by two forces, which is closer to the reality. The computation is performed on a Microsoft Windows 10 enterprise 64-bit operation system, and the simulations are run using 40 cores with 192 GB RAM machine.

3. Experimental Verification and Comparison

In order to verify the correctness of the calculation method, the result in this paper is compared with the relevant test in literature [16]. In the literature, the Phase Doppler Particle Analyzer (PDPA) is used to sample 3000 oil droplet diameters in the aero-engine bearing chamber, and the histogram of oil diameter distribution is obtained. The distribution of droplet diameter is in a range of 14 to 120 μm. The operating condition in literature is air inflow m_g = 10 g/s and oil inflow V_L = 100 L/h. The oil droplet diameter distribution is calculated under these conditions. It should be noted here that in the calculation, the oil droplets larger than 120 μm in diameter are not considered, so only the oil droplet diameter distribution in a range of 10 to 120 μm is calculated. The calculation results are compared with the results in the literature in Figure 4. Through comparison, the calculation results are consistent with the experimental results in the literature, which verifies the rationality of the research method in this paper. The calculation results given in this paper can provide more accurate initial conditions for the analysis of the flow and heat transfer, and also can provide reference for the further test of oil mist concentration in the bearing chamber.

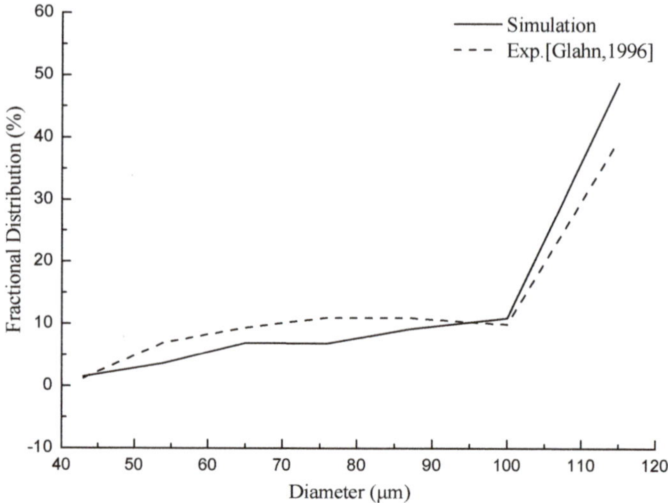

Figure 4. Comparison and verification of oil drop size distribution.

4. Results and Discussion

In this paper, the distribution of air phase velocity field and the change of oil droplet volume fraction with time in the bearing chamber at different times are analyzed when the air inflow is 12 g/s, oil inflow is 15 g/s and initial oil droplet diameter is 28 μm. Moreover, the influences of different initial oil droplet diameter, air inflow, and oil inflow on the oil droplet diameter distribution are further analyzed. The operating condition parameters are shown in Table 1.

Table 1. Operating condition.

Parameters	Value
Air mass inflow rate m_g (g/s)	8, 12, 14, 16
Oil mass inflow rate m_L (g/s)	15, 20, 25, 30
Initial diameter D_0 (μm)	20, 28, 40, 56

4.1. The Distribution of Air Phase Velocity Field and the Volume Fraction of Different Oil Droplet Diameter at Different Time

According to the mechanism of oil droplet coalescence and breakup, the turbulence of air phase flow field has a great influence on the coalescence and breakup of oil droplet. Therefore, the velocity of air phase in the bearing chamber is firstly analyzed. Figure 5 shows the distribution of air phase velocity in the bearing chamber at different time when the air inflow is 12 g/s, oil inflow is 15 g/s, and the initial oil droplet diameter is 28 μm. As can be seen from the figure, the velocity near the outlet of the bearing chamber is largest, and the other region is smaller. Because the seal air enters into the bearing chamber from the air inlet, it first moves along the axial direction. When the air is blocked by the wall surface, the velocity changes, which results in an uneven velocity distribution of the flow field. The air velocity in the middle region is significantly lower than that in the inner wall surface and the outer wall surface. Meanwhile, many vortices with different size existed in the bearing chamber. With the passage of time, the flow field structure tends to stable.

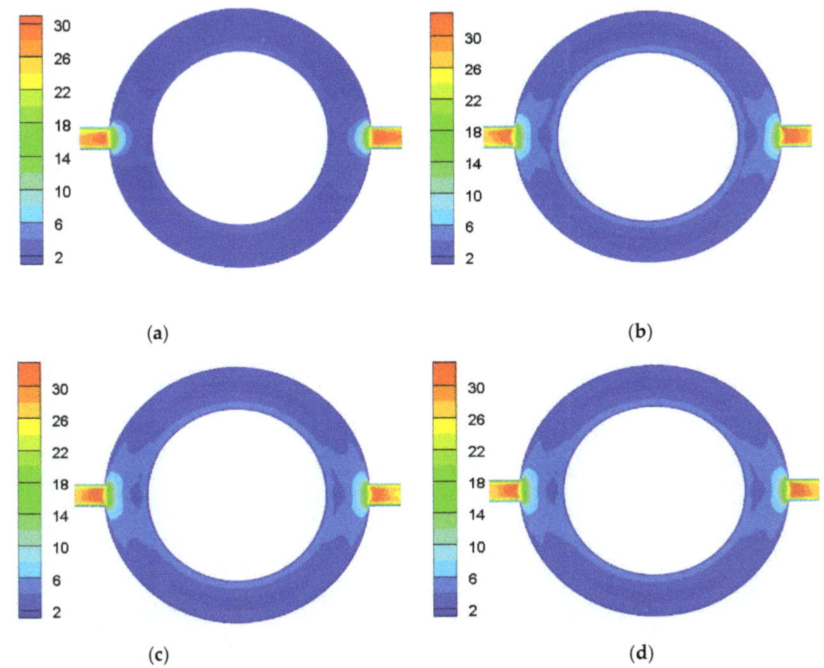

Figure 5. The velocity (m/s) distribution of air at different time when the air inflow is 12 g/s, oil inflow is 15 g/s, and the initial oil droplet diameter is 28 μm at $z = 0$ plane (the far left is vent; the far right is scavenge). (**a**) $t = 50$ ms; (**b**) $t = 100$ ms; (**c**) $t = 200$ ms; (**d**) $t = 400$ ms.

Figures 6 and 7 show the volume fraction of oil diameter in the range of 56 to 80 μm and 28 to 40 μm when the air inflow is 12 g/s, oil inflow is 15 g/s, and initial oil droplet diameter is 28 μm, respectively. In Figure 6, the volume fraction of the large diameter oil droplet on the outer wall surface of the bearing chamber is large, that near the inlet is relatively low, and gradually increases along the z axial direction. The volume fraction of large diameter oil droplet near the outer wall of the bearing chamber increases gradually with the passage of time. The reason is that the oil droplet flow with the air phase, and enter the bearing chamber from inlet. During the flow process, the oil droplets with smaller diameter constantly gather. When the air phase flow field reaches the wall that is opposite to the inlet, the flow direction of air phase changes and flows along the wall. During the movement of the oil droplet with the air phase flow field, the oil droplets constantly gather again. Therefore, the volume fraction of oil droplets at the external wall surface of the bearing chamber is large. It can also be seen from the figure that when the oil droplet with larger diameter runs to the entrance with larger velocity, the breakup will occur. The oil droplets are coalesced and broken in the bearing chamber, and finally reach a steady state. In Figure 7, the small diameter oil droplet is mainly distributed near the inlet, and this distribution gradually decreases along the axial direction. It can be seen that the size of the oil droplet in the bearing chamber is different because of the coalescence of oil droplets. The small diameter oil droplet is mainly concentrated near the inlet, while the large diameter oil droplet is mainly concentrated on the outer wall of the bearing chamber.

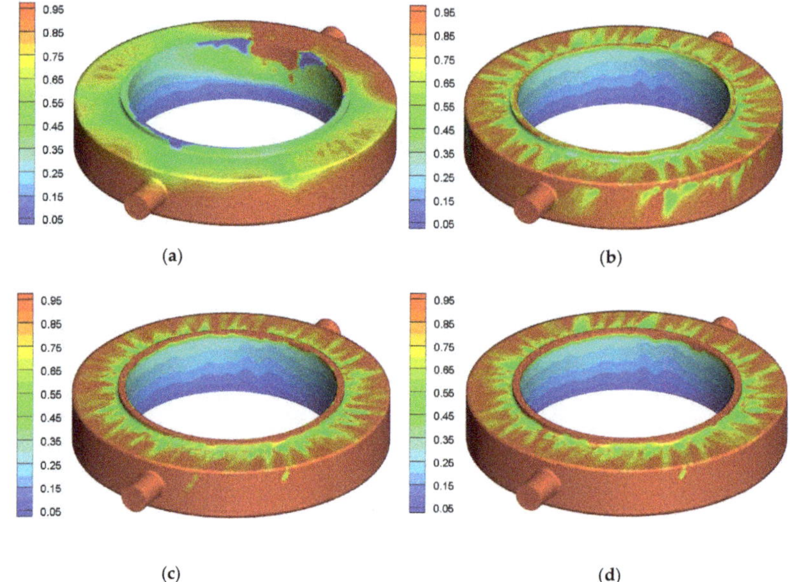

Figure 6. Oil droplet volume fraction (%) at different time with diameter in the range of 56 to 80 μm when the air inflow is 12 g/s, oil inflow is 15 g/s, and initial oil droplet diameter is 28 μm (the oil inlet is on the bottom, the vent is in front and the z axial direction is up). (**a**) t = 50 ms; (**b**) t = 100 ms; (**c**) t = 200 ms; (**d**) t = 400 ms.

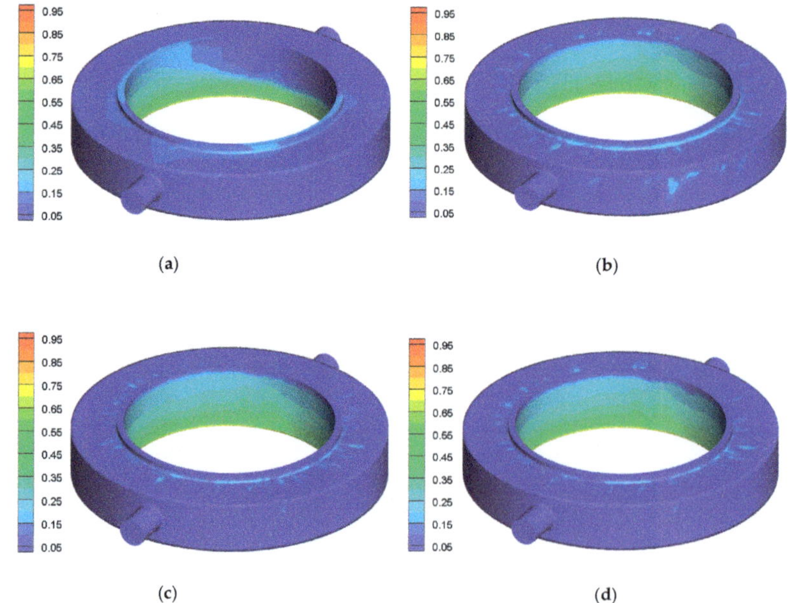

Figure 7. Oil droplet volume fraction (%) at different time with diameter in the range of 28 to 40 μm when the air inflow is 12 g/s, oil inflow is 15 g/s, and initial oil droplet diameter is 28 μm (the oil inlet is on the bottom, the vent is in front and the z axial direction is up). (**a**) t = 50 ms; (**b**) t = 100 ms; (**c**) t = 200 ms; (**d**) t = 400 ms.

4.2. The Change Rule of Droplet Size Distribution with Time

The change of oil droplet diameter distribution in the bearing chamber with time when the air inflow is 12 g/s, the oil inflow is 15 g/s, and the initial oil droplet diameter is 28 μm is shown in Figure 8. The oil droplets enter into the bearing chamber, due to the flow field development is not sufficient oil droplets coalesce firstly in the initial stage. Moreover, the volume fraction of large diameter droplets increases simultaneously. Then, the breakup of the oil droplet emerges prominently under the action of gas phase turbulence, and the volume fraction of big diameter droplets decreases. At last oil droplet diameter distribution tends to stable and the size of oil droplet mainly concentrates in the vicinity of 50 μm and 80 μm. From another angle, it shows that the oil droplet size distribution in the bearing chamber is not a simple R–R distribution

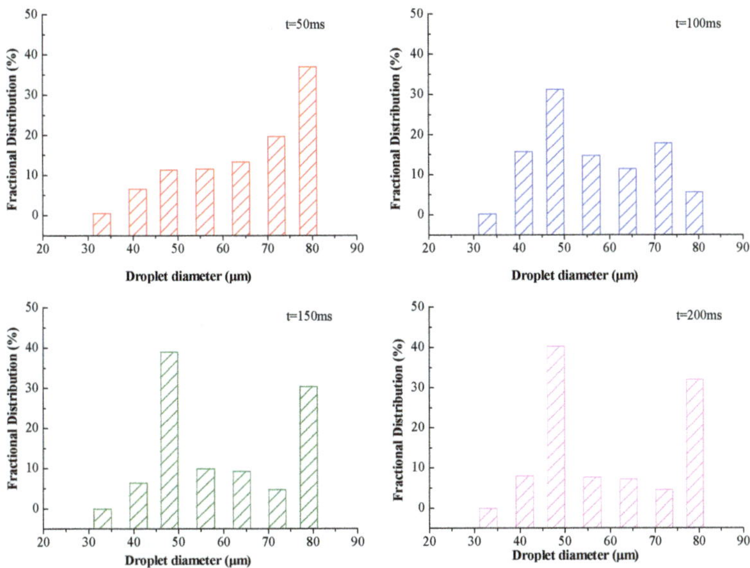

Figure 8. Droplet size distribution at different time when the air inflow is 12 g/s, the oil inflow is 15 g/s, and the initial oil droplet diameter is 28 μm.

4.3. The Influence of Difference Initial Diameter of Oil Droplet on the Size Distribution

Figure 9 shows the influence of different initial diameter on the oil droplet size distribution in the bearing chamber when the air inflow is 12 g/s and the oil inflow is 15 g/s at $t = 250$ ms. It can be seen that the influence of the initial diameter of the oil droplet on the volume fraction of different diameter oil droplet is obvious, and the peak diameter moves in the direction of large diameter. When the initial diameter increases to 56 μm, the peak value of oil droplet diameter increases to 70–80 μm. The size of the oil droplet diameter determines the value of the inertia force. The small diameter oil droplet is greatly affected by the air flow field, while the large diameter oil droplet is the opposite. The increasing of the inertia force of oil droplet weakens the moving velocity of the oil droplet, so the coalescence of oil droplet strengthens. Therefore, the proportion of large oil droplet is larger with the initial diameter of oil droplet increasing in the bearing chamber.

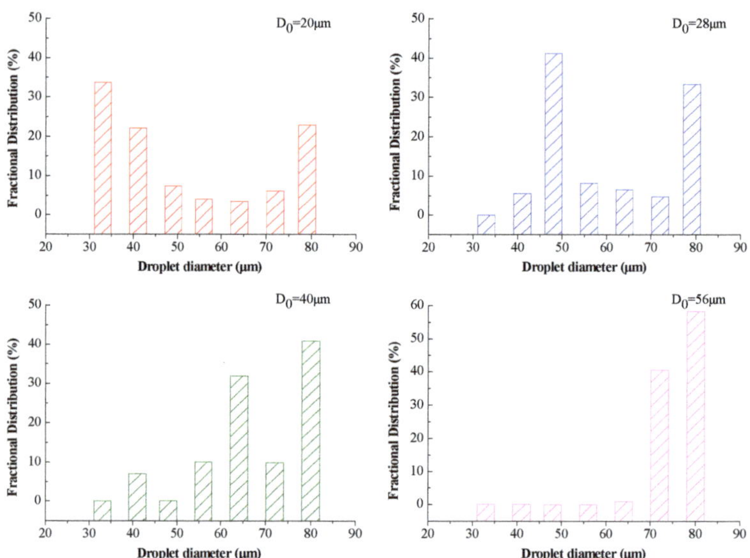

Figure 9. The influence of initial diameter on oil droplet size distribution when the air inflow is 12 g/s and the oil inflow is 15 g/s at $t = 250$ ms.

4.4. The Influence of Air Inflow on the Oil Droplet Size Distribution

The change of oil droplet size distribution under different air inflow is shown in Figure 10; when the oil inflow is 15 g/s, the initial oil droplet diameter is 28 μm at $t = 250$ ms. It can be seen that the peak value of oil droplet diameter is mainly concentrated around 50–80 μm. When the air flow is small, the proportion of oil droplet with 50 μm is larger than that with diameter of 80 μm. With air inflow increasing, the effect of turbulence intensity increases, and the strong disturbance in the flow field and the entrainment of vortices enhance the mutual collision between oil droplets. After the oil droplets collide with each other, the proportion of large diameter oil droplets increases in the oil droplet group due to the adhesion. Meanwhile, the increase of air inflow leads to the increase of oil droplet velocity, which promotes the coalescence between oil droplets. The smaller diameter oil droplet coalesces into larger diameter oil droplets. When the air inflow is 20 g/s, the peak value of oil droplet diameter distribution is mainly around 80 μm. This shows that the air inflow has a significant effect on the oil droplet diameter distribution.

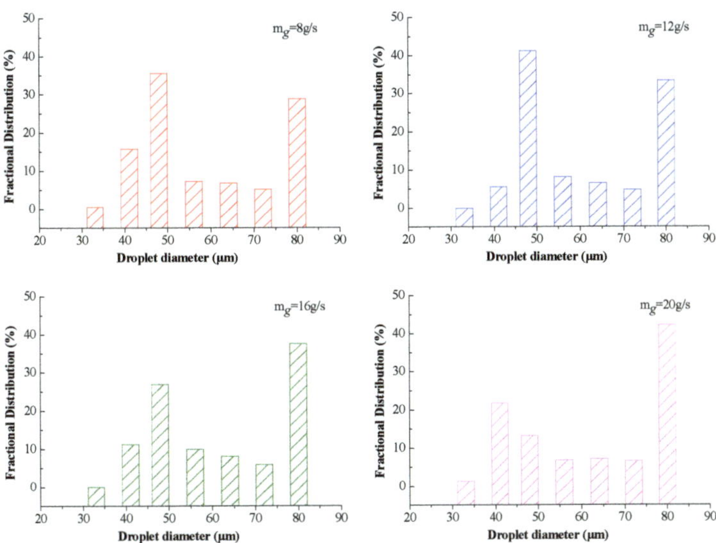

Figure 10. The influence of air inflow on the oil droplet size distribution when the oil inflow is 15 g/s, the initial oil droplet diameter is 28 µm at t = 250 ms.

4.5. The Influence of Oil Inflow on the Oil Droplet Size Distribution

The oil droplet diameter distribution under different oil inflow when the air inflow is 12 g/s, the initial oil droplet diameter is 28 µm at t = 250 ms is shown in Figure 11. It can be seen that the peak value of oil droplet diameter distribution is concentrated at 45–80 µm. The effect of oil inflow on the distribution is not obvious because the increasing oil inflow only increases the volume fraction of oil in the bearing chamber, but does not change the size distribution of the oil droplets.

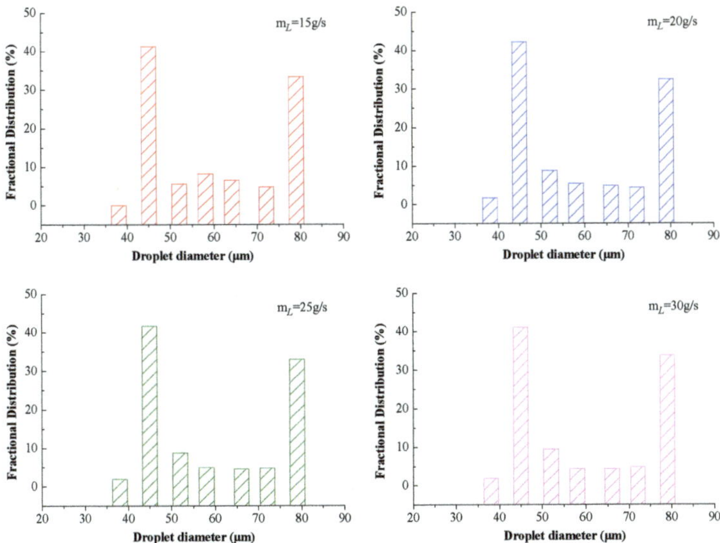

Figure 11. The influence of oil inflow on the oil droplet diameter distribution when the air inflow is 12 g/s, the initial oil droplet diameter is 28 µm at t = 250 ms.

5. Conclusions

(1). In the initial stage, the oil droplets were mainly coalesced, which makes the volume fraction of the oil droplets with large diameter increase. However, with time elapsing, the oil droplet with large diameter breaks up again, which reduces the volume fraction of the large oil droplet, and the oil droplet size distribution finally tends to stable.

(2). With initial oil droplet diameter and air flow increasing, the enhancement of coalescence and breakup between oil droplets has a significant effect on the oil droplet size distribution. However, the coalescence and breakup between oil droplets did not change significantly with the oil inflow increasing, and the effect on the oil droplet diameter distribution is not notable.

(3). Compared with the experimental results in the literature, the rationality of considering the influence of oil droplet coalescence and breakup on the oil droplet size distribution in the bearing chamber is verified.

The oil droplet size distribution directly affects the mixture state of oil and gas in the bearing chamber. Compared with the R–R distribution, the oil droplet size distribution obtained in this paper is more consistent with the real situation by considering the influence of coalescence and breakup of oil droplets. It can provide more accurate initial conditions for the future calculation of heat transfer coefficient internal and external the bearing chamber.

Author Contributions: Conceptualization, F.W.; Methodology, F.W.; Software, F.W.; Validation, F.W.; Formal Analysis, F.W. Investigation, F.W. and L.W. Writing—original draft preparation, F.W.; Writing—review and editing, F.W., L.W., G.C., and D.Z.; Supervision, G.C.; Project Administration, G.C.; Funding Acquisition, L.W. All authors have read and agreed to the published version of the manuscript.

Funding: This research was funded by the National Natural Science Foundation of China, grant number 51975475, and the Fundamental Research Funds for the Central Universities, grant number 31020200503002.

Conflicts of Interest: The authors declare no conflict of interest.

Nomenclature

ρ_l	oil density
ρ_g	gas density
μ_l	oil dynamic viscosity
σ_l	oil surface tension coefficient
μ_g	air dynamic viscosity
d_s	diameter of oil scavenge
d_v	diameter of air vent
h_b	height of the bearing chamber
w_b	width of the bearing chamber
r_s	rotor radius
α_q	volume fraction of the q phase
V_q	volume of the q phase
R_q	the Reynolds- stress tensors
τ_q	the viscous- stress tensors of the q phase
k	turbulent kinetic energy
ε	turbulent energy dissipation term
ν_t	turbulent viscosity
G_k	generating term of turbulent kinetic energy
n	number distribution function
a	coalescence rate function
b	breakup rate function
β	probability distribution density function
N_i	number of oil droplets in unit volume
M_q	the interfacial force term between the gas phase and liquid phase
U_r	the slip velocity
C_d	the drag coefficient

Abbreviations

PBM	Population Balance Model
PDPA	Phase Doppler Particle Analyzer
PBE	Population Balance Equation
QMOM	Quadrature Method of Moments

References

1. Chen, G.D.; Chen, B.; Wang, J. Research into configuration and flow of wall oil film in bearing chamber based on droplet size distribution. *Chin. J. Aeronaut.* **2011**, *24*, 355–362. [CrossRef]
2. Hulburt, H.; Katz, S. Some problems in particle technology, A statistical mechanical formulation. *Chem. Eng. Sci.* **1964**, *19*, 555–574. [CrossRef]
3. Liao, Y.X.; Lucas, D. A literature review of theoretical models for drop and bubble breakup in turbulent dispersions. *Chem. Eng. Sci.* **2009**, *64*, 3389–3406. [CrossRef]
4. Liao, Y.X.; Lucas, D. A literature review on mechanisms and models for the coalescence process of fluid particles. *Chem. Eng. Sci.* **2010**, *65*, 2851–2864. [CrossRef]
5. Kumar, S.; Ramkrishna, D. On the solution of population balance equations by discretization—I. A fixed pivot technique. *Chem. Eng. Sci.* **1996**, *51*, 1311–1332. [CrossRef]
6. Coulaloglou, C.A.; Tavlarides, L.L. Description of interaction processes in agitated liquid-liquid dispersions. *Chem. Eng. Sci.* **1977**, *32*, 1289–1297. [CrossRef]
7. Tsouris, C.; Tavlarides, L.L.; Bonnet, J.C. Application of the ultrasonic technique for real-time holdup monitoring for the control of extraction columns. *Chem. Eng. Sci.* **1990**, *45*, 3055–3062. [CrossRef]
8. Luo, H.; Svendsen, H.F. Theoretical model for drop and bubble breakup in turbulent dispersions. *AIChE J.* **1996**, *42*, 1225–1233. [CrossRef]
9. Li, D.; Gao, Z.; Buffo, A.; Podgorska, W.; Marchisio, D.L. Droplet breakage and coalescence in liquid–liquid dispersions: Comparison of different kernels with EQMOM and QMOM. *AIChE J.* **2017**, *63*, 2293–2311. [CrossRef]
10. Li, S.-W.; Jing, S.; Zhang, Q.; Wu, Q.-L. Advances in simulation of liquid-liquid two-phase flow in extraction columns with CFD-PBM. *Chin. J. Process Eng.* **2012**, *12*, 702–711. (In Chinese)
11. Xing, C.; Wang, T.; Wang, J. Experimental study and numerical simulation with a coupled CFD–PBM model of the effect of liquid viscosity in a bubble column. *Chem. Eng. Sci.* **2013**, *95*, 313–322. [CrossRef]
12. Wang, T.F.; Wang, J.F.; Jin, Y. A novel theoretical breakup kernel function for bubbles/droplets in a turbulent flow. *Chem. Eng. Sci.* **2003**, *58*, 4629–4637. [CrossRef]
13. Wang, T.F.; Wang, J.F.; Jin, Y. Population balance model for gas-liquid flows, Influence of bubble coalescence and breakup models. *Ind. Eng. Chem. Res.* **2005**, *44*, 7540–7549. [CrossRef]
14. Xu, Z.; Zhao, H.; Zhao, H. CFD-population balance Monte Carlo simulation and numerical optimization for flame synthesis of TiO2 nanoparticles. *Proc. Combust. Inst.* **2015**, *36*, 1099–1108. [CrossRef]
15. Chen, G.D.; Chen, B.; Wang, J. Oil droplet size distribution and deposition properties in bearing chamber. *J. Mech. Eng.* **2011**, *47*, 139–144. (In Chinese) [CrossRef]
16. Glahn, A.; Kurreck, M.; Willman, M.; Wittig, S. Feasibility study on oil droplet flow investigation inside aero-engine bearing chambers-pdpa techniques in combination with numerical approaches. *J. Eng. Gas Turbines Power* **1996**, *118*, 749–755. [CrossRef]
17. LIN, J.H. *Aero Engine Design Manual-Transmission and Lubrication Systems*; Aviation Industry Press: Beijing China, 2002. (In Chinese)
18. Drew, D. Mathematical modeling of two–phase flow. *Annu. Rev. Fluid Mech.* **1982**, *15*, 261–291. [CrossRef]
19. Gosman, A.; Lekakou, C.; Politis, S.; Issa, R.; Looney, M. Multidimensional modeling of turbulent two–phase flows in stirred vessels. *AIChE J.* **1992**, *38*, 1946–1956. [CrossRef]
20. Behzadi, A.; Issa, R.; Rusche, H. Modelling of dispersed bubble and droplet flow at high phase fractions. *Chem. Eng. Sci.* **2004**, *59*, 759–770. [CrossRef]

Appl. Sci. **2020**, *10*, 5648

21. Yeoh, G.H.; Cheung, C.P.; Tu, J. *Multiphase Flow Analysis Using Population Balance Modeling: Bubbles, Drops and Particles*; Butterworth-Heinemann, Elsevier: Oxford, UK, 2014; pp. 69–90.
22. Laakkonen, M.; Alopaeus, V.; Aittamaa, J. Validation of bubble breakage, coalescence and mass transfer models for gas-liquid dispersion in agitated vessel. *Chem. Eng. Sci.* **2006**, *61*, 218–228. [CrossRef]

© 2020 by the authors. Licensee MDPI, Basel, Switzerland. This article is an open access article distributed under the terms and conditions of the Creative Commons Attribution (CC BY) license (http://creativecommons.org/licenses/by/4.0/).

Article

A New Vector-Based Signal Processing Method of Four-Sensor Probe for Measuring Local Gas–Liquid Two-Phase Flow Parameters Together with Its Assessment against One Bubbly Flow

Xiaohang Qu [1], Qianjian Guo [2,*], Yi Zhang [1,3,*], Xiaoni Qi [1] and Lei Liu [3]

[1] Department of Energy and Power Engineering, Shandong University of Technology, Zibo 255000, Shandong, China; qxh@sdut.edu.cn (X.Q.); xiaoniqi@sdut.edu.cn (X.Q.)
[2] School of Mechanical Engineering, Shandong University of Technology, Zibo 255000, Shandong, China
[3] School of Energy and Power Engineering, Shandong University, Jinan 250061, Shandong, China; 201814133@mail.sdu.edu.cn
* Correspondence: guoqianjian@163.com (Q.G.); zhy20360@mail.sdu.edu.cn (Y.Z.)

Received: 15 July 2020; Accepted: 5 August 2020; Published: 7 August 2020

Abstract: A multiphase flow measurement technique plays a critical role in the studies of heat and mass transfer characteristics and mechanism of the gas–liquid two-phase, the practical measurement of the gas–liquid flow and the improvement of multiphase theoretical models. The four-sensor electrical probe as an emerging measurement method has been proved to be able to get the local flow parameters of multi-dimensional two-phase flow. However, few studies have been reported using the four-sensor probe to obtain the interface information (e.g., the interface direction and velocity). This paper presents a new signal processing method by which the interface direction and velocity can be obtained, besides void fraction, interfacial area concentration (IAC) and bubble chord length. The key solution is to employ the vector-based calculating method, which possesses the merits of simplicity and efficiency, to gain the interface velocity vector through legitimately assuming a direction of the interface velocity. A miniaturized four-sensor electrical probe was made and a gas–liquid two-phase flow experiment was performed to test the proposed signal process scheme. The two-phase flow was controlled to be in cap-bubble flow regime. To validate the availability and reliability of the proposed method, the local flow parameters obtained by the probe measurement were compared with the results from visual measurement technique in the same flow conditions. The comparison indicates that the above local flow parameters from four-sensor probe measurement are in good agreement with the visual measurement results, with maximum deviations of chord length of 8.7%, thereby proving the correctness of the proposed method.

Keywords: gas–liquid two-phase flow; local flow parameters measurement; four-sensor probe; vector-based signal processing; visual measurement experiment

---

1. Introduction

Gas–liquid two-phase flow is a common phenomenon occurring in petroleum, chemical, refrigeration and power generation industries [1–3]. Due to the unstable flow, heat and mass transfer process, the flow pattern and interface structure of two-phase flow are usually complex [4–6]. Therefore, the two-phase flow parameters are difficult to measure quickly and accurately [7]. However, the knowledge of local parameters of two-phase flow plays a critical role in the studies of the heat and mass transfer characteristics and mechanism of the gas–liquid two-phase, the development of two-phase modeling research, the optimization of two-phase flow patterns and the safety and stability of equipment operation [8–10]. The theoretical models, including the two-fluid model and interfacial area transportation model, rely heavily on the advanced measurement techniques to provide benchmark database and possibility of inspecting new phenomenon and physical laws. The flow parameters including phase distribution, void fraction, interfacial area concentration (IAC), bubble size, velocity, etc. of two-phase flow have great impacts on the heat and mass transfer characteristics, the reaction efficiency and operation safety of multiple chemical applications [11–15]. Hence, it is particularly necessary to develop and utilize accurate, fast and convenient methods to explore the two-phase flow details.

Currently, the measurement methods of local two-phase flow parameters can be divided into the following two categories: (1) the photography and image process techniques, having the advantage of lack influence on the flow, usually called visual measurement [16–18] and (2) the point by point measurement [3,19,20], in which optical or electrical signal that can be altered by involved phases. However, the first kind of methods is only applied to systems where the flow channel or vessel is transparent, unless the photography employs high energy ray to penetrate metal walls [21,22]. Besides, it is hard for visual measurements to distinguish bubbles at different depths when bubbles are overlapped in complex two-phase flow [23]. Nevertheless, this method has contributed greatly in the identification of flow regimes, track of simple bubbly flow and determination of bubble condensation rate in phase changing flows [23–28]. For the second category of methods, one of its important kinds is the wire-sensor mesh [29–33] where a matrix of measurement points is created at the cross points of two arrays of parallel mesh wires. Electrical signals from the wire-sensor mesh are collected and then analyzed by a program to get void fraction, bubble size, bubble velocity and so on across the whole flow field. Another kind is the sensor probe including single-sensor probe, double-sensor probe, four-sensor probe and other multiple-sensor probes [34–42]. Moving the probe across the two-phase flow field, local flow parameters at different positions can be measured. Although both the wire-sensor mesh and sensor probe interfere measurements affect the original two-phase flow field, this intrusion can be limited by a great extent with very thin wires or probes. Moreover, compared with the visual measurement, the point-to-point sensors are easier to be adopted online for practical opaque tube and vessel and more complex two-phase flow.

In the above-mentioned method, only two sensors apart from the probe shell are used in the double-sensor probe, and thus two series of signals can be collected. In the simple measurement theorem of double-sensor probe, it is assumed that the direction of interface and velocity of bubbles (or interface) are both in parallel to the connection between the two sensors' tips. As a result, the double-sensor probe can be applied to measure local flow void fraction, IAC and magnitude of velocity when the two-phase flow is very consistently stable bubbly flow, and the accuracy can be very high without too much care to the signal process and correction algorithm [34]. However, more assumptions and more complicated signal processing and correction algorithms are necessary if the flow becomes complicated or more parameters, e.g., bubble size and interface direction, have to be detected. Inevitably, more uncertainties would be produced in the application of single-sensor probe or double-sensor probe if employing these assumptions [43–46]. As uncertain assumptions must be incorporated, it is hard for the double-sensor probe to obtain real local flow parameters in multi-dimensional two-phase flow.

Meanwhile, the typical multiple-sensor probe, namely the four-sensor probe, as a promising alternative compared with single-sensor probe and double-sensor probe, is developed by a few researchers to omit the involvement of so many assumptions. According to measurement theorem of four-sensor probe, it is unnecessary to coordinate the connections of the sensors' tips to the direction of interface or direction of interface displacement velocity, while the void fraction, IAC and chord length can still be calculated. In addition, the interface direction can be obtained by the proper processing and trigonometric operation of the four series of signals from the four sensors [47–51]. However, the bubble velocity or interface displacement velocity is still unknown, except its component in the normal direction of interface [36]. It has been proved that this problem can be compensated by employing several four-sensor probes or multiple-sensor probe, five-sensor probe, six-sensor probe, etc. [52,53]. With a much simpler assumption of the two-phase flow, obtaining the full velocity vector by using merely one four-sensor probe can also be realized. For instance, assuming a spherical or symmetric bubble shape and a bubble velocity perpendicular to the symmetric plane has been demonstrated to be a concise way to get the full components of bubble velocity [54–56]. The above simplifications when calculating the total interface velocity vector only apply to special cases. Considering this fact, the local interface velocity directions can be practically provided by prior measurement, flow simulation and even other legitimate assumption in many two-phase flows. It is necessary and significant for obtaining interface details to propose a new method to get the total interface velocity vector based on the known interface velocity directions.

Besides the development of algorithm for four-sensor probe, the miniaturization of the probe and sensors are essential to guarantee the accuracy of measurement. According to the published literature [47,50], both the diameter of sensor wire and the total front area formed by all four sensors have great influence on the bubbles' behavior. Thus, it is recommended to use as small as possible sensors and probe of cross-sectional area. With the assumption of a much smaller probe than the size of bubble, the error produced during the application of the probe stems mainly from the bouncing away of bubble from the probe and the slipping away of interface through the gap between sensors. Their errors or uncertainties are acceptable [36,47,49]. Therefore, with the solution of the many issues of four-sensor probe, including its fabrication and further miniaturization, the correction for its disturbance to original flow and the improvement of signal process algorithm, it is promising to be widely applied in multiphase flow measurement inside chemical reactor, oil piping, power generation facility and heat exchanger. In addition, the method will be easy to be adopted for different combinations of fluid components. Even if the flow were experiencing phase changing, including boiling and condensation, the method would still be applicable with proper algorithm improvement and correction.

However, the bubbles in two-phase flow keep deforming and are hard to be taken as sphere or symmetric, resulting in few studies on the accurate, convenient and efficient measurement of local flow parameters having been published thus far. Therefore, a new signal processing method for four-sensor probe to get the bubble velocity vector was developed in this study and the vector-based calculation was used for the first time to deduct the local flow parameters. Besides, the interface direction obtained from the probe were for the first time validated against a visual experiment that was also performed. The void fraction, IAC, bubble velocity, bubble chord length and interface direction resulting from the probe measurement were compared with the visual measurement. The application perspectives of this method in the field of mass and heat transfer of gas–liquid two-phase flow is discussed. The proposed methods in the present paper are expected to be useful in the heat and mass transfer characteristics and mechanism studies of the gas–liquid two-phase and the direct measurement of two-phase flow. Meanwhile, it can able provide significant database for the improvement of two-phase models.

2. Measurement Principles of Four-Sensor Probe

2.1. Electrical Circuit of Four-Sensor Probe

As shown in Figure 1, an electrical circuit can be adopted in the four-sensor probe measurement. Since signal filtering and noise reduction can be easily realized by the appropriate signal pre-processing MATLAB codes (MATLAB 2017b, MathWorks, Inc., Natic, MA, USA), the circuit elements responsible for these functions were not necessary and thereby only four electrical resistances were employed in the circuit. The probe contains four sensors denoted s_0, s_1, s_2 and s_3, which connect to the negative electrode of the DC power supply through their respective resistance. The four sensors are covered by a rigid stainless-steel shell, which connects to the positive electrode of the DC power and functions as the common high voltage pole.

The shell of the probe made of metal was not insulated and thus always in contact with water. The four sensors are well insulated except at their very ends where the sensors can contact the water and thus get through to the positive pole of DC supplier. The signals from s_0 to s_3 are either high level or low level, depending on whether the sensor tips are submerged into water or exposed to the bubble air. A high voltage level indicates that a particular sensor is submerged in water and a low voltage level indicates that a particular sensor is exposed to air. Finally, the signals of high or low voltage from four sensors were transmitted to data collection system composed by a data acquisition unit (Art Technology, Beijing, China) and a PC. It is worth noting that the sensors have to be connected to the negative electrode of the DC power to avoid electrochemical corrosion and expand the probe's lifespan.

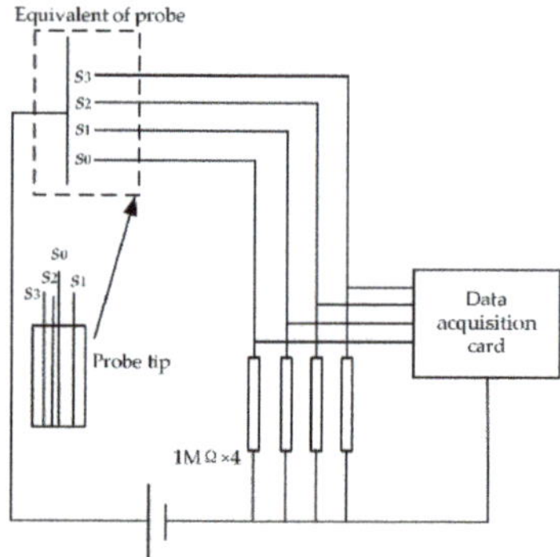

Figure 1. Schematic diagram for the measurement circuit of the four-sensor probe.

For the data acquisition system, it is suggested in the literature that a sample frequency higher than 10 kHz is required to guarantee resolution. Therefore, a sample frequency of 10 kHz for each of the four sensors was employed in this study.

2.2. Fabrication of the Probe

As the impact of probe on bubble only becomes negligible with small probe size, the accuracy of the measurement improves with the downsizing of the four-sensor probe, thus the size of the probe should be miniaturized as much as possible. The four-sensor probe used in the present study was hand-made using stainless steel tube with internal diameter 1.2 mm and outer diameter 1.5 mm as the positive electrode and four copper wires of diameter 0.1 mm as the negative electrodes. The specific dimensions of the probe and its picture in reality are shown in Figure 2.

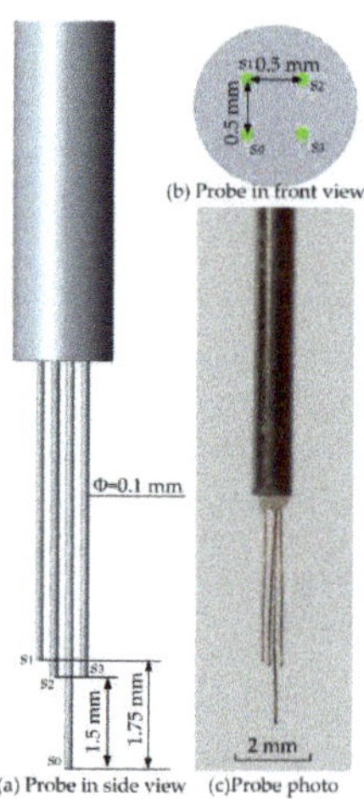

Figure 2. Dimensions of the four-sensor probe. (**a**) Probe in side view; (**b**) Probe in front view; (**c**) Probe photo.

The dimensions of the four-sensor probe and the relative position of sensor tips are shown in Figure 2a,b, respectively. As shown in Figure 2c, to provide support for the probe, the four sensors made of copper wire were fixed inside a stainless tube by epoxy resin, and the stainless tube with a length of 300 mm was then fixed by its end inside a short tube of internal dimeter of 5 mm with resin, before mounting it in a slide module, as shown in Section 3.1. Each copper wire was covered by electrically insulation material except the tip was polished by sandpaper for electrical contact with water. Choosing the wire tip as coordinate center and the axis of s_0 as the y-coordinate, the three vectors formed by the probe are S_1 (0, 1.75, 0.5), S_2 (0.5, 1.5, 0.5) and S_3 (0.5, 1.5, 0), which are used in Section 2.4.

2.3. Signal Pre-Processing

After data collection, the signals were processed by a MATLAB program consisting of mainly two functions, the pre-processing and deduction of local flow parameters. The signal pre-processing was designed to obtain the time instants when each of the four sensors penetrate or recede a bubble.

In an ideal case, the signals should be square wave (see Figure 3d) with the high-level representing sensor contact with water and the low-level representing passing-by of a particular bubble. However, mainly due to the delaying of data collection system and electromagnetic interference, the practically collected signal demonstrates noising inclining and fluctuating features, as shown in Figure 3a. Average filtering was applied to attenuate the noise signal firstly, with the result shown in Figure 3b. Then, a threshold voltage was chosen for signal binarization, as shown in Figure 3c. The signal inversion so that the high level corresponds to air is shown in Figure 3d. The threshold value adopted should be slightly higher than the noise voltage, to avoid the influence of noise while ensuring the accuracy of instants when sensor penetrates or recedes a bubble. After that, the resulted square wave was extracted to separate the rising edge corresponding to bubble approaching a sensor and falling edge corresponding to bubble leaving a senor, as shown in Figure 3e.

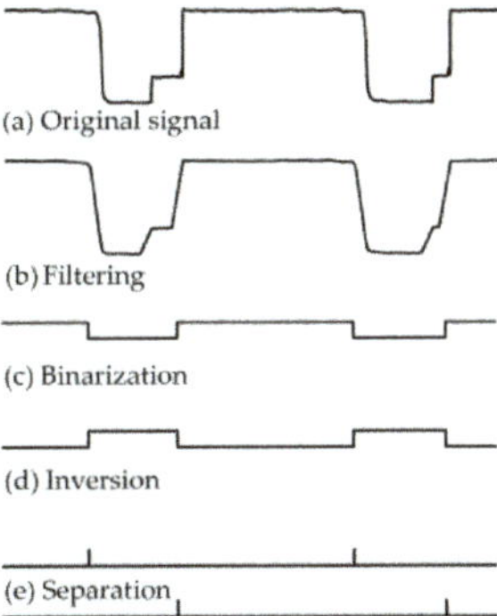

Figure 3. A piece of signal during pre-processing. (**a**) Original signal; (**b**) Signal filtering processing; (**c**) Binarization processing; (**d**) Inversion processing; (**e**) Separation processing.

The above procedures shown in Figure 3a–e are the same for s_0 to s_3, and thus eight sets of signal are obtained. Since one bubble has two interfaces passing-by one sensor, eight rising and falling instants are produced in total (four rising edges and four falling edges), as shown in Figure 4, where the eight instants produced by one bubble are noted by t_i' and t_i'' ($i = 0, 1, 2, 3$) with ′ noting rising instant and ″ noting falling instant. To make sure each of the eight-instant group belong to the same bubble (effective bubble), the collected signals are screened by cross-checking every bubble using the method described by Equation (1):

$$\max\left(\left|t_i'' - t_j''\right|, \left|t_i' - t_j'\right|\right) < \min\left(\left(t_i'' - t_i'\right), \left(t_j'' - t_j'\right)\right), i, j = 0, 1, 2, 3 \text{ and } i \neq j \tag{1}$$

which guarantees the time delay when bubble approaches (or leaves) one sensor and another must be smaller than the retaining duration of one bubble. This is in the first place required by the assumption that the four-sensor probe is much smaller than the measured bubbles.

The above procedures allow omitting two kinds of ineffective bubbles. The first kind is those bubbles for which eight-instant signal is incomplete, which means that only part of the four sensors penetrate them. This means bubbles that slide or bounce off the probe. Another kind of ineffective bubbles are those whose eight-edge group does not fit the conditions of Equation (1). These bubbles can be either very small or highly deformed and are also considered as ineffective bubble. Neglecting ineffective bubbles when calculating the local flow parameters might cause error because these bubbles still have contribution to IAC. To counteract the error, the ineffective bubbles were kept counted in the program and the IAC was corrected using the average contribution of effective bubbles. The details are discussed in Section 2.4.

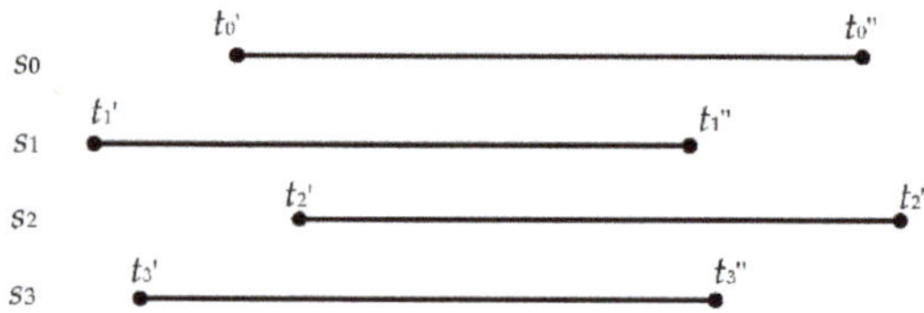

Figure 4. A group of signals produced by one bubble.

2.4. Deduction of Local Flow Parameters from Electrical Signals

Before calculation of the local flow parameters, the following assumptions should be made: (1) The probe containing the four sensors is very small in size in comparison to the bubble diameter, indicating that all measured bubbles are so-called large bubbles. Small bubbles can be detectable by one or a few sensors of the four-sensor probe, but they are neglected during the signal pre-processing based on the fact that a few rising or falling edges are missing, or no edge is missing but the eight-instant does not fit Equation (1). (2) The magnitudes of interface velocity and its direction remain unchanged when an interface passes by the sensors of probe. This is true as long as the probe size is small compared to the bubbles.

The void fraction equals the ratio of the duration of the sensor contact with air to the total measurement duration. It should be noted that as there are four sensors in the probe, thus the final void fraction can be determined by their average value, as shown by Equation (2).

$$\overline{V_f} = \frac{1}{4}\left(\sum_{i=0,1,2,3} \frac{\Sigma_{bubbles}\left(t_i'' - t_i'\right)}{t_{total}} \right) \tag{2}$$

where t_{total} denotes the total measurement time.

With the eight-instant (edge) of each effective bubble and known size and positions of the four sensors in the probe, the local flow parameters contributed by each effective bubble can be obtained. The measurement principles of other local flow parameters are schematically presented in Figure 5, which shows the relationships between different vectors.

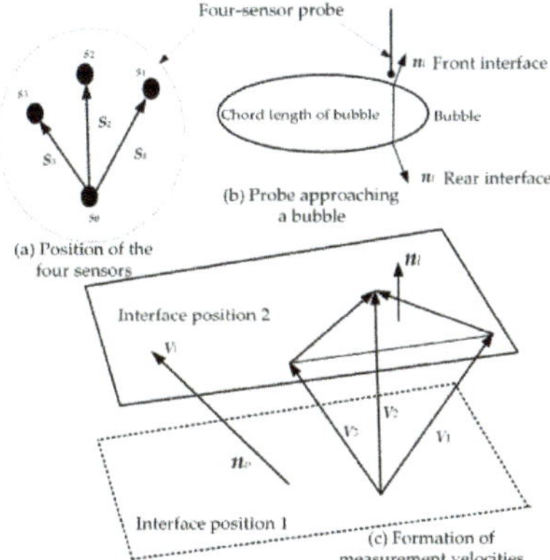

Figure 5. Measurement principle of four-sensor probe. (**a**) Position of the four sensors; (**b**) Probe approaching a bubble; (**c**) Formation of measurement velocities.

Local time-averaged IAC was predicted by Ishii [57] to be related to the interface velocity projected in the normal direction of particular plane:

$$\bar{a} = \frac{1}{t_{total}}\left(\sum_l \frac{1}{|V_l \cdot n_l|}\right) \tag{3}$$

where l, V_l and n_l denote the lth interface, the vector of interface velocity and the unitary vector normal to the lth interface, respectively, at a particular measurement point. The short line above a indicates a time-averaged value.

It is worth noting that, as mentioned in Section 2.3, since the neglected ineffective bubbles also contribute the IAC, Equation (3) must be corrected by multiplication factor, as shown in Equation (4):

$$\bar{a}' = \bar{a}f = \bar{a}\left(\frac{N_{\text{eff}} + N_{\text{ineff}}}{N_{\text{eff}}}\right) \tag{4}$$

where N_{eff} and N_{ineff} are the number of effective and ineffective bubbles, respectively. This correction to IAC has taken the contribution of each ineffective bubbles to be equal to the average contribution of effective bubbles. A bubble is treated as ineffective mainly due to its small size and continually deforming feature. However, because of its higher surface to volume ratio, its contribution to IAC is usually larger than a bubble of large size and regular shape. Therefore, the result from Equation (4) is still expected to be lower than the true IAC.

As the methods proposed by previous researchers who have employed trigonometric functions to determine the direction of interface are not conducive to comprehensible and fast calculation, a distinct and brief vector-based calculation is proposed and performed to obtain the interface direction here in this section. The cross-product of two vectors is also a vector and its direction is perpendicular to the plane formed by the original two vectors, hence the normal vector of one interface can be determined by three velocities measurable by the probe, as shown in Figure 5c and mathematically by Equation (5):

$$n_l = \frac{(V_2 - V_1) \times (V_2 - V_3)}{\left|(V_2 - V_1) \times (V_2 - V_3)\right|} \tag{5}$$

where V_1, V_2 and V_3 are the three measured velocities of a particular interface, respectively. They, respectively, have the same directions with the three position vectors formed by the probe, S_1, S_2 and

S_3, and their magnitudes are obtained by dividing each position vector by the time delay from the probe signals. Mathematically, it can be expressed by:

$$V_i = \frac{S_i}{(t_i - t_0)}, \quad i = 1, 2, 3 \tag{6}$$

where t_i ($i = 1, 2, 3$) and t_0 denote the time-instant of rising or falling edges of s_i and s_0, respectively. ($t_i - t_0$) are the time-instant delays between s_i and s_0. When t_i and t_0 correspond to rising edges, n_l and V_i are the unitary normal vector and measured velocities of the front interface of a bubble, respectively; and, when t_i and t_0 corresponds to falling edges, n_l and V_i are the unitary normal vector and measured velocities of the rear interface of a bubble, respectively.

The interfacial measurement theorem proposed by Shen [36] indicates the projections on the n_l of the interface displacement velocity vector V_l and the measured velocities V_i are the same, and the theorem can be expressed as:

$$V_l \cdot n_l = V_i \cdot n_l, \quad i = 1, 2, 3 \tag{7}$$

from which the velocity component into the normal direction of interface can be easily obtained. In contrast, obtaining the three components of V_l requires more assumptions or measuring parameters, for instance a further assumption of sphere-shape bubble [51] or symmetric bubble [56]. However, the application of these assumptions is only suitable when bubbles encountered in two-phase are not highly distorted or deforming.

A new method to get the whole components of V_l is proposed in this study by using the known or legitimately assumed velocity direction, i.e., the unit vector of V_l. Although the shape and size of bubbles keep changing in two-phase flow, the direction of bubbles velocity is usually constant and thus the direction of V_l often remains constant or constant on an averaged level at one fixed position of the flow field. As a result, by a known or assumed interface velocity direction n_v (Figure 5), the magnitude of V_l can be obtained as follows:

$$V_l \cdot n_v \cdot n_l = V_i \cdot n_l, \quad i = 1, 2, 3 \tag{8}$$

The above equation applies well when the two-phase flow is limited internal flow, where the direction of the interface velocity can be regarded as parallel to the channel axis. For arbitrary multi-dimensional two-phase flow, the equation and the resulted method still apply, as long as the local flow direction can be provided by legitimate assumption, flow simulation or prior measurement. For a fixed location, the local time-averaged magnitude of interface velocity can be obtained by averaging through numerous bubbles and denoted by V_l.

As shown by Equation (9), the local time-averaged chord length of a bubble at a fixed position is an averaged product of the V_l and the averaged time duration of the four signals at high voltage.

$$\overline{C} = \frac{1}{N} \sum_N \overline{V_l} \cdot \frac{\sum_{i=0}^{3} \left(t_i'' - t_i' \right)}{4} \tag{9}$$

where N denotes the number of bubbles during the measurement duration.

The direction of interface n_l can be shown more explicitly by the angle between n_l and the axis of the flow by Equation (10).

$$\theta = \cos^{-1} \frac{n_l \cdot n_{axis}}{|n_l \cdot n_{axis}|} \tag{10}$$

where n_{axis} is the direction of the channel axis.

2.5. Innovations of the Present Probe Algorithm

In the pioneering literature related to four-sensor probe, the interfacial velocity can only be obtained for the component that is vertical to the interface itself. If the full interfacial velocity vector is intended to be obtained, assumptions must be made. For instance, bubbles formed in the flow field in [51] were so small that they were assumed to be spherical. In [53], the authors made an assumption that the interface is very large and it moves only in its normal direction. In [56], the flattened bubbles were regarded as symmetrical to a center plane. Although these assumptions help ascertain the full components of the interfacial velocity, they only apply in special cases since the interface in practical gas–liquid flow is usually quite complicated.

Considering the fact that the explicit expression of bubble or interfacial velocity for four-sensor probe has not been completely developed, this paper proposes that, if the moving direction of interface can be ascertained prior to probe measurement, then the above assumptions will be unnecessary and the full interfacial velocity can be acquired. Fortunately, the fields of averaged moving directions of the interface, for so many quasi-steady two-phase flows, are actually the flow fields for the two-phase mixture and are easy to make certain through methods of prior measurement, simulation or even legitimate assumptions. These are the primary innovations of the present algorithm for four-sensor probe, as shown in Section 2.4.

Besides, the signal deduction process is all vector-based, which is different from the complex matrix, trigonometric functions and tensors. Although the basic rules are the same in essence and the results are expected to be same, no matter vector-based or trigonometric function-based algorithms are used, the method proposed in this paper has the merits of easy to read, clear and efficient to modify or improve.

3. Experimental Facilities

To validate the availability and correctness of the probe measurement methods, an experiment was performed to compare the local flow parameters obtained by the four-sensor probe and visual measurement, including void fraction, IAC, velocity, chord length and interface direction. An air–water two-phase flow with bubbles approximately the same size injected vertically up in a transparent tube was adopted. There is no doubt that such flow properties can be easily determined by senor probe or visual techniques with high accuracy. Therefore, it was chosen as a validation case for the proposed four-sensor probe measurement method.

Although the proposed probe and algorithm have not yet been validated, the existing fundamental measurement principles described in Section 2 can also apply to micro and conventional large systems. For fierce two-phase flow scenarios, corrosive fluid, high pressure system, high flow rate and cases of flow experiencing heat and mass transfer, the measurement methods still apply as long as the probe is fixed strongly and prevented from damage. The experiment in this study was only designed for the purpose of primary method validation, considering that it is convenient to be measured by visual technique and easy to be replicated.

3.1. Bubbly Flow in Vertical Pipe

An air–water two-phase bubbly flow system in a vertical tube was built to get the local flow parameter by both four-sensor probe and visual techniques. The test facility and flow rate were chosen for obtaining a simple and steady two-phase flow pattern. As shown in Figure 6, a transparent glass tube with the length of 0.5 m and internal diameter of 8 mm was used as the test section, and water was filled up to a height of 0.4 m during the experiment. Air produced by air compressor and regulated by surge tank and control valve was injected from the bottom of the tube. The air flow rate was maintained at 0.1 L/min during the test. As a result, it was found that a steady series of cap bubbles was produced inside the tube.

The top of the transparent tube was open to the atmosphere, and the four-sensor probe was vertically mounted in a one-dimensional sliding module with its tip pointing downside, so the probe could move horizontally to measure parameters across the tube diameter transversely. Only the 4.5 mm in the middle part of the 8-mm-diameter tube was accessible for measurement, resulting in a range from −2.5 to 2 mm with the interval of 0.5 mm.

The DC power supply of the probe was from a 9-V battery to avoid voltage fluctuation characterizing AC power supplier. The probe was connected to a data acquisition card which then transferred the collected data to a laptop. The data collection system was able to collect and transform the analog electrical signal to digital form at frequency of 10 kHz for each of the four sensors. For every transverse position of the probe in the tube, the data collecting persisted for 80 s and thus 800,000 data points were obtained for each sensor.

A high-speed camera was employed to record the images of bubble in a system without a probe. Images of 1262-pixel vertically and 710-pixel horizontally with a frequency of 50 fps were shot for 80 s. Through the image processing and analysis, the local flow parameters of the two-phase flow could be obtained, which is typical for the so-called visual measurement.

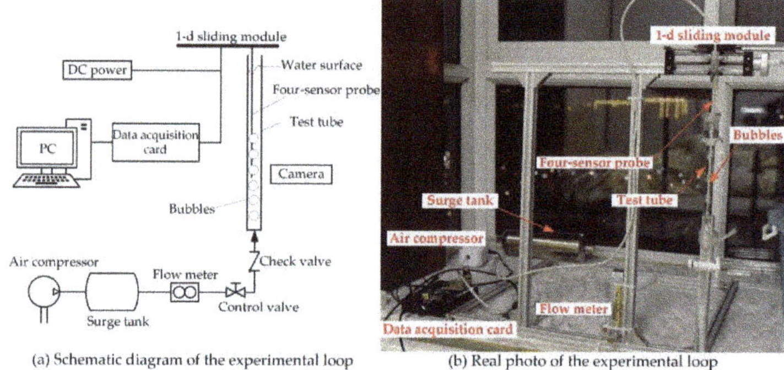

(a) Schematic diagram of the experimental loop   (b) Real photo of the experimental loop

Figure 6. Experiment loop of the present study. (**a**) Schematic diagram of the experimental loop; (**b**) Real photo of the experiment loop.

3.2. Visual Measurement Techniques

Besides the probe measurement described in Section 2, the flow parameters can also be obtained by the visual measurement which contains image recording and processing.

A series of continuous captured images is shown in Figure 7, with time intervals between each of 20 ms. It can be seen that the recorded bubbles are roughly in cap shape, and the space intervals between bubbles are roughly constant. When a bubble reached the location where the four-sensor probe was located, as shown by the red frame of Figure 7, the image was taken as one of the images constituting the visual measurement.

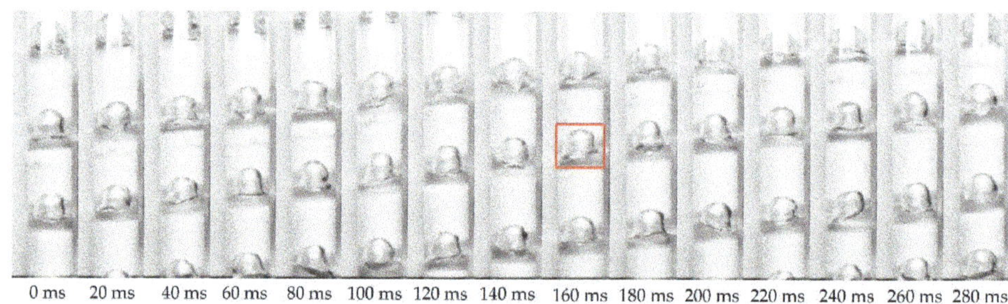

Figure 7. A series of images showing continuous captured bubbles.

For each of the chosen images, the subsequent bubble image processing is shown in Figures 8 and 9. For the first step of image processing (Figure 8a as an example), each bubble was cropped out, according to marked edges and converted into binary black and white image (matrix), as shown in Figure 8b–d, respectively. Five successive chosen bubbles are shown in Figure 9a, and it can be seen that, although the flow conditions remain unchanged during the test, the shape of cap bubbles change continuously. To obtain the interface direction and bubble chord length, the images of bubble were added up in MATLAB code and divided by the number of bubbles to get the averaged bubble shape. The resulted image (matrix) shown in Figure 9b stands for the probability of a pixel occupied by gas phase (void fraction). As the number of images n increases, the difference of the resulted images reduces to minor, and it was found that $n = 100$ is enough in this research. By binarization of the last image of Figure 9b taken with 0.5 as the threshold value, the averaged bubble and its edge are shown in Figure 9c,d, respectively. Based on Figure 9d, the time-averaged bubble interface direction θ and bubble chord length C at different radial location can be read by MATLAB.

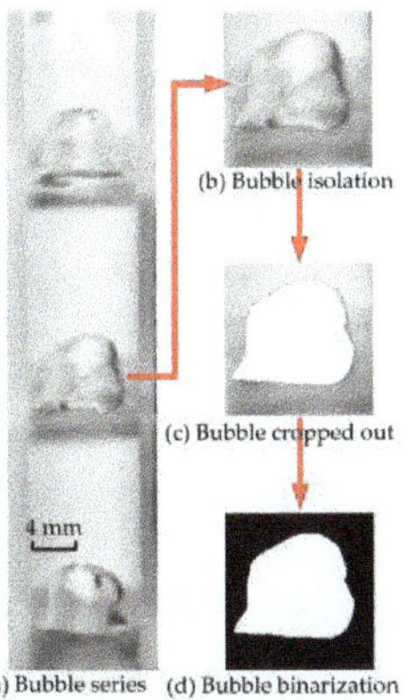

Figure 8. Cut out bubbles and its binarization. (**a**) Bubble series; (**b**) Bubble isolation; (**c**) Bubble cropped out; (**d**) Bubble binarization.

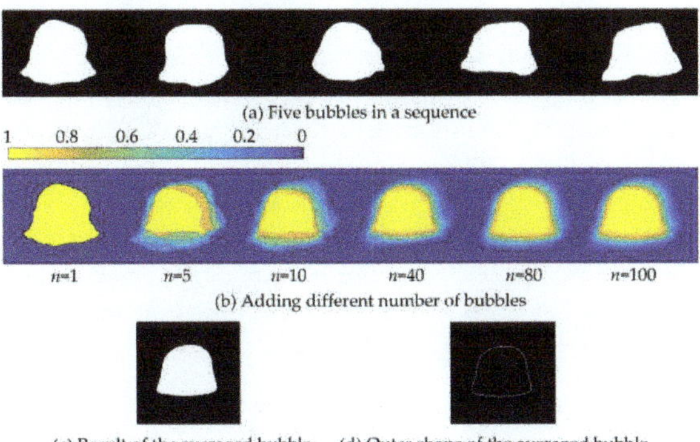

Figure 9. Averaging of bubbles. (**a**) Five bubbles in a sequence; (**b**) Adding different number of bubbles; (**c**) Result of the averaged bubble; (**d**) Outer shape of the averaged bubble.

Assuming the averaged bubble in Figure 9c is axisymmetric, the bubble volume can be obtained by integrating the bubble's cross area at each horizontal layer throughout the bottom to top of the bubble, as shown in Equation (11). The total area of the interface can be obtained by integrating the bubble's interface area at each horizontal layer throughout the bottom to top of the bubble, as shown in Equation (12):

$$\overline{B_V} = \int_{bottom}^{top} \pi r(y)^2 dy \tag{11}$$

$$\overline{B_a} = \int_{bottom}^{top} 2\pi r(y)dy \tag{12}$$

where r, a function of vertical location, is the radius of bubble at each horizontal layer.

The time-averaged void fraction across the whole diameter can be obtained through dividing B_V by the average interval volume between two successive bubbles, as shown in Equation (13):

$$\overline{\overline{V_f}} = \frac{\overline{B_V}}{A\overline{L}} \tag{13}$$

where the two short lines above V_f indicate time-averaged and space-averaged value for the same time, A is the cross area of the test tube and L is the average distance between two successive bubbles.

In a similar manner, the time-averaged IAC across the whole diameter can be obtained through dividing B_a by the average interval volume between two successive bubbles, as shown in Equation (14):

$$\overline{\overline{a}} = \frac{\overline{B_a}}{A\overline{L}} \tag{14}$$

The time-averaged bubble velocity equals the bubble production frequency multiplied by the averaged bubble distance and can be expressed by:

$$\overline{V_l} = \frac{N}{t_{total}}\overline{L} \tag{15}$$

where N and t_{total} are the total number of bubbles produced and the total measurement time, respectively.

4. Results and Discussion

It is worth noting that the four-sensor probe can only give out local flow parameters of two-phase flow. It is unable to discriminate different two-phase flow regimes of stratified, slug, wavy, etc. Extra correlation research is required to make connections between local flow parameters to global flow regimes. Measuring local flow parameters at multiple locations is the purpose of the four-sensor probe and the validations against visual measurement are hence extended below.

For cap bubble occupying almost the tube diameter from −4 to 4 mm, ten radial locations ranging from −2.5 to 2 mm are measured by the four-sensor probe. The numbers of total detectable bubbles and the numbers and ratios of the effective bubbles in the 80-s measurement duration at each radial location are given in Table 1. It can be seen that 350 bubbles on average are produced in 80 s, resulting in a bubble frequency of 4.375 per second. The uncertainty of the counted bubble number changes between −5 to 6, indicating that the steady and uniform features of the bubbly flow, which is necessary for this verification test.

Table 1. Number and ratio of the effective bubbles to total detected bubbles.

Location (mm)	−2.5	−2	−1.5	−1	−0.5	0	0.5	1	1.5	2
N_{total}	349	356	351	347	347	353	345	355	352	350
N_{eff}	107	202	252	291	317	313	309	302	271	140
N_{eff}/N_{total}	0.307	0.567	0.718	0.839	0.914	0.887	0.897	0.851	0.770	0.400

As expected, the effective number of bubbles recognized by the MATLAB code is lower than their total number. Meanwhile, both the effective number and the ratio of the effective number to the total number decrease, and this decrease becomes faster toward the ends of the test range. For the test location beyond 2 mm, the ratio can be well below 0.5. This is because, as the probe moves away from tube axis, the bubble interfaces become more inclined and it becomes easy for the bubble to slip away from the probe. It should be noted that the errors of the obtained local flow parameters increase with the decrease of ratio of number of effective bubbles to their total number.

The void fraction from four-sensor probe by Equation (2) and visual measurement by Equation (13) are compared in Figure 10. Since the void fraction from Equation (13) is an average void fraction across the whole tube diameter, it appears as a horizontal straight line in Figure 10. In view of the bubble shape shown in Figure 9c, the bubbly flow should produce a void fraction distribution with a peak value in the tube axis and decrease towards the tube wall. This trend was successfully reflected by the probe measurement, as shown by the black square dots in Figure 10. Its arithmetic mean value shown by the dashed line agrees well with the visual measurement, with an overestimation of 8.4%.

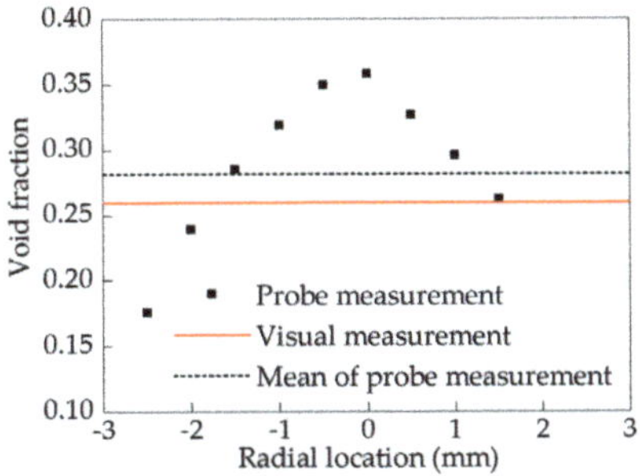

Figure 10. Void fraction from probe and visual measurement.

The IAC from four-sensor probe by Equation (4) and visual measurement by Equation (14) are compared in Figure 11. Since the IAC from Equation (14) is an average void fraction across the whole tube diameter, it appears as a horizontal straight line in Figure 11. In view of the bubble shape shown in Figure 9c, the normal vector of bubble interface changes from vertical near the tube axis to pointing right-upwards near the tube wall, thus the angle between bubble velocity which is parallel to the tube and the normal vector of interface increase and the denominator of Equation (3) decreases. As a result, a larger IAC should be observed near the tube wall, with a minimum value emerging at the tube axis. This trend was successfully captured by the probe measurement, as shown by the black square dots in Figure 11. Its arithmetic mean value shown by the dashed line agrees well with the visual measurement, with an underestimation of 1.7%.

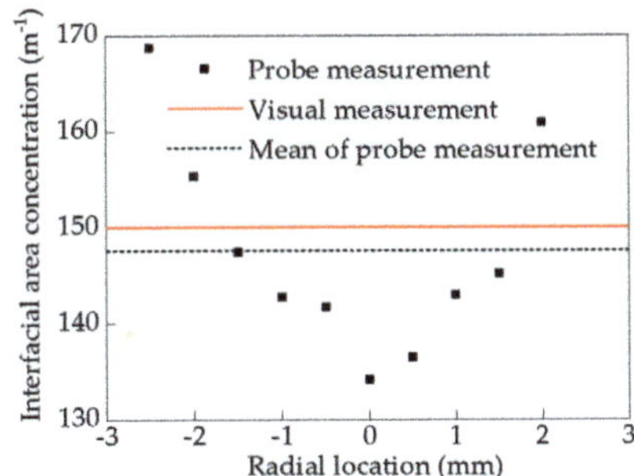

Figure 11. IAC from probe and visual measurement.

The bubble velocity from four-sensor probe by Equation (8) and visual measurement by Equation (15) are compared in Figure 12. Because bubble moves as an entity, it is worth noting that the bubble velocity measured at different radial locations should remain constant when the probe moves from tube axis to tube wall. However, the black square dots in Figure 12 representing probe measured velocity demonstrate fluctuation feature. This is mainly caused by error and it can be seen the error increases towards the tube wall. Nevertheless, its arithmetic mean value shown by the dashed line agrees well with the visual measurement, with an overestimation of 9.3%.

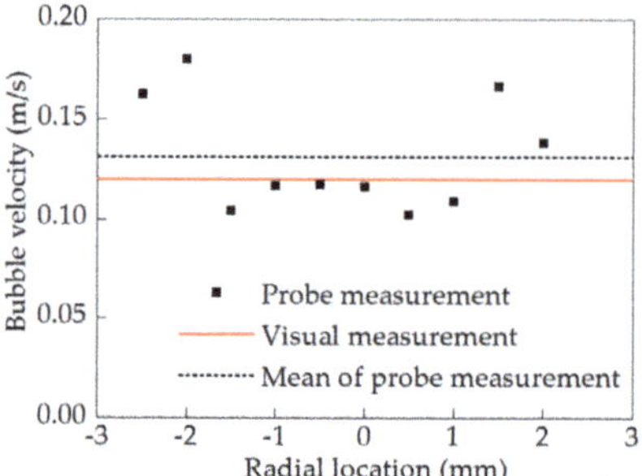

Figure 12. Bubble velocity from probe and visual measurement.

The chord length from four-sensor probe by Equation (9) and visual measurement are compared in Figure 13. Both the probe measurement and visual measurement show the same trend with the chord length emerging in the middle with maximum value and decreasing towards both sides. The largest deviation between them is 8.7% at −2.5 mm, and the deviation decreases towards the pipe axis.

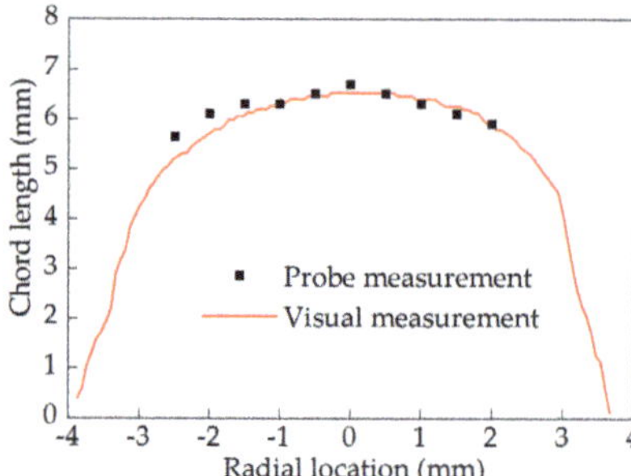

Figure 13. Bubble chord length from probe and visual measurement.

Although the interface direction is required to calculate the interface displacement velocity in the interfacial measurement theorem expressed by Equation (7), it is rarely verified against other measurement techniques in the accessible literature. Instead of directly using of the interface normal vector, the interface direction can be represented more conveniently by the angle between the interface normal vector and the tube axis. The angle from four-sensor probe by Equation (10) and visual measurement are compared, as shown in Figure 14. Figure 14a,b shows the comparisons at front interface and rear interface, respectively. As can be seen, both measurement methods show very close results with the deviation between them growing distinct towards the tube wall, and the largest errors are 22.4% and 3.1% for the front and rear interfaces, respectively.

The quantitative comparisons of the measured parameters, the deviation of the probe measurement from the visual measurement are shown in Tables 2 and 3. From the above, it is demonstrated that the measurement of bubbly flow by the four-sensor probe can give agreeable flow parameters with visual measurement techniques in aspects of void fraction, IAC, bubble velocity, bubble chord length and interface direction. Thereby, it also proves the correctness of the proposed method.

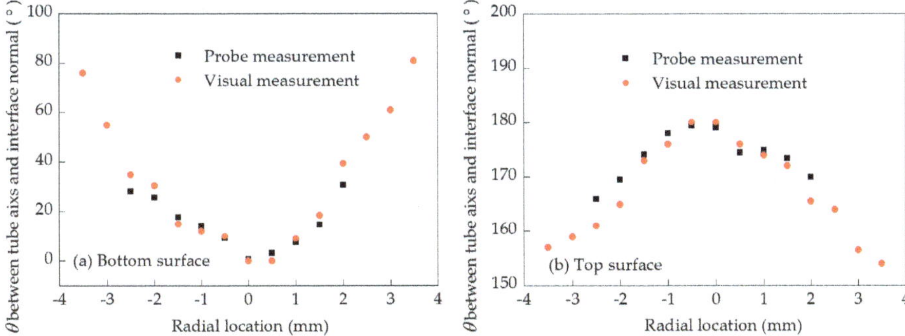

Figure 14. Angle between interface normal and tube axis by probe and visual measurement. (**a**) Bottom surface direction of the bubble; (**b**) Top surface direction of the bubble.

Table 2. Mean probe and visual measurement of void fraction, IAC and bubble velocity.

Items	Void Fraction	IAC (m^{-1})	Bubble Velocity (m/s)
Probe measurement	0.28	147.54	0.13
Visual measurement	0.26	150.15	0.12
Deviation (%)	8.42	−1.74	9.25

It is common knowledge that the bubble condensation is a kind of typical enhanced heat and mass transfer method encountered in numerous industrial processes such as steam–air mixture injected into subcooled zone, subcooled flow boiling and direct contact condensation [27]. Further knowledge of the bubble interface phenomena and the accurate measurement of the local flow parameters at the multi-scale interfaces, such as the void fraction, IAC, bubble velocity, bubble chord length and interface direction, are important parameters to study the process and mechanism of heat and mass transfer in gas–liquid two-phase flow. As previously reported in the literature [27,45,46,58], during the gas–liquid two-phase heat and mass transfer, the interface structure, void fraction, IAC, bubble equivalent diameter, their velocity, etc. show nonlinear variations, resulting in that they are difficult to be measured quickly and accurately. Although the double-sensor probe with multiple assumptions can also be employed to measure these parameters, it is hard to obtain realistic local data in multi-dimensional two-phase flow. As a result, the method proposed in the present study is expected to promote the solution of this problem.

Table 3. Probe and visual measurement of bubble chord length and interface direction.

Radial Location (mm)	Chord Length (mm)			Angle at Front Interface (°)			Angle at Rear Interface (°)		
	Probe	Visual	Deviation (%)	Probe	Visual	Deviation (%)	Probe	Visual	Deviation (%)
−2.5	5.65	5.20	8.65	28.1	35.0	−19.74	166.0	161.0	3.10
−2.0	6.10	5.73	6.46	25.7	30.5	−15.83	169.5	165.0	2.73
−1.5	6.30	6.09	3.45	17.7	15.0	18.27	174.1	173.0	0.62
−1.0	6.30	6.40	−1.56	14.0	12.0	16.40	177.9	176.0	1.10
−0.5	6.50	6.47	0.46	9.3	10.0	−6.94	179.4	180.0	−0.34
0.0	6.70	6.53	2.60	0.5	0.0	-	179.1	180.0	−0.52
0.5	6.50	6.53	−0.46	3.2	0.0	-	174.5	176.0	−0.87
1.0	6.30	6.40	−1.56	7.5	9.0	−17.05	174.9	174.0	0.49
1.5	6.10	6.27	−2.71	14.6	18.5	−21.21	173.3	172.0	0.77
2.0	5.90	5.87	0.51	30.7	39.5	−22.39	170.0	165.5	2.70

The interface velocity and its direction obtained using the brief vector-based calculation method proposed in this study are beneficial to further master the heat and mass transfer process at the interface from a macro viewpoint. Meanwhile, the IAC is an important parameter to characterizing the interface transfer phenomenon (heat and mass transfer). Compared with the double-sensor probe, the four-sensor probe is used to measure the IAC to be able to solve the detection error caused by the escape and retreat of bubbles. On this basis, by combining the proposed vector-based signal processing method, a variety of assumptions can be avoided, which will be of great significance to further study the internal relationship between IAC and bubble size, bubble deformation and condensation effect, as well as to modify the calculation model of IAC. This shows that the proposed method can be widely applied in the future research of heat and mass transfer characteristics.

Combined with several groups of four-sensor probes and the new vector-based signal processing method, the variation laws of the void fraction, the bubble size and number of bubbles in the condensation area can be measured accurately, quickly and conveniently, and thereby the phase distribution characteristics in the condensation area can be known. Especially during the air–steam mixture bubble condensation, the phase distribution characteristics, the interface velocity and direction can reveal the mechanism affecting the interface heat and mass transfer. Furthermore, some small bubbles with high non-condensable gas concentration appear at the later stage of the air–steam mixture bubbles condensation [58]. According to the previous study [59], when the content of non-condensable air is constant, the volume change rate of bubble increases with the decrease of bubble diameter. This means that the process of small bubble condensation is of great significance for studying the enhancement of condensation heat transfer of air–steam mixture bubbles. Therefore, the proposed method in this study will also be promising for investigating the influence mechanism of non-condensable air at the interface on the tiny steam bubble condensation enhancement.

5. Conclusions

The void fraction, IAC, bubble size, velocity, etc. of two-phase flow have great influences on the heat and mass transfer characteristics, the reaction efficiency and the operation safety of multiple chemical applications. To obtain detailed knowledge of two-phase flow, a miniaturized four-sensor probe was made firstly in this study. Then, a new method based on vector calculation to get the interface direction and a new method to get the bubble velocity magnitude were proposed, by which other local flow parameters can be obtained from the signals produced by four-sensor probe. For the purpose of verification of the probe made and the elaboration algorithm developed, an experimental facility was also built and air–water two-phase bubbly flow was tested. Besides the measurement by probe, the flow parameters were also obtained by video recording and image process techniques. For the first time, the direction of bubble interface was compared with visual image of bubble.

The calculation of interface velocity vector was shown to be realizable using legitimate interface velocity direction. The comparison between the results obtained from probe measurement and visual measurement indicates their good agreement. However, since the number of ineffective bubbles increases moving towards tube wall, the measurement error by probe also increases while the probe moves away from tube axis. The averaged values across the tube are over predicted by 8.4%, under predicted by 1.7% and over predicted by 9.3% with probe measurement for void fraction, IAC and bubble velocity, respectively, in comparison to the visual measurement counterpart. The chord length and angle between tube axis and normal to the interface show great consistency with visual results, with maximum deviations of 8.7% and 22.4%, respectively.

After more validation tests against multiple kinds of two-phase flow parameters, it can be concluded that the methods proposed in this study are promising in the characteristics and mechanism studies of the gas–liquid heat and mass transfer, the direct measurement of gas–liquid two-phase flow and providing significant database for the improvement of two-phase flow models.

Author Contributions: Conceptualization, X.Q. (Xiaohang Qu) and Y.Z.; methodology, X.Q. (Xiaohang Qu), Q.G. and Y.Z.; software, X.Q. (Xiaohang Qu) and X.Q. (Xiaoni Qi); validation, X.Q. (Xiaohang Qu), Q.G. and Y.Z.; formal analysis, X.Q. (Xiaoni Qi), Q.G., L.L. and Y.Z.; investigation, X.Q. (Xiaohang Qu) and Y.Z.; data curation, X.Q. (Xiaohang Qu), Y.Z. and L.L.; writing—original draft preparation, X.Q. (Xiaohang Qu), and Y.Z.; writing—review and editing, Q.G. and X.Q. (Xiaoni Qi); visualization, X.Q. (Xiaohang Qu), Y.Z., X.Q. (Xiaoni Qi) and L.L.; and supervision, Q.G. and Y.Z. All authors have read and agreed to the published version of the manuscript.

Funding: This research was funded by the National Natural Science Foundation of China (Grant Nos. 51806128 and 51879154) and the Natural Science Foundation of Shandong Province (Grant Nos. ZR2019BEE008 and ZR2018LE011).

Acknowledgments: The authors gratefully acknowledge financial support from the National Natural Science Foundation of China (Grant Nos. 51806128 and 51879154) and the Natural Science Foundation of Shandong Province (Grant Nos. ZR2019BEE008 and ZR2018LE011).

Conflicts of Interest: The authors declared that there is no conflict of interest.

Nomenclature

The following abbreviations are used in this manuscript:

A	tube cross section area
a	interfacial area concentration
B_a	bubble surface area
B_v	bubble volume
C	bubble chord length
f	ratio of effective to total number
L	distance between bubbles
N	number
$\mathbf{n}_l$	interface normal direction
$\mathbf{n}_v$	interface velocity direction
r	radius
S_1, S_2, S_3	vectors formed by sensors
s_0, s_1, s_2, s_3	sensors
t	time
V_1, V_2, V_3	measurable velocity vectors
V_f	void fraction
$\mathbf{V}_l$	interface velocity vector
V_l	interface velocity magnitude
x, y	coordinates
Greek letters	
θ	angle
Subscripts and Superscripts	
axis	tube axis
eff	effective value
i, j	denoting 0,1,2,3
l	interface
v	interface velocity
-	averaged value
′	when bubble approaches probe
″	when bubble leaves probe

References

1. Zhang, J.-G.; Xu, K.-J.; Dong, S.; Liu, Z.; Hou, Q.-L.; Fang, Z.-Y. Mathematical model of time difference for Coriolis flow sensor output signals under gas-liquid two-phase flow. *Measurement* **2017**, *95*, 345–354. [CrossRef]

2. Yeo, D.Y.; No, H.C. A zero-dimensional dryout heat flux model based on mechanistic interfacial friction models for two-phase flow regimes with channel flow in a packed bed. *Int. J. Heat Mass Transf.* **2019**, *141*, 554–568. [CrossRef]

3. Wang, C.; Zhao, N.; Chen, C.; Sun, H. A method for direct thickness measurement of wavy liquid film in gas-liquid two-phase annular flow using conductance probes. *Flow Meas. Instrum.* **2018**, *62*, 66–75. [CrossRef]

4. Wang, C.; Zhao, N.; Fang, L.; Zhang, T.; Feng, Y. Void fraction measurement using NIR technology for horizontal wet-gas annular flow. *Exp. Therm. Fluid Sci.* **2016**, *76*, 98–108. [CrossRef]

5. Wang, Y.; Liu, Z.; Chang, Y.; Zhao, X.; Guo, L. Experimental study of gas-liquid two-phase wavy stratified flow in horizontal pipe at high pressure. *Int. J. Heat Mass Transf.* **2019**, *143*, 118537. [CrossRef]

6. Zhu, H.; Duan, J.; Cui, H.; Liu, Q.; Yu, X. Experimental research of reciprocating oscillatory gas-liquid two-phase flow. *Int. J. Heat Mass Transf.* **2019**, *140*, 931–939. [CrossRef]

7. Liang, F.; Sun, Y.; Fang, Z.; Sun, S. Application of multi-slot sampling method for gas-liquid two-phase flow rate measurement. *Exp. Therm. Fluid Sci.* **2016**, *79*, 213–221. [CrossRef]

8. Cui, F.; Hong, F.; Li, C.; Cheng, P. Two-phase flow instability in distributed jet array impingement boiling on pin-fin structured surface and its affecting factors. *Int. J. Heat Mass Transf.* **2019**, *143*, 118495. [CrossRef]

9. Zhai, L.-S.; Bian, P.; Gao, Z.-K.; Jin, N.-D. The measurement of local flow parameters for gas–liquid two-phase bubbly flows using a dual-sensor probe array. *Chem. Eng. Sci.* **2016**, *144*, 346–363. [CrossRef]

10. Zhou, Y.; Chang, H.; Lv, Y. Gas-liquid two-phase flow in a horizontal channel under nonlinear oscillation: Flow regime, frictional pressure drop and void fraction. *Exp. Therm. Fluid Sci.* **2019**, *109*, 109852. [CrossRef]

11. Pietrzak, M.; Witczak, S. Flow patterns and void fractions of phases during gas–liquid two-phase and gas–liquid–liquid three-phase flow in U-bends. *Int. J. Heat Fluid Flow* **2013**, *44*, 700–710. [CrossRef]

12. Guan, X.; Gao, Y.; Tian, Z.; Wang, L.; Cheng, Y.; Li, X. Hydrodynamics in bubble columns with pin-fin tube internals. *Chem. Eng. Res. Des.* **2015**, *102*, 196–206. [CrossRef]

13. Besagni, G.; Inzoli, F. Influence of internals on counter-current bubble column hydrodynamics: Holdup, flow regime transition and local flow properties. *Chem. Eng. Sci.* **2016**, *145*, 162–180. [CrossRef]

14. Wu, H.; Buschle, B.; Yang, Y.; Tan, C.; Dong, F.; Jia, J.; Lucquiaud, M. Liquid distribution and hold-up measurement in counter current flow packed column by electrical capacitance tomography. *Chem. Eng. J.* **2018**, *353*, 519–532. [CrossRef]

15. He, H.; Pan, L.-M.; Ren, Q.-Y.; Ye, T.-P.; Zhang, D.-F. Local liquid film behavior of annular two-phase flow on rod-bundle geometry-I. Experimental phenomenon and analysis. *Int. J. Heat Mass Transf.* **2019**, *141*, 58–70. [CrossRef]

16. Zheng, S.Q.; Yao, Y.; Guo, F.F.; Bi, R.S.; Li, J.Y. Local bubble size distribution, gas–liquid interfacial areas and gas holdups in an up-flow ejector. *Chem. Eng. Sci.* **2010**, *65*, 5264–5271. [CrossRef]

17. Ahmadi, R.; Ueno, T.; Okawa, T. Visualization study on the mechanisms of net vapor generation in water subcooled flow boiling under moderate pressure conditions. *Int. J. Heat Mass Transf.* **2014**, *70*, 137–151. [CrossRef]

18. Fu, Y.; Liu, Y. Development of a robust image processing technique for bubbly flow measurement in a narrow rectangular channel. *Int. J. Multiph. Flow* **2016**, *84*, 217–228. [CrossRef]

19. Zheng, G.-B.; Jin, N.-D.; Jia, X.-H.; Lv, P.-J.; Liu, X.-B. Gas–liquid two phase flow measurement method based on combination instrument of turbine flowmeter and conductance sensor. *Int. J. Multiph. Flow* **2008**, *34*, 1031–1047. [CrossRef]

20. He, D.; Chen, S.; Bai, B. Void fraction measurement of stratified gas-liquid flow based on multi-wire capacitance probe. *Exp. Therm. Fluid Sci.* **2019**, *102*, 61–73. [CrossRef]

21. Zeitoun, O.; Shoukri, M. Axial void fraction profile in low pressure subcooled flow boiling. *Int. J. Heat Mass Transf.* **1996**, *40*, 869–879. [CrossRef]

22. Hampel, U.; Speck, M.; Koch, D.; Menz, H.J.; Mayer, H.G.; Fietz, J.; Hoppe, D.; Schleicher, E.; Zippe, C.; Prasser, H.M. Experimental ultra fast X-ray computed tomography with a linearly scanned electron beam source. *Flow Meas. Instrum.* **2005**, *16*, 65–72. [CrossRef]

23. Qu, X.H.; Tian, M.C. Acoustic and visual study on condensation of steam–air mixture jet plume in subcooled water. *Chem. Eng. Sci.* **2016**, *144*, 216–223. [CrossRef]

24. Petrovic de With, A.; Calay, R.K.; de With, G. Three-dimensional condensation regime diagram for direct contact condensation of steam injected into water. *Int. J. Heat Mass Transf.* **2007**, *50*, 1762–1770. [CrossRef]

25. Kim, S.J.; Park, G.C. Interfacial heat transfer of condensing bubble in subcooled boiling flow at low pressure. *Int. J. Heat Mass Transf.* **2011**, *54*, 2962–2974. [CrossRef]

26. Tang, J.G.; Yan, C.Q.; Sun, L.C. A study visualizing the collapse of vapor bubbles in a subcooled pool. *Int. J. Heat Mass Transf.* **2015**, *88*, 597–608. [CrossRef]

27. Qu, X.H.; Tian, M.C.; Zhang, G.M.; Leng, X.L. Experimental and numerical investigations on the air–steam mixture bubble condensation characteristics in stagnant cool water. *Nucl. Eng. Des.* **2015**, *285*, 188–196. [CrossRef]

28. Tang, J.G.; Yan, C.Q.; Sun, L.C.; Li, Y.; Wang, K.Y. Effect of liquid subcooling on acoustic characteristics during the condensation process of vapor bubbles in a subcooled pool. *Nucl. Eng. Des.* **2015**, *293*, 492–502. [CrossRef]

29. Prasser, H.M.; Scholz, D.; Zippe, C. Bubble size measurement using wire-mesh sensors. *Flow Meas. Instrum.* **2001**, *12*, 299–312. [CrossRef]

30. Prasser, H.M.; Beyer, M.; Carl, H.; Gregor, S.; Lucas, D.; Pietruske, H.; Schütz, P.; Weiss, F.-P. Evolution of the structure of a gas–liquid two-phase flow in a large vertical pipe. *Nucl. Eng. Des.* **2007**, *237*, 1848–1861. [CrossRef]

31. Frank, T.; Zwart, P.J.; Krepper, E.; Prasser, H.M.; Lucas, D. Validation of CFD models for mono- and polydisperse air–water two-phase flows in pipes. *Nucl. Eng. Des.* **2008**, *238*, 647–659. [CrossRef]

32. Lucas, D.; Beyer, M.; Szalinski, L. Experimental database on steam–water flow with phase transfer in a vertical pipe. *Nucl. Eng. Des.* **2013**, *265*, 1113–1123. [CrossRef]

33. Hernández Cely, M.M.; Baptistella, V.E.C.; Rodriguez, O.M.H. Study and characterization of gas-liquid slug flow in an annular duct, using high speed video camera, wire-mesh sensor and PIV. *Exp. Therm. Fluid Sci.* **2018**, *98*, 563–575. [CrossRef]

34. Revankar, S.T.; Ishii, M. Local interfacial area measurement in bubble flow. *Int. J. Heat Mass Transf.* **1992**, *35*, 913–925. [CrossRef]

35. Revankar, S.T.; Ishii, M. Theory and measurement of local interfacial area using a four sensor probe in two-phase flow. *Int. J. Heat Mass Transf.* **1993**, *36*, 2997–3007. [CrossRef]

36. Shen, X.; Saito, Y.; Mishima, K.; Nakamura, H. Methodological improvement of an intrusive four-sensor probe for the multi-dimensional two-phase flow measurement. *Int. J. Multiph. Flow* **2005**, *31*, 593–617. [CrossRef]

37. Xue, J.; Al-Dahhan, M.; Dudukovic, M.P.; Mudde, R.F. Four-point optical probe for measurement of bubble dynamics: Validation of the technique. *Flow Meas. Instrum.* **2008**, *19*, 293–300. [CrossRef]

38. Mizushima, Y.; Sakamoto, A.; Saito, T. Measurement technique of bubble velocity and diameter in a bubble column via single-tip optical-fiber probing with judgment of the pierced position and angle. *Chem. Eng. Sci.* **2013**, *100*, 98–104. [CrossRef]

39. Besagni, G.; Brazzale, P.; Fiocca, A.; Inzoli, F. Estimation of bubble size distributions and shapes in two-phase bubble column using image analysis and optical probes. *Flow Meas. Instrum.* **2016**, *52*, 190–207. [CrossRef]

40. Liu, F.; Wang, X.L.; Ye, S.; Hang, T.Y.; Zheng, J.J.; Wang, H.D.; Chen, X.Y. A model for measuring the velocity vector of bubbles and the pierced position vector in breaking waves using four-tip optical fiber probe, Part I: Computational method. *Ocean Eng.* **2018**, *161*, 384–392. [CrossRef]

41. Han, Y.F.; Jin, N.D.; Yin, Z.Q.; Ren, Y.Y.; Gu, Y. Measurement of oil bubble size distribution in oil-in-water emulsions using a distributed dual-sensor probe array. *Exp. Therm. Fluid Sci.* **2017**, *86*, 204–223. [CrossRef]

42. Gurau, B.; Vassallo, P.; Keller, K. Measurement of gas and liquid velocities in an air–water two-phase flow using cross-correlation of signals from a double sensor hot-film probe. *Exp. Therm. Fluid Sci.* **2004**, *28*, 495–504. [CrossRef]

43. Wu, Q.; Ishii, M. Sensitivity study on double-sensor conductivity probe for the measurement of interfacial area concentration in bubbly flow. *Int. J. Multiph. Flow* **1999**, *25*, 155–173. [CrossRef]

44. Wu, Q.; Welter, K.; McCreary, D.; Reyes, J.N. Theoretical studies on the design criteria of double-sensor probe for the measurement of bubble velocity. *Flow Meas. Instrum.* **2001**, *12*, 43–51. [CrossRef]

45. Wang, D.; Liu, Y.; Talley, J.D. Numerical evaluation of the uncertainty of double-sensor conductivity probe for bubbly flow measurement. *Int. J. Multiph. Flow* **2018**, *107*, 51–66. [CrossRef]

46. Cartellier, A. Simultaneous void fraction measurement, bubble velocity, and size estimate using a single optical probe in gas–liquid two-phase flows. *Rev.Sci. Instrum.* **1992**, *63*, 5442–5453. [CrossRef]

47. Kim, S.; Fu, X.Y.; Wang, X.; Ishii, M. Development of the miniaturized four-sensor conductivity probe and the signal processing scheme. *Int. J. Heat Mass Transf.* **2000**, *43*, 4101–4118. [CrossRef]

48. Manera, A.; Ozar, B.; Paranjape, S.; Ishii, M.; Prasser, H.M. Comparison between wire-mesh sensors and conductive needle-probes for measurements of two-phase flow parameters. *Nucl. Eng. Des.* **2009**, *239*, 1718–1724. [CrossRef]

49. Lucas, G.; Zhao, X.; Pradhan, S. Optimisation of four-sensor probes for measuring bubble velocity components in bubbly air–water and oil–water flows. *Flow Meas. Instrum.* **2011**, *22*, 50–63. [CrossRef]

50. Pradhan, S.R.; Mishra, R.; Ubbi, K.; Asim, T. Optimal design of a four-sensor probe system to measure the flow properties of the dispersed phase in bubbly air–water multiphase flows. *Sens. Actuators A Phys.* **2013**, *201*, 10–22. [CrossRef]

51. Shen, X.; Nakamura, H. Spherical-bubble-based four-sensor probe signal processing algorithm for two-phase flow measurement. *Int. J. Multiph. Flow* **2014**, *60*, 11–29. [CrossRef]

52. Shen, X.; Mishima, K.; Nakamura, H. A method for measuring local instantaneous interfacial velocity vector in multi-dimensional two-phase flow. *Int. J. Multiph. Flow* **2008**, *34*, 502–509. [CrossRef]

53. Shen, X.; Nakamura, H. Local interfacial velocity measurement method using a four-sensor probe. *Int. J. Heat Mass Transf.* **2013**, *67*, 843–852. [CrossRef]

54. Mishra, R.; Lucas, G.P.; Kieckhoefer, H. A model for obtaining the velocity vectors of spherical droplets in multiphase flows from measurements using an orthogonal four-sensor probe. *Meas. Sci. Technol.* **2002**, *13*, 1488–1498. [CrossRef]

55. Luther, S.; Rensen, J.; Guet, S. Bubble aspect ratio and velocity measurement using a four-point fiber-optical probe. *Exp. Fluids* **2004**, *36*, 326–333.

56. Tian, D.; Yan, C.; Sun, L. Model of bubble velocity vector measurement in upward and downward bubbly two-phase flows using a four-sensor optical probe. *Prog. Nucl. Energy* **2015**, *78*, 110–120. [CrossRef]

57. Ishii, M.; Hibiki, T. *Thermo-Fluid Dynamics of Two-Phase Flow*; Springer: New York, NY, USA, 2010.

58. Qu, X.; Revankar, S.T.; Tian, M. Numerical simulation of bubble formation and condensation of steam air mixture injected in subcooled pool. *Nucl. Eng. Des.* **2017**, *320*, 123–132. [CrossRef]

59. Tang, L. *Numerical Analysis on Condensing Process of the Steam and the Air-Steam Mixture Bubble*; Shandong University: Jinan, China, 2018.

 © 2020 by the authors. Licensee MDPI, Basel, Switzerland. This article is an open access article distributed under the terms and conditions of the Creative Commons Attribution (CC BY) license (http://creativecommons.org/licenses/by/4.0/).

Article

CFD Analysis of Subcooled Flow Boiling in 4 × 4 Rod Bundle

Ye-Bon Seo, SalaiSargunan S Paramanantham, Jin-Yeong Bak, Byongjo Yun and Warn-Gyu Park *

School of Mechanical Engineering, Pusan National University, Busan 46241, Korea;
seoyebon308@pusan.ac.kr (Y.-B.S.); salaisargunansp@pusan.ac.kr (S.S.P.); jinyb0309@pusan.ac.kr (J.-Y.B.);
bjyun@pusan.ac.kr (B.Y.)
* Correspondence: wgpark@pusan.ac.kr

Received: 10 June 2020; Accepted: 24 June 2020; Published: 30 June 2020

Abstract: Rod bundle flow is an important research field related to reactor cooling in nuclear power plants. Owing to the rapid development of computerized performance assessments, interest in coolant flow analysis using computational fluid dynamics has garnered research interest. Rod bundle flow research data compared with experimental results under various conditions are thus needed. To address this, a boiling model verification study was conducted with reference to experiments. This study adopts the Reynolds-averaged Navier–Stokes equation, a practical analysis method compared to direct numerical simulation and large eddy simulation, including turbulence modeling, to predict the flow of coolant inside a rod bundle. This study also investigates void behavior in low-pressure subcooled flow boiling using a Eulerian approach (two-fluid model). The rod bundle has a length of 0.59 m and a hydraulic diameter of approximately 14.01 mm. At the cross-section at a height of 0.58 m, near the exit, numerical results were compared with the experimental values of the volume fraction of vapor and interfacial area concentration. The simulation results showed good agreement with the experimental data for six different initial conditions with constant density.

Keywords: subcooled flow boiling; two-fluid model; vapor volume fraction; rod bundle; STAR-CCM+

---

1. Introduction

Computational fluid dynamics (CFD) is being increasingly used in the analysis of three-dimensional two-phase flow for the safe operation of a nuclear power plant. In recent years, subcooled boiling in the rod bundle is one of the important phenomena observed in small break loss of coolant accidents (SBLOCA) for pressurized water reactor (PWR) and is also expected in the loss of heat sink accident of nuclear spent fuel pool (SFP) as a safety issue. Thus, experiments and measurements on multiphase turbulent flow in pipes or rod bundles have been conducted, and numerous fluid properties have been studied. In recent years, various simulations have been conducted in various papers [1–5]. Numerical analysis of subcooled two-phase flow at high pressure has confirmed that the Eulerian multiphase model including a Reynolds stress turbulence model is suitable for predicting the actual phenomenon [6–10]. Single-phase convective heat transfer and vapor–water two-phase boiling heat transfer experiments have been conducted in a vertical seven-rod bundle at a low flow rate near atmospheric pressure. In this case, bubble generation is dominantly determined by the enhancement factor of forced convection by the inlet mass flux and the suppression factor of nucleate boiling. Thus, it was necessary to react sensitively to the fluid velocity, and in the next step, excessive bubble prediction is prevented through a nuclear boiling suppression factor. By taking in account the enhancement factor of forced convection and the suppression factor of nucleate boiling, a new correlation in terms of boiling number, Reynolds number, and Martinelli number was found to be effective in predicting two-phase-flow boiling heat transfer coefficients [11]. Analytical studies using various turbulence models for rod bundles in various shapes have also been conducted. A 3 × 3 rod bundle flow was analyzed using the k–ε model, shear stress transport k-ω model, and Reynolds stress model; and the pressure rate, flow velocity, and temperature were predicted [12]. For supercritical pressure conditions, the importance of the turbulence mixing factor was studied thermal hydraulics characteristics of 2×2 rod bundle by numerical analyzing [13]. In total of 2000 mm length of the 4×4 rod bundle with a mixing grid installed at 2/3 height, turbulence was processed in a Reynolds-averaged Navier–Stokes method using a nonlinear and viscous model to simulate CFD [14]. A study on the effect of spacer grid distance considered the influence of spacer grid blockage ratio and spacer grid spacing on heat transfer in a vertical rod bundle [15]. Reynolds number of the working fluid in this paper is 24,800 to 30,100 to investigate the flow effect on heat transfer. As such, this study was carried out with reference to other studies on the two-phase model, turbulence model, and various rod bundle shapes and grids.

The present study addresses the local flow inside the rod bundle. As the inside of a bundle of nuclear fuel rods has a very narrow flow path, a two-phase flow study using three-dimensional heat transfer analysis and a boiling model is required for reactor safety design and analysis. For two-phase flow analysis, the bubble size is a very important factor for accurately predicting bubble behavior in the flow path. In line with this focus, this study analyzes the results of the numerical analysis of the volume fraction of vapor (VF) and interfacial area concentration (IAC) compared with experimental values. VF is the volume of vapor in volume rate and IAC is defined by VF and interaction length scale. It focuses on the analysis of coolant boiling in a narrow flow path inside a nuclear fuel reactor 4×4 rod bundle. This phenomenon is solved by the Eulerian multiphase model as a segregated flow. The important elements to accurately predict the behavior of vapor are the interfacial momentum transfer force and wall boiling model of heat and mass transfer force. The usability of the detailed model used in this study has been evaluated by researchers as follows.

Tomiyama (1998) studied simple but reliable correlations for a drag coefficient by comparing experimental data. The correlations were obtained for the drag coefficient of single bubbles under a wide range of fluid properties, bubble diameters, and acceleration of gravity. These were developed according to the balance of forces acting on the bubbles in stagnant liquid, and the available empirical correlations of the terminal rising velocities of single bubbles [16].

It has been shown that in many flows, the inertial force can simply be added to the generalization of the lift force introduced by Auton (1987) [17]. The lift force of Auton (1988) consists of the rotational force and inertial force or additional mass [18].

Owing to the neglect of the wall area force, the predicted pore profile peaked instead of approaching zero at the wall. Therefore, Achard & Cartellier (1985) concluded that a different kind of transverse force is needed to break the bubbles away from the wall [19]. Later, Antal (1991) developed a two-fluid model of multidimensional laminar bubble two-phase flow to analyze vertical pipe flow. In this study, the wall lift force for the dispersed phase (i.e., bubbles) and the wall lubrication force about the wall force were calculated. The results showed good agreement with the data when the considered model was used [20].

Lahey et al. (1993) used $C_{TD} = 1$ to predict the lateral phase distribution of a three-dimensional water/air bubble upflow or downflow in a pipe using the k–ε turbulence model and an algebraic stress law. It can be confirmed [21]. These results are very interesting because a properly formulated multidimensional two-fluid model has the inherent ability to predict the lateral phase distribution phenomena in simple and complex geometric conduits for bubbles of size 1 mm $< D_b <$ 6 mm.

The model of Kurul and Podowski (1990) was used to determine the influence of bubble wall area fraction properties [22]. This model can estimate the fraction of the wall area affected by the sweeping of the liquid inflow under the bubble departure.

In the two-fluid model, active nucleation site density is an important factor for estimating the IAC. Thus, the model by Hibiki–Ishii (2003) proposed that the active nucleation site density is a function of the critical cavity size and contact angle [23]. This model clearly demonstrated the dependence of active nucleation site density on the superheated wall by various studies.

Kocamustafaogullari (1983) reported that the bubble departure diameter was associated with water over a wide range of pressures [24]. Using the correlation proposed by Kocamustafaogullari, the mean deviation of the experimental data is approximately ±33%. It accurately simulated the available experimental water data, considering various free surfaces.

Cole (1960) demonstrated through experimental measurements that the main forces acting on bubbles leaving the surface at high heat flows are the buoyancy and drag forces [25]. The dimensionless relationship was developed by the bubble velocity, bubble diameter, and contact angle at breakoff.

There are many other studies. However, the boiling phenomenon of the subchannel is still a field that requires various studies depending on the total length, hydraulic diameter, flow rate, and heat transfer rate. This study analyzes the coolant flow in 4 × 4 rod bundles with applied symmetrical conditions. The diameter, pitch, and hydraulic diameter of the rods are 16, 21, and 14.01 mm respectively, and the total length is 590 mm. The cross-sectional area of the coolant is 0.004008 m^2. The grid size has three prism layers total of 0.8 mm near the wall, and 1 mm in other areas. Since the fluid belongs to high-Re, High-y+ wall treatment (applicable to $30 \leq y+ \leq 100$) was applied. This simulation is calculated with steady state and constant density. Initial condition covered a mass flux ranging from 250 to 310 kg/m^{2}s, heat flux ranging from 148 to 200 kW/m^2, and inlet subcooled temperature ranging from 20.2 °C to 25.2 °C. Numerical analysis is performed by STAR-CCM+ ver. 12.06.011 considering three-dimensional turbulence.

2. Subcooled Boiling Experiment in 4 × 4 Rod Bundle

In this study, an experiment was conducted in the subcooled boiling water flow with a 4 × 4 rod bundle heater. Instruments such as Coriolis flow meters, static pressure gauges, differential pressure gauges, thermal couples, and power meters were equipped in the facility, and control valves for precisely controlling the flow rate and preheater to maintain the target inlet temperature were installed. The measurement instrument used in the experiment and the uncertainty are listed in Table 1. The pressure could be adjusted with a pressurizer connected to the water storage tank consisting of a test loop. A separator for separating steam and water was installed at the outlet of the test section, and a condenser and heat exchanger for heat removal from steam and water, respectively, were installed at the rear end of the water separator. An additional heat exchanger was installed at the outlet of the test section to prevent the flashing of high-temperature water. To minimize heat loss from the flow test channel during the experiment, insulating material was applied to the outside wall of the flow channel. The test section was designed as a square channel of which the length of the sides was 85 mm. Each heater rod was 16 mm in outer diameter and 590 mm in heated length (Z_h). The pitch-to-diameter ratio was 1.3 in the bundle arrangement; therefore, the hydraulic diameter (D_H) of the whole subchannel was 14 mm.

Table 1. Measurement instrument with each uncertainty.

Parameter	Measurement Instrument	Measurement Uncertainty
Mass flow rate	Coriolis mass flow meter (Micromotion CMF100S)	0.25 g/s
Temperature	K-type thermocouple (OMEGA)	0.66 K
Pressure	Pressure transmitter (Rosemount 3051S)	0.8 kPa
Power	Power meter & Shunt (Yokogawa WT330)	0.57 kW

A two-sensor optical fiber probe (2S-OFP) consisting of one front and one rear sensor was applied to measure local two-phase flow parameters in the rod bundle. The local two-phase flow parameters measured in the experiment were the void fraction ($\overline{\alpha}^t$), interfacial area concentration ($\overline{a_i}^t$, IAC), interfacial velocity ($\overline{v_{ij}}^t$), and bubble diameter ($\overline{D_{sm}}^t$). The local void fraction is the ratio of passage time of bubbles per unit time at a local point in the two-phase flow condition. The IAC, defined as the interfacial area per unit volume, is a parameter of prime importance in terms of mass, momentum, and energy transfers between phases. In this study, the interfacial velocity is the velocity of the actual velocity of the bubbles along the axial direction of a flowing channel. The detailed equations for each parameter are as follows.

$$\overline{\alpha}^t = \frac{1}{T} \sum_j^{N_b} (t_{F2} - t_{F1}) \tag{1}$$

$$\overline{a_i}^t = 2N_b \overline{\frac{1}{\overrightarrow{v_{ij}}}}^t I(\phi_{j,max}) \tag{2}$$

$$\overline{v_{ij}}^t = \frac{N_b}{\sum_j^{N_b} \frac{(t_{R1} - t_{F1})}{S}} \tag{3}$$

where T is the total measurement time, N_b is the number of bubbles in contact with local optical fiber sensor, t_{F2} is the time when the bubble completely passes through the front sensor, t_{F1} is the time when the bubble starts to pass through the front sensor, t_{R1} is the time when the bubble starts to pass through the rear sensor, $I(\phi_{j,max})$ is the angle between the mean flow direction vector and the velocity vector of the interface, proposed by Hibiki et al. (1998) [26], and superscript t is the time average. The bubble diameter is six times the void fraction over the interfacial area concentration. Shen and Nakamura (2013) [27] showed a measurement error of ±7.8% when measuring the void fraction of small bubbles with an optical fiber probe technique. The interfacial velocity for small bubbles estimated with error less than ±2.3% when comparing the photographic and double sensor probe method used in our laboratory.

The measurement was performed on a total of 26 points in the area corresponding to 1/8 of the center subchannel at $Z_h/D_H = 41.8$ downstream from the inlet of the test section. The flow condition covered a mass flux ranging from 250 to 310 kg/m^{2}s, heat flux ranging from 148 to 200 kW/m^2, and an inlet subcooled temperature ranging from 20.2 °C to 25.2 °C at a 200 kPa inlet pressure.

3. Numerical Methods

For the analysis performed in this study, a two-fluid analysis method based on the Eulerian equation and a standard k–ε turbulence model were used [28,29]. In particular, this study considers the drag force of Tomiyama et al. (1998) [16], the lift force of Auton et al. (1988) [18], the wall lubrication force of Antal et al. (1991) [20], the turbulent dispersion force of Lahey et al. (1993) [21], and the virtual mass force of Auton et al. (1988). [18] The wall heat distribution model of Kurul and Podowski (1990) [22] was used to calculate the boiling flow, considering the interfacial length scale from Bak (2015) [30] of the bubble diameter model. Hibiki–Ishii (2003) [23] nucleation density, Kocamustafaogullari (1983) [24] bubble departure diameter, and the Cole (1960) [25] bubble departure frequency model were used to calculate the wall bubble generation.

3.1. Turbulence Model

Prior to the numerical analysis, the standard k–ε and realized k–ε turbulence models were compared to confirm the turbulence model sensitivity [31]. Figure 1 shows that there is no significant difference between the results of the two models. However, when calculations were made using the standard k–ε model under the same conditions, the results were closer to the experimental values. Therefore, the standard k–ε model was adopted for all cases.

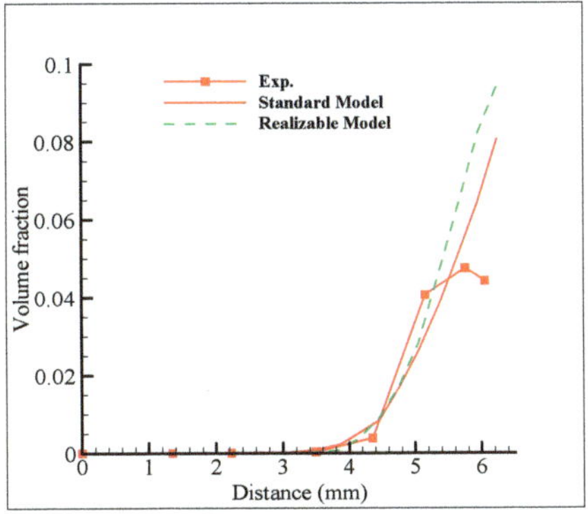

Figure 1. Turbulence model test.

3.2. Two-Phase Interaction Models

The drag force and coefficient of Tomiyama et al. (1998) [16] are given by

$$F_{ij}^D = C_{ij}^D \frac{1}{2}\rho_c|u_j - u_i|(u_j - u_i)\left(\frac{a_{ij}}{4}\right)$$
(4)

$$C_{ij}^D = max\left[\left(\frac{24}{Re}\left(1 + 0.15Re^{0.687}\right)\right), \frac{8Eo}{3(E0 + 4)}\right]$$
(5)

In Equation (5), *Re* is the Reynolds number and *Eo* is the Eotvos number, expressed as

$$Re = \frac{\rho_c|u_r|l}{\mu_c}$$
(6)

$$Eo = \frac{|\rho_c - \rho_d|gl^2}{\sigma}$$
(7)

This method is suitable for liquid-gas bubble systems with a low Reynolds number and low Morton number flow; for example, small bubbles in an air–water system. It is only available when the dispersed phase is a gas. The lift force and coefficient of Auton et al. (1988) [18] are given by

$$F_{ij}^L = C_{ij}^L \rho_c \alpha_d [u_r \times (\nabla \times u_r)]$$
(8)

$$C_{ij}^L = Const.(-0.5 \sim 0.5)$$
(9)

The wall lubrication force that prevents bubbles from touching the wall, and the coefficient of Antal et al. (1991) [20] are given by

$$F_{ij}^{WL} = -C_{ij}^{WL}\frac{\alpha_d\rho_c u_r^2}{d}n$$
(10)

$$C_{ij}^{WL} = max\left[C_1 + C_2\frac{d}{y_w}, 0\right]$$
(11)

In Equation (11), C_1 and C_2 are constant. The turbulence dispersion force and coefficient of Lahey et al. (1993) [21] are given by

$$F_{ij}^{TD} = C_{ij}^{TD}\frac{1}{2}\rho_c(u_j - u_i)\left(\frac{a_{ij}}{4}\right)\frac{v_c^t}{\sigma}\left(\frac{\nabla a_j}{a_j} - \frac{\nabla a_i}{a_i}\right)$$
(12)

$$C_{ij}^{TD} = 1 \tag{13}$$

The virtual mass force and coefficient of Auton et al. (1988) [18] are given by

$$F_{ij}^{VM} = C_{ij}^{VM} \rho_c \alpha_d \left(a_j - a_i\right) \tag{14}$$

$$C_{ij}^{VM} = 0.5 \tag{15}$$

The nucleation site number density determines the number of locations on the heated surface where bubbles form, per unit area. The nucleation site number density of Hibiki–Ishii (2003) [23] was used as follows

$$d_\omega = d_1 \theta \left(\frac{\sigma}{g\Delta\rho}\right)^{0.5} \left(\frac{\sigma}{\rho_g}\right)^{0.9} \tag{16}$$

where d_1 is 2.64×10^{-5} m/deg, and θ is $41.36°$ for the active nucleation site number density as a function of critical cavity size and contact angle. It is designed for use with the Kocamustafaogullari (1983) [24] model for bubble departure diameter as follows

$$n'' = \overline{n''}\left(1 - \exp\left[-\frac{\theta^2}{8\mu^2}\right]\right)\left(\exp\left[f\left(p^+\right)\frac{\lambda'}{R_e}\right] - 1\right) \tag{17}$$

The bubble departure frequency model of Cole (1960) [25] was used as follows

$$f = \sqrt{\frac{4}{3}\frac{g\left(\rho_l - \rho_g\right)}{d_\omega \rho_l}} \tag{18}$$

3.3. Bubble Diameter Model

In this three-dimensional two-phase flow analysis, the bubble size is a very important factor for accurate prediction of bubble behavior in the flow path. Yao and Morel (2004) and Yeoh and Tu (2005) developed a kinematic model based on the IAC and bubble number density transport equations for accurate bubble size prediction. The model has been developed and requires subsidiary equations [32,33].

The bubble size is known to be related to the hydraulic diameter of the flow path, Laplace length scale, volume fraction, and Reynolds number of the bubble through the previously developed models of Zeitoun et al. (1994) [34], Zeitoun and Shoukri (1996) [35], and Hibiki et al. (2006) [36]. Thus, bubble size model is formed by

$$d_{sm} = \alpha^A N_{Reb}^B L_O^C \tag{19}$$

where α is the volume fraction. The bubble Reynolds number (N_{Reb}) to consider the turbulent dissipation term (ε) and the Laplace length scale (L_O) are given by

$$N_{Reb} = \frac{\varepsilon^{1/3} L_O^{4/3}}{v_f} \tag{20}$$

$$L_O = \sqrt{\frac{\sigma}{g\Delta\rho}} \tag{21}$$

In Equation (20), the subscript f means liquid phase. In Equation (21), σ is the surface tension, g is the gravitational acceleration, and $\Delta\rho$ is the density difference between phases.

Based on Equation (19), Hibiki realized a bubble size model applicable to the bubble flow as

$$d_{sm} = 4.34\alpha^{0.36} N_{Reb}^{-0.696} L_O^{0.665} / D_H \tag{22}$$

A simple form of Equation (20) is derived by adding the density ratio and excluding the hydraulic diameter included in the bubble Reynolds number [37]. Bak derived a bubble diameter model from Yun by applying DEDALE, Hibiki, and DEBORA experimental data for both low- and high-pressure conditions. The bubble size model of Bak (2015) [30] used in this study is given by

$$d_{sm} = 32.39\alpha^{0.289} N_{Reb}^{-0.529} N_{\rho}^{-0.06} L_{O}$$ (23)

where the density ratio (N_{ρ}) to reflect pressure is given by

$$N_{\rho} = \frac{\rho_f}{\rho_g}$$ (24)

and the subscripts f and g are the liquid phase and gas phase, respectively.

4. Operating Conditions

The numerical analysis of the rod bundle has six initial conditions. It also proves that the grid test was conducted before showing the numerical analysis results.

4.1. Initial and Boundary Conditions

The total length of the rod bundles was 590 mm. The diameter, pitch, and hydraulic diameter are 16, 21, and 14.02 mm, respectively. The cross-sectional area through which the coolant flows is approximately 0.00401 m^2. The experimentally measured data on Lines 1–5 comparing the numerical analysis results at the cross-section of 580 mm in height are shown in Figure 2.

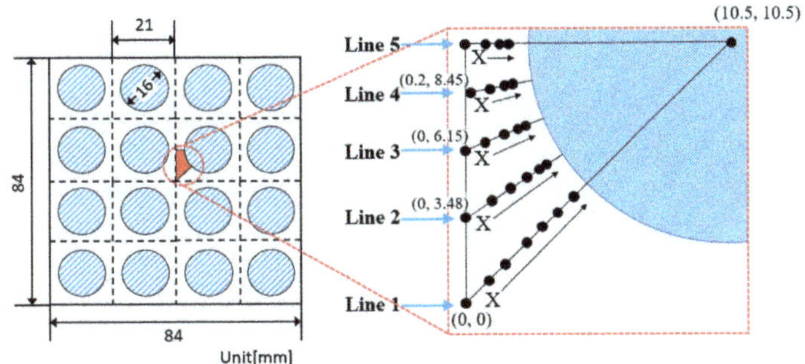

Figure 2. Measured points on Lines 1–5 for numerical analysis.

The initial conditions for the six experimental cases used in the calculations are listed in Table 2. At the outlet of all cases, the fluid has a pressure of 194.45 kPa. This shows a decrease of about 5 kPa from the inlet. Calculation was performed with constant density.

Table 2. Initial conditions of six cases.

Case No.	Mass Flux (kg/m^{2}s)	Velocity (m/s)	Heat Flux (kW/m^2)	Sub-Cooled Temp. (°C)	Inlet Temp. (°C)
1	310	0.322917	168	22.7	97.5
2	310	0.322917	168	20.2	100.
3	310	0.322917	168	25.2	95.0
4	310	0.322917	148	22.7	97.5
5	310	0.322917	200	22.7	97.5
6	250	0.260417	168	25.2	95.0

In Table 2, the different conditions are highlighted in gray. Cases 2 and 3 have different inlet temperatures compared with Case 1. Cases 4 and 5 have different heat flux compared with Case 1. Case 6 has a different inlet velocity from that of Case 3.

4.2. Grid Test

Given that the fluid does not have different results at a centerline of 4×4 and 2×2, the numerical analysis of the rod bundle was modeled by constructing a shape with one 2×2 side and by applying a symmetrical condition twice. The geometry part is the gray area in Figure 3b.

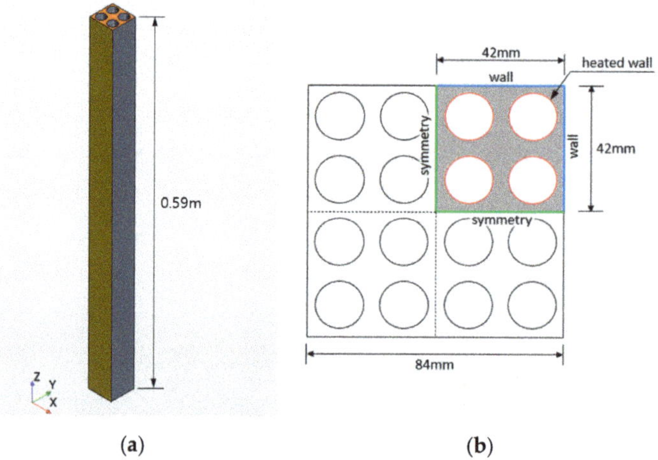

Figure 3. Shape of the rod bundle: (**a**) three-dimensional model viewed in the x–z plane and (**b**) cross-section of 0.58 m in height and wall conditions.

The analysis results are shown in Figure 4, with approximately one million, two million, and four million grids. A comparison of the analysis values was made between Line 1 and Line 4.

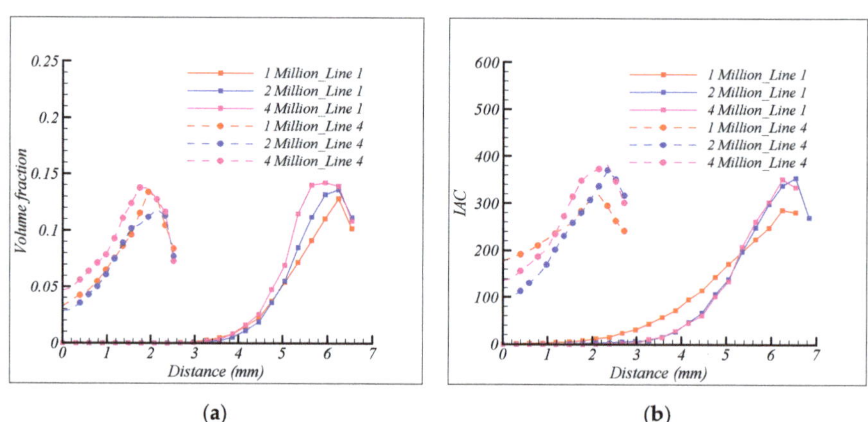

Figure 4. Comparison of one million, two million, and four million grids for tests on Lines 1 and 4: (**a**) volume fraction and (**b**) interfacial area concentration.

The grid test showed no significant difference between the approximately one million, two million, and four million grids. Therefore, numerical analysis was performed with one million grids, which had the shortest calculation time. The cross-sectional shape of the one million grid is shown in Figure 5.

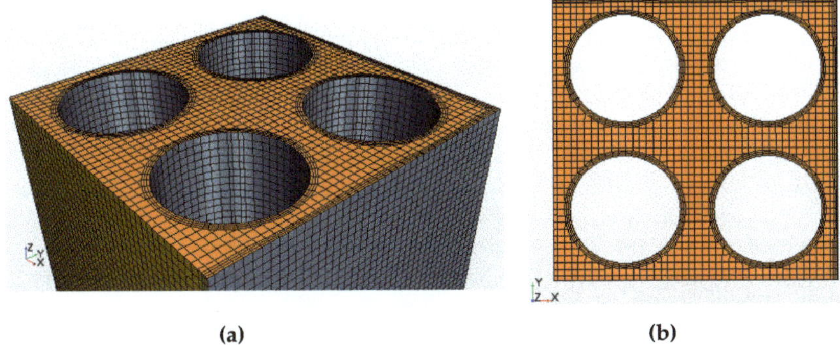

(a)   **(b)**

Figure 5. One-million-grid rod bundle: (**a**) 3-D near the outlet and (**b**) cross-section of the outlet.

The one million grid consists of hexahedron grids. By applying three prism layers to the wall surface, the spacing is approximately 0.2667 mm near the wall, for a total of 0.8 mm. In other areas, the grid spacing was 1 mm. This grid method calculates three important measures of mesh quality at the start of a run: aspect ratio—maximum 0.99: good; skewness angle—maximum 4.79: good; volume change—maximum 0.98: good. Mean aspect ratio is 0.81, mean skewness is 4.75, mean volume change is 0.8. In terms of overall mesh quality minimum quality is 0.53, maximum 0.99, and mean 0.97.

5. Results

5.1. Results

Before the results of the six cases are shown, Figure 6 illustrates a stable flow falling by approximately 5.53 kPa of pressure from about 200 kPa from the inlet to the outlet. The pressure distribution is shown in Figure 6.

Figure 6. Pressure distribution of numerical analysis results in Case 1.

For the parameters of the results, the meanings of the volume fraction and IAC are as follows.

$$Volume\ fraction = \frac{Volume\ of\ vapor}{Volume} \tag{25}$$

$$IAC = 6 \frac{Volume\ fraction}{Interaction\ length\ scale} \tag{26}$$

In Equation (26), the interaction length scale is defined as the bubble diameter.

Figures 7–12 show the comparison of the results of Lines 1–5 in Cases 1–6. In each figure, (a) shows the volume fraction and (b) shows the IAC.

In Case 1, it is considered that the VF and IAC were accurately predicted. Cases 2 and 3 (subcooled temperature of 20.2 °C or 25.2 °C) have different inlet temperature from Case 1 (subcooled temperature of 22.7 °C). The axis shows numerical values similar to experimental values, but has smaller values to the rod surface. There was no difficulty in interpreting the same tendency, such as the slight decrease in numerical results near the rod surface, as the experimental results. The result of Case 4 shows accurate prediction. Case 4 also indicates low VF and IAC, the same as the experiments, because of the low heat flux of 148 kw/m^2. In Case 5, which had a high heat flux of 200 kw/m^2, the IAC was low near the rod surface, but it can be said that it measures some reasonable value in the VF. Lastly, Case 6, which had a low flow rate, was under-interpreted near the rod surface except on Line 1, but it shows that it interprets the appropriate values near the axis and across Line 1.

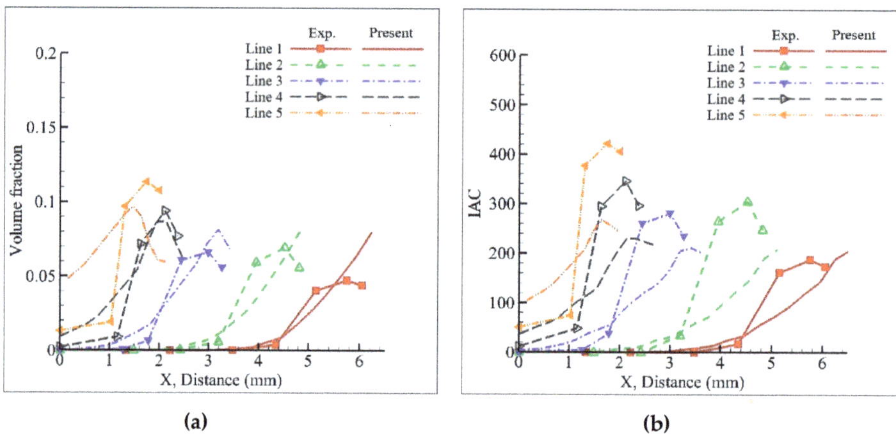

(a)   **(b)**

Figure 7. Case 1 numerical results of Lines 1–5 compared with experiment results: (**a**) volume fraction and (**b**) interfacial area concentration.

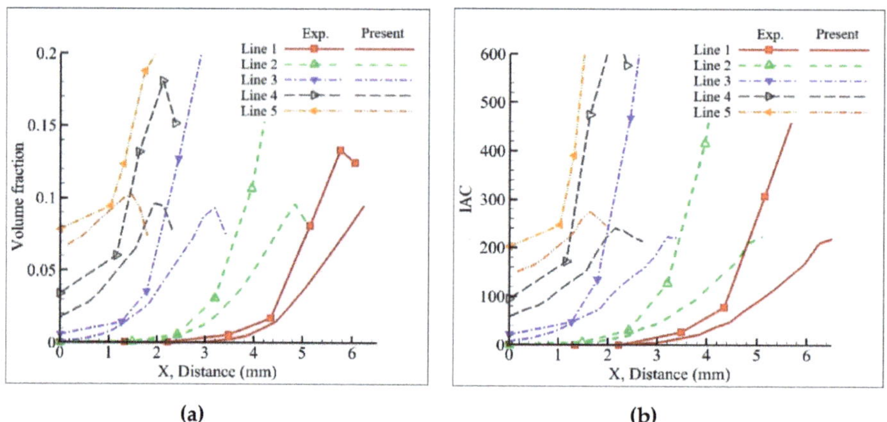

(a)   **(b)**

Figure 8. Case 2 numerical results of Lines 1–5 compared with experiment results: (**a**) volume fraction and (**b**) interfacial area concentration.

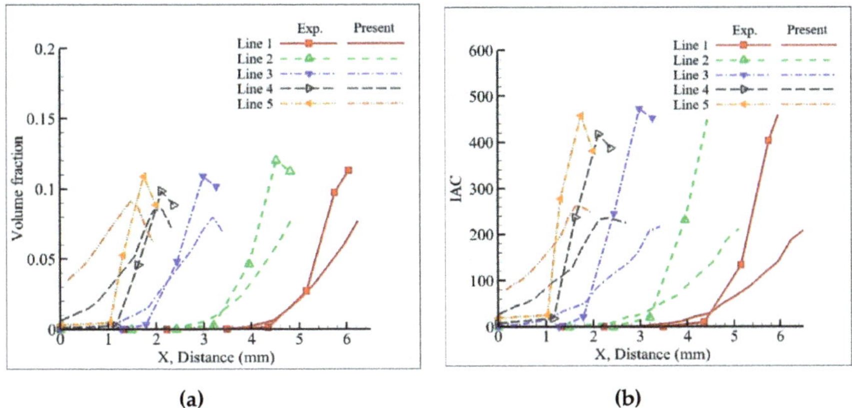

(a)   **(b)**

Figure 9. Case 3 numerical results of Lines 1–5 compared with experiment results: (**a**) volume fraction and (**b**) interfacial area concentration.

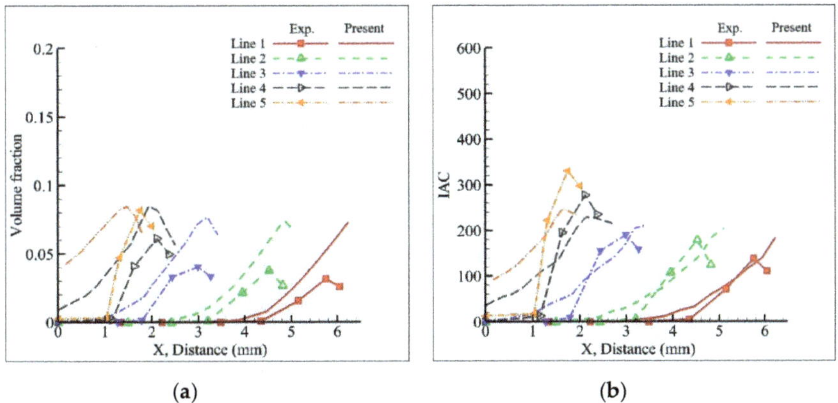

(a)   **(b)**

Figure 10. Case 4 numerical results of Lines 1–5 compared with experiment results: (**a**) volume fraction and (**b**) interfacial area concentration.

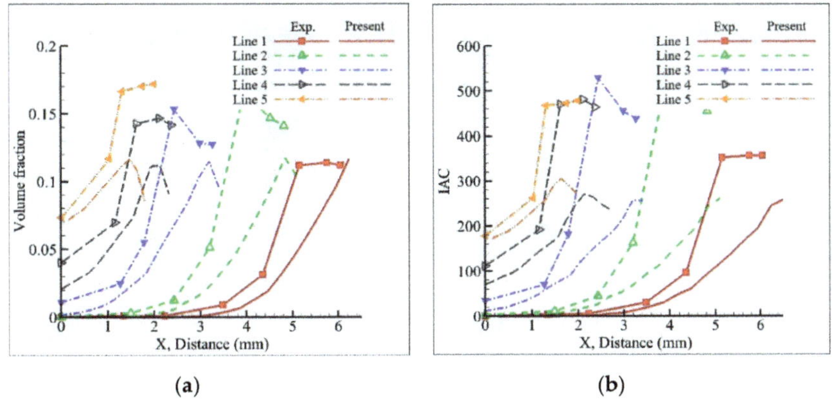

(a)   **(b)**

Figure 11. Case 5 numerical results of Lines 1–5 compared with experiment results: (**a**) volume fraction and (**b**) interfacial area concentration.

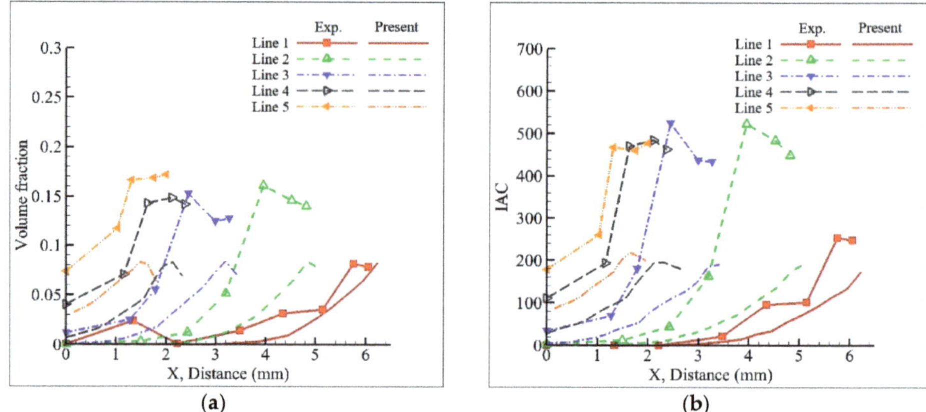

Figure 12. Case 6 numerical results of Lines 1–5 compared with experiment results: (**a**) volume fraction and (**b**) interfacial area concentration.

Figure 13 shows the numerical analysis results for the volume fraction distribution contour of all cases for the cross-section with a height of 580 mm. In Figure 13c,d, the volume fraction distributions are characterized by a low value between the rod surface and the wall. However, in Figure 13a,b,e,f, the volume fraction distributions are characterized by a high value between the rod surface and the wall.

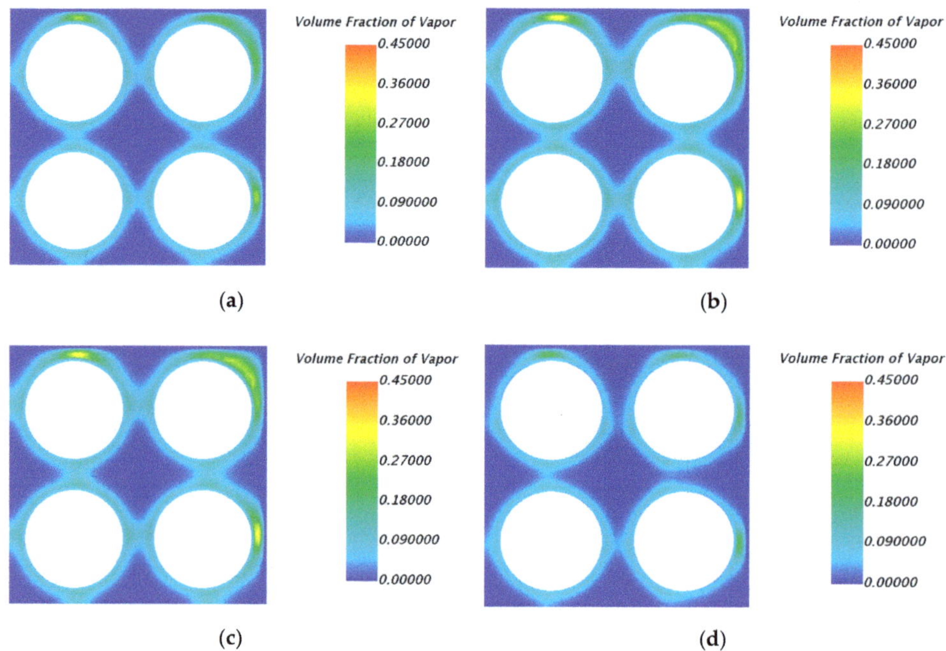

Figure 13. *Cont.*

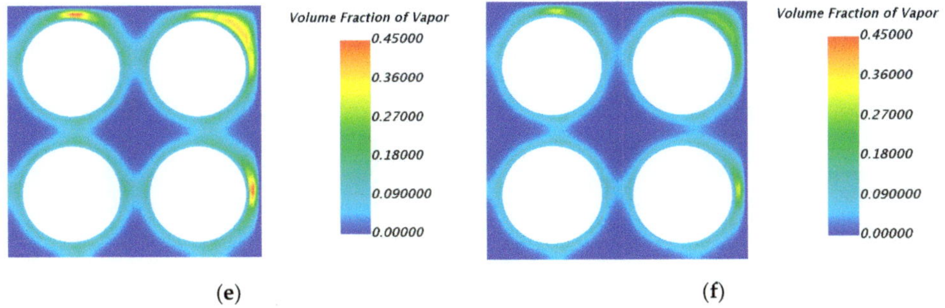

Figure 13. Volume fraction distribution on the cross-section with a height of 580 mm: (**a**) Case 1, (**b**) Case 2, (**c**) Case 3, (**d**) Case 4, (**e**) Case 5, and (**f**) Case 6.

This can be explained by the velocity and temperature of the liquid. Under the same conditions, Case 4 had the smallest heat flux (148 kW/m^2) and Case 5 had the largest heat flux (200 kW/m^2). The detailed analysis of this is shown in Figures 14 and 15, taking Cases 4 and 5 as examples. As shown in Figure 14a, the liquid velocity decreased more as it approached the outlet in the wall edge. Therefore, the liquid temperature increased because of heat transfer. However, because Case 4 was a case with very low heat flux, the temperature at the wall edge excluding the rod surface did not rise sufficiently to the boiling point, as shown in Figure 14b. Therefore, as shown in Figure 14c, air bubbles were not generated at the corners. Because no bubbles were generated, the velocity of the bubbles remained low, as shown in Figure 14d.

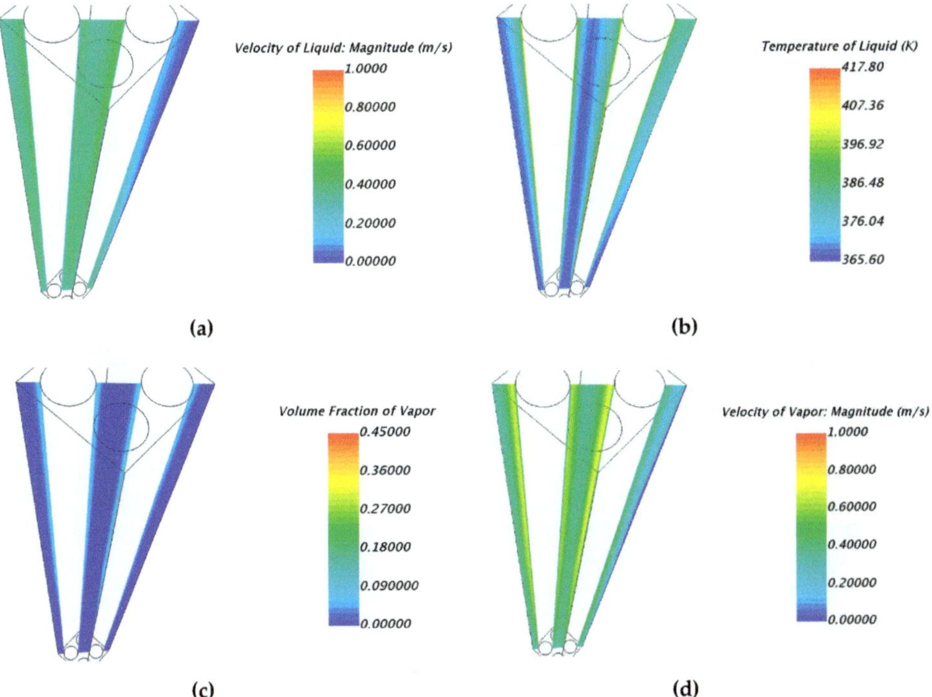

Figure 14. Distribution contour in Case 4: (**a**) velocity of liquid, (**b**) temperature of liquid, (**c**) volume fraction, and (**d**) velocity of vapor on the vertical plane at the wall edge.

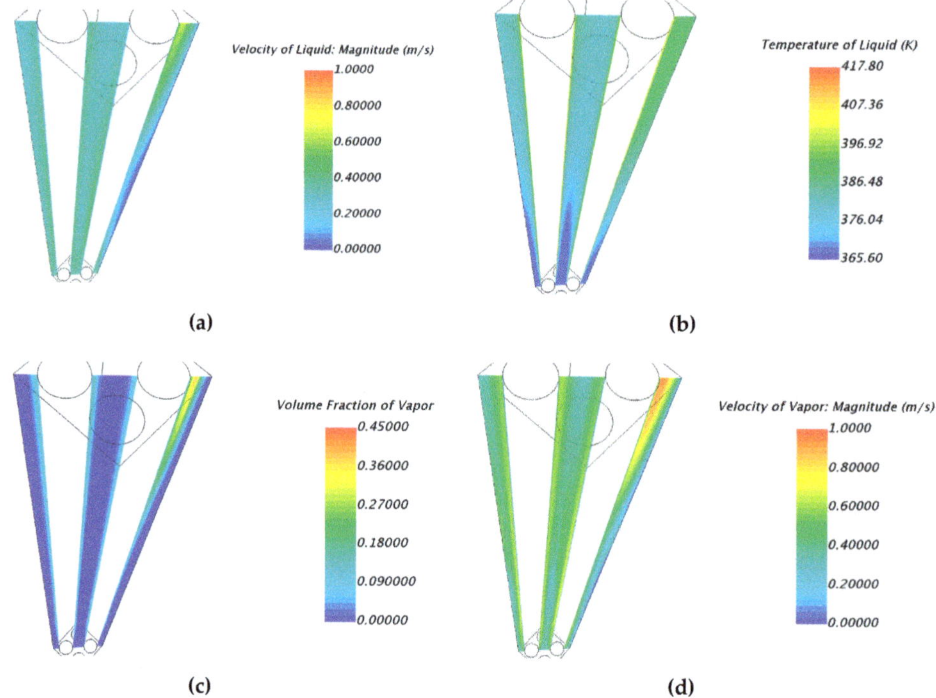

Figure 15. Distribution contour in Case 5: (**a**) velocity of liquid, (**b**) temperature of liquid, (**c**) volume fraction, and (**d**) velocity of vapor on the vertical plane at the wall edge.

Figure 15a, however, shows that the liquid flowed with a low velocity at the wall edge from the inlet to the middle height of the rod bundle. Therefore, as the liquid velocity decreased, the temperature increased enough to reach the boiling point from the middle height, as shown in Figure 15b. Consequently, more bubbles were generated, as shown in Figure 15c. Figure 15d shows that the velocity of the liquid increased more as the generated vapor velocity increased.

5.2. Model Validation

Various initial conditions were used for the numerical analysis of the six cases, with inlet fluid velocity of approximately 0.260417 or 0.322917 m/s, heat flux of 148–200 kw/m^2, and inlet subcooled temperature of 20.2–25.2 °C. The graphs on the cross-section near the outlet from the rod can confirm the validity of the model in Figures 7–12. In particular, this model predicts the reasonable value on center of rod bundles in a wide flow path and the high inlet temperature or the large heat flux; the sub-cooled temperature is 20.2 °C or the heat flux is 200 kw/m^2. All the volume fractions and IACs in Figures 7–12 tend to be similar to the experimental values, which indicates that this study will be useful for understanding the vapor behavior of the entire steady-state flow in the simulations for predicting the boiling phenomenon. The results shown in Figures 14 and 15 are expected to aid in predicting the correlation between liquid and volume fraction, which is the most important factor considered in bubbly flow CFD research. These CFD results generally reproduced the trend of experimental values. However, this simulation has the same kind of error in all cases. Thus, we must describe the error. Numerical analysis results in higher errors as the mass flux is higher and the heat flux and temperature are lower. Moreover, it can be confirmed that the VF is smaller than the experimental value. The analysis of this cause from the model aspect shows that is due to the use of a constant value for each coefficient of the interface momentum transfer force models in this paper.

In each model, the coefficient was inputted as a certain and reasonable constant with reference to the papers, but it is not a value derived from the exact same experiment and the same working fluid, so an error occurs. In the interfacial momentum transfer force model that predicts the movement of bubbles, the most dominant force is the drag force, and then the lift force. When looking at the horizontal cross section of a vertical pipe, the lift force, which determines the location of the bubble, is one of the important factors in determining how well the bubble moves after generation and departure from the heating surface [38–40].

6. Conclusions

In this study, numerical analysis of two-phase flow boiling by heat transfer was conducted. The experimental values and CFD results showed good agreement in all six cases considered. The analysis was a steady state of 3D and the calculated model was performed at constant density. The fluid was calculated using the Eulerian multiphase model including a Reynolds stress turbulence model. The interfacial momentum transfer force model and wall boiling model are important models in two-phase flow. The momentum transfer force was calculated by considering the drag force, lift force, wall lubrication force, turbulent dispersion force, and virtual mass force. The nucleation site number density, bubble departure diameter, and bubble departure frequency of the wall boiling model were considered in the heat and mass transfer force. This means that the model considered the behavioral characteristics of the fluid to be as realistic as possible. Thus, it is possible to sufficiently interpret the code. In particular, it shows stable analysis results in all cases with different conditions. The main contributions of this study are as follows.

1. It enables various water–air flow analyses of low- and high-pressure conditions of the model used in numerical analysis.
2. When the error rate is very small in a wide flow path, and the inlet temperature is high or the heat flux is large, the flow boiling phenomenon is properly predicted.
3. Through an analysis of vertical plane flow, which cannot be measured by experiments, the cause of the high-volume fraction at the edge was investigated by using the liquid velocity and correlation.

Author Contributions: Conceptualization, W.-G.P.; formal analysis, Y.-B.S. and S.S.P.; funding acquisition, W.-G.P.; investigation, Y.-B.S. and S.S.P.; methodology, W.-G.P.; project administration, W.-G.P. and B.Y.; Supervision, W.-G.P. and B.Y.; writing—original draft, Y.-B.S. and J.-Y.B.; writing—review & editing, Y.-B.S., W.-G.P. and B.Y.; All authors have read and agreed to the published version of the manuscript.

Funding: This research was funded by "Human Resources Program in Energy Technology" of the Korea Institute of Energy Technology Evaluation and Planning (KETEP), granted financial resource from the Ministry of Trade, Industry & Energy, Republic of Korea. (No. 20184010201660) and the Nuclear Research & Development Program of the NRF (National Research Foundation of Korea) grant funded by the MSIT (Ministry of Science and ICT). (NRF-2019M2D2A1A03056998).

Conflicts of Interest: The authors declare no conflict of interest.

References

1. Hosokawa, S.; Ogawa, Y.; Tomiyama, A. Liquid velocity distribution in turbulent bubbly flow in a 2 × 2 rod bundle. In Proceedings of the Ninth Korea-Japan Symposium on Nuclear Thermal Hydraulics and Safety, Buyeo, Korea, 16–19 November 2014.
2. Hosokawa, S.; Yamamoto, T.; Okajima, J.; Tomiyama, A. Measurements of turbulent flows in a 2 × 2 rod bundle. *Nucl. Eng. Des.* **2012**, *249*, 2–13. [CrossRef]
3. Hosokawa, S.; Tomiyama, A. Multi-fluid simulation of turbulent bubbly pipe flows. *Chem. Eng. Sci.* **2009**, *64*, 5308–5318. [CrossRef]
4. Hayashi, K.; Shigeo, H.; Tomiyama, A. Void Distribution bubble motion in bubbly flows in a 4 × 4 rod bundle. Part II: Numerical simulation. *J. Nucl. Sci. Technol.* **2014**, *51*, 580–589. [CrossRef]
5. Kamei, A.; Hosokawa, S.; Tomiyama, A.; Kinoshita, I.; Murase, M. Void Fraction in a Four by Four Rod Bundle under a Stagnant Condition. *J. Power Energy Syst.* **2010**, *4*, 315–326. [CrossRef]

6. Colombo, M.; Fairweather, M. Accuracy of Eulerian–Eulerian, two-fluid CFD boiling models of subcooled boiling flows. *Int. J. Heat Mass Transf.* **2016**, *103*, 28–44. [CrossRef]

7. Murallidharan, J.S.; Prasad, B.V.S.S.S.; Patnaik, B.S.V.; Hewitt, G.F.; Badalassi, V. CFD investigation and assessment of wall heat flux partitioning model for the prediction of high pressure subcooled flow boiling. *Int. J. Heat Mass Transf.* **2016**, *103*, 211–230. [CrossRef]

8. Nemitallah, M.A.; Habib, M.A.; Ben Mansour, R.; El Nakla, M. Numerical predictions of flow boiling characteristics: Current status, model setup and CFD modeling for different non-uniform heating profiles. *Appl. Therm. Eng.* **2015**, *75*, 451–460. [CrossRef]

9. Yan, B.H.; Yu, Y.Q.; Gu, H.Y.; Yang, Y.H.; Yu, L. Simulation of turbulent flow and heat transfer in channels between rod bundles. *Heat Mass Transf.* **2010**, *47*, 343–349. [CrossRef]

10. Zhang, K.; Fan, Y.Q.; Tian, W.X.; Guo, K.L.; Qiu, S.Z.; Su, G.H. Pressure drop characteristics of two-phase flow in a vertical rod bundle with support plates. *Nucl. Eng. Des.* **2016**, *300*, 322–329. [CrossRef]

11. Zhang, K.; Hou, Y.D.; Tian, W.X.; Fan, Y.Q.; Su, G.H.; Qiu, S.Z. Experimental investigations on single-phase convection and steam-water two-phase flow boiling in a vertical rod bundle. *Exp. Therm. Fluid Sci.* **2017**, *80*, 147–154. [CrossRef]

12. Jin, H.; Lee, Y.; Park, N. CFD analysis on a three by three nuclear fuel bundle with five spacer grid assemblies. *J. Mech. Sci. Technol.* **2012**, *26*, 3965–3972. [CrossRef]

13. Zang, J.; Yan, X.; Du, D.; Zeng, X.; Huang, Y. Numerical Simulations of turbulent mixing factor in 2×2 rod bundles at supercritical pressures. In Proceedings of the NURETH-16, Chicago, IL, USA, 30 August–4 September 2015.

14. Zuber, N. On the dispersed two-phase flow in the laminar flow regime. *Chem. Eng. Sci.* **1964**, *19*, 897–917. [CrossRef]

15. Sohag, F.A.; Mohanta, L.; Cheung, F.B. CFD analyses of mixed and forced convection in a heated vertical rod bundle. *Appl. Therm. Eng.* **2017**, *117*, 85–93. [CrossRef]

16. Tomiyama, A.; Kataoka, I.; Zun, I.; Sakaguchi, T. Drag coefficients of single bubbles under normal and micro gravity conditions. *JSME Int. J. Ser. B Fluids Therm. Eng.* **1998**, *41*, 472–479. [CrossRef]

17. Auton, T.R. The lift force on a spherical body in a rotational flow. *J. Fluid Mech.* **1987**, *183*, 199. [CrossRef]

18. Auton, T.R.; Hunt, J.C.R.; Prudhomme, M. The Force Exerted on a Body in Inviscid Unsteady Non-Uniform Rotational Flow. *J. Fluid. Mech.* **1988**, *197*, 241–257. [CrossRef]

19. Achard, J.-L.; Cartellier, A. Local characteristics of upward laminar bubbly flows. *PCH. Physicochem. Hydrodyn.* **1985**, *6*, 841–852.

20. Antal, S.P.; Lahey, R.T., Jr.; Flaherty, J.E. Analysis of phase distribution in fully developed laminar bubbly two-phase flow. *Int. J. Multtphase Flow* **1991**, *17*, 635–652. [CrossRef]

21. Lahey, R.T.; Debertodano, M.L.; Jones, O.C. Phase Distribution in Complex-Geometry Conduits. *Nucl. Eng. Des.* **1993**, *141*, 177–201. [CrossRef]

22. Kurul, N.; Podowski, M.Z. Multidimensional effects in forced convection subcooled boiling. *Ihtc-9* **1990**, 21–26. Available online: http://www.ihtcdigitallibrary.com/conferences/6ec9fdc764f29109,3b6668e93c4d03e5,1968e4d71e4e45b1.html (accessed on 10 December 2019). [CrossRef]

23. Hibiki, T.; Ishii, M. Active nucleation site density in boiling systems. *Int. J. Heat Mass Transf.* **2003**, *46*, 2587–2601. [CrossRef]

24. Kocamustafaogullari, G. Pressure-Dependence of Bubble Departure Diameter for Water. *Int. Commun. Heat Mass Transf.* **1983**, *10*, 501–509. [CrossRef]

25. Cole, R. A photographic study of pool boiling in the region of the critical heat flux. *AIChE J.* **1960**, *6*, 533–538. [CrossRef]

26. Hibiki, T.; Hogsett, S.; Ishii, M. Local measurement of interfacial area, interfacial velocity and liquid turbulence in two-phase flow. *Nucl. Eng. Des.* **1998**, *184*, 287–304. [CrossRef]

27. Shen, X.; Nakamura, H. Local interfacial velocity measurement method using a four-sensor probe. *Int. J. Heat Mass Transf.* **2013**, *67*, 843–852. [CrossRef]

28. Jones, W.P.; Launder, B.E. The Prediction of Laminarization with a Two-Equation Model of Turbulence. *Int. J. Heat Mass Tranf.* **1972**, *15*, 301–314. [CrossRef]

29. Launder, B.E.; Sharma, B.I. Application of the Energy Dissipation Model of Turbulence to the Calculation of Flow Near a Spinning Disc. *Lett. Heat Mass Transf.* **1974**, *1*, 131–138. [CrossRef]

30. Bak, J.Y.; Yun, B.J.; Jeong, J.J. Bubble Size Models for the Prediction of Bubbly Flow with CMFD Code. In Proceedings of the KNS 2015 spring meeting, Daejeon, Korea, 6–8 May 2015.

31. Shih, T.-H.; Liou, W.W.; Shabbir, A.; Yang, Z.; Zhu, J. *A New k-e Eddy Viscosity Model for High Reynolds Number Turbulent Flows: Model Development and Validation*; Technical Report for NASA Lewis Research Cente: Cleveland, OH, USA, 1994 August.

32. Yao, W.; Morel, C. Volumetric interfacial area prediction in upward bubbly two-phase flow. *Int. J. Heat Mass Transf.* **2004**, *47*, 307–328. [CrossRef]

33. Yeoh, G.H.; Tu, J.Y. A unified model considering force balances for departing vapour bubbles and population balance in subcooled boiling flow. *Nucl. Eng. Des.* **2005**, *235*, 1251–1265. [CrossRef]

34. Zeitouna, O.; Shoukria, M.; Chatoorgoon, V. Measurement of interfacial area concentration in subcooled liquid-vapour flow. *Nucl. Eng. Des.* **1994**, *152*, 243–255. [CrossRef]

35. Zeitoun, O.; Shoukri, M. Bubble behavior and mean diameter in subcooled flow boiling. *J. Heat Transf. Trans. ASME* **1996**, *118*, 110–116. [CrossRef]

36. Hibiki, T.; Lee, T.-H.; Lee, J.-Y.; Ishii, M. Interfacial area concentration in boiling bubbly flow systems. *Chem. Eng. Sci.* **2006**, *61*, 7979–7990. [CrossRef]

37. Yun, B.J.; Splawski, A.; Lo, S.; Song, C.H. Prediction of a subcooled boiling flow with advanced two-phase flow models. *Nucl. Eng. Des.* **2012**, *253*, 351–359. [CrossRef]

38. Sivalingam, S.; Paramanantham, S.S.; Park, W.G. Numerical simulation of vapor volume fraction in a vertical channel under low-pressure conditions. *J. Mech. Sci. Technol.* **2018**, *32*, 4657–4664. [CrossRef]

39. Marta, S.; Marek, J. Multiphase model of flow and separation phases in a whirlpool: Advanced simulation and phenomena visualization approach. *J. Food Eng.* **2020**, *274*, 109846.

40. Tomiyama, A.; Tamai, H.; Zun, I.; Hosokawa, S. Transverse migration of single bubbles in simple shear flows. *Chem. Eng. Sci.* **2002**, *57*, 1849–1858. [CrossRef]

 © 2020 by the authors. Licensee MDPI, Basel, Switzerland. This article is an open access article distributed under the terms and conditions of the Creative Commons Attribution (CC BY) license (http://creativecommons.org/licenses/by/4.0/).

MDPI
St. Alban-Anlage 66
4052 Basel
Switzerland
Tel. +41 61 683 77 34
Fax +41 61 302 89 18
www.mdpi.com

Applied Sciences Editorial Office
E-mail: applsci@mdpi.com
www.mdpi.com/journal/applsci

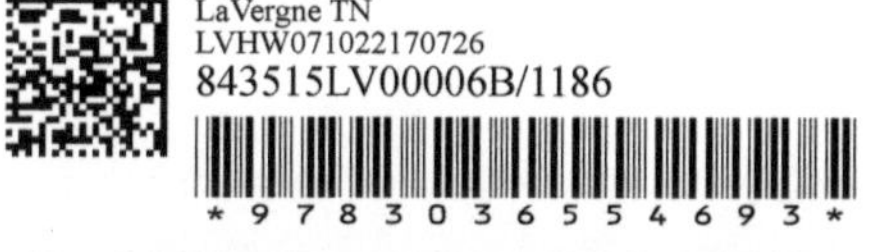

www.ingramcontent.com/pod-product-compliance
Lightning Source LLC
LaVergne TN
LVHW071022170726
843515LV00006B/1186